W0079614

NUCLEAR PHYSICS, LARGE AND SMALL

Related Titles from AIP Conference Proceedings

To learn more about these titles, or the AIP Conference Proceedings Series, please visit
the webpage **http://proceedings.aip.org**

NUCLEAR PHYSICS, LARGE AND SMALL

International Conference on
Microscopic Studies of Collective Phenomena

Morelos, México   19 – 22 April 2004

EDITORS

Roelof Bijker
ICN-UNAM, México

Richard F. Casten
Yale University, New Haven, Connecticut

Alejandro Frank
ICN-UNAM, México

SPONSORING ORGANIZATIONS
Consejo Nacional de Ciencia y Tecnología (CONACyT)
División de Física Nuclear de la SMF (DFN-SMF)
Instituto de Ciencias Nucleares (ICN-UNAM)
National Science Foundation
University of Delaware

Melville, New York, 2004
AIP CONFERENCE PROCEEDINGS ■ VOLUME 726

Editors:

Roelof Bijker
Instituto de Ciencias Nucleares
UNAM
Apartado Postal 70-543
04510 México, D. F.
MEXICO
E-mail: bijker@nuclecu.unam.mx

Richard F. Casten
Yale University
272 Whitney Avenue
New Haven, CT 06520
USA
E-mail: rick@riviera.physics.yale.edu

Alejandro Frank
Instituto de Ciencias Nucleares
UNAM
Apartado Postal 70-543
04510 México, D. F.
MEXICO
E-mail: frank@nuclecu.unam.mx

Authorization to photocopy items for internal or personal use, beyond the free copying permitted under the 1978 U.S. Copyright Law (see statement below), is granted by the American Institute of Physics for users registered with the Copyright Clearance Center (CCC) Transactional Reporting Service, provided that the base fee of $22.00 per copy is paid directly to CCC, 222 Rosewood Drive, Danvers, MA 01923. For those organizations that have been granted a photocopy license by CCC, a separate system of payment has been arranged. The fee code for users of the Transactional Reporting Service is: 0-7354-0207-8/04/$22.00.

© 2004 American Institute of Physics

Individual readers of this volume and nonprofit libraries, acting for them, are permitted to make fair use of the material in it, such as copying an article for use in teaching or research. Permission is granted to quote from this volume in scientific work with the customary acknowledgment of the source. To reprint a figure, table, or other excerpt requires the consent of one of the original authors and notification to AIP. Republication or systematic or multiple reproduction of any material in this volume is permitted only under license from AIP. Address inquiries to Office of Rights and Permissions, Suite 1NO1, 2 Huntington Quadrangle, Melville, N.Y. 11747-4502; phone: 516-576-2268; fax: 516-576-2450; e-mail: rights@aip.org.

L.C. Catalog Card No. 2004111398
ISBN 0-7354-0207-8
ISSN 0094-243X
Printed in the United States of America

Contents

*Italicized name indicates author who presented the paper.

*Italicized name indicates author who presented the paper.

POSTERS

*Italicized name indicates author who presented the paper.

Preface

This volume contains the Proceedings of the *Stufiesta*, an International Conference on *Nuclear Physics, Large and Small: Microscopic Studies of Collective Phenomena* which was held on April 19-22, 2004 at the Hacienda Cocoyoc, Morelos, Mexico. The meeting celebrated the distinguished career of our esteemed colleague and dear friend, Stu Pittel, on the occasion of his 60th birthday.

Stuart Pittel has made essential contributions to nuclear physics. Among the most important is his pioneering work on the relation between the neutron-proton interaction and nuclear deformation, carried out in the 70's in collaboration with Pedro Federman. These ideas were fundamental to establish a bridge between the collective and single-particle views of the nucleus, a connection which is now considered classical. The influence of this work has carried over to current frontier research, such as its relevance to neutron-proton pairing in exotic nuclei and as a mechanism for shape-phase transitions. In recent work on facilitating the theoretical description of weakly bound nuclei, his work on the Transformed Harmonic Oscillator (THO) basis illustrates the breadth of his achievements. A proper litany of his achievements could extend for many pages. In addition to his exceptional scientific work, Stuart has had an important role in establishing high level scientific collaborations with many different countries, including Mexico, Argentina, and Bulgaria. As a recognition of his many contributions to physics, he was elected as a fellow of the American Physical Society in 1990 and as a corresponding member of the Mexican Academy of Sciences in 1995.

At the meeting, recent developments in nuclear structure physics were discussed from both the experimental and theoretical points of view. The program consisted of 31 invited plenary talks and a poster session. These contributions are included in this Proceedings, as well as short comments by some of the session chairs. The high level of the talks and the posters and the active discussions bear witness to the enormous activity in the field, as well as to the many aspects of Nuclear Physics where Stu's influence is enduring.

The meeting was held in the beautiful surroundings of the Hacienda Cocoyoc, the same location as that of the Nuclear Physics Conference *Contemporary Topics in Nuclear Structure Physics*, organized by Rick Casten, Alejandro Frank, Marcos Moshinsky and Stu Pittel sixteen years ago in 1988. Unfortunately, as that conference drew near, Stu fell ill and could not attend the meeting he had helped to organize. The present meeting gave Stu a second chance to spend time in wonderful Cocoyoc, accompanied by a remarkable group of nuclear physicists and friends from all over the world, linked by their common love for our beautiful subject and their admiration for Stu's contributions.

We particularly enjoyed the variety of topics discussed, as well as the participation of many of our colleagues and friends at the Conference's banquet, where poetry and music accompanied the many anecdotes attesting to Stu's wonderful sense of humor and love for diversity and human values.

Finally, we would like to acknowledge the generous support from a joint grant from the Consejo Nacional de Ciencias y Tecnología in Mexico and the National Science Foundation, which made it possible to help many young scientists to attend and participate in this meeting. We are also very grateful to the University of Delaware, the Instituto de Ciencias Nucleares of the Universidad Nacional Autónoma de México and the Sociedad Mexicana de Física for their support. Our special thanks go to Jackie Mooney and Raquel Polo, the conference secretaries, who made life easy for the organizers during the meeting, and to Paula Fox whose artistic and technical expertise was crucial to preparing these Proceedings for publication.

Roelof Bijker
Rick Casten
Alejandro Frank

May, 2004

International Conference
Nuclear Physics, Large and Small
Microscopic Studies of Collective Phenomena

Hacienda Cocoyoc, Morelos, Mexico
April 19-22, 2004

In honor of Stuart Pittel
on the occasion of his 60th birthday

StuFiesta
Nuclear Physics, Large and Small
Hacienda Cocoyoc April 19 - 22, 2004

Happy Birthday Stuart

Somebody we know has become sixty of late
So we have all come here to celebrate.

A guy who has had a career so great
And even more has a wonderful mate.

His name is Professor Stuart Pittel
And whatever he does, he does it well.

We thought doing a Conference for you would be so cool
Because we knew if we didn't, you'd push us in the pool.

Stu is a physicist with achievements pioneering
There's no problem that he's a-fearing.

Happy birthday to Stu tonight
You may be old but you're just right.

Not always was Stu the Stu we know today
He grew up in his own strange way.

Stu as a boy of six was a mischievous young pup
In fact, he had already appeared in a police lineup.

His crime was subtle, like much of his work would later be
He cut up the Venetian blinds so no one could see.

He grew up being a baseball fan
Wanting to impress the girls as best he can.

He was good at math and science in school
But the girls thought a jock was far more cool.

To be a ball player would increase the babe attraction
Even more than knowing the pair interaction.

He so wanted to wear a Major League hat
But he had (has?) two main problems–short and fat.

So he settled to be a big baseball fan
'Cause almost everyone can play better than he can.

He loves to root for the underling
That has always been his favorite thing.

So he supports the Phils, who haven't won yet
What's more they just blew up the Vet.

This year, however, I have to say
It won't be the Phils' (or the Red Sox') day.

They will all get thoroughly beat
By A-Rod and Jeter and Mariano's heat.

Stu has a prodigious memory
He thinks it's the best in history.

Yet, to be fair, it really is true
He can remember almost anything right out of the blue.

This could be an ancient physics discussion
Or the name of an obscure scientist Russian.

Don't ever challenge him on sports trivia
He knows more facts than there are nuclei at RIA.

He can tell you how many homers Duke Snider got
That is, how many times he did the home run trot.

He can give the definition of a hit or an error
Or how many runners were nailed by Berra.

He knows so much he recently got the call
To give an invited talk on baseball.

To enliven a conference of RIA physics
With many arcane sports statistics.

Because of the great physics Stu has done
He has the worldwide respect of everyone!

He links nuclear structure from the big to the small
All the while having a worldwide ball.

To understand nuclear deformation
He used his incredible imagination.

Putting protons and neutrons in special orbits you see
And letting them interact proved to be the key.

Known as the Federman-Pittel mechanism
So pioneering that today it's a truism.

When he said shells were cracking
This was an idea we had been lacking.

This was something you didn't hear people say
Yet it's become a mantra today.

He loves bosons and fermions
From which he makes pairons and orbitons.

Working now with scientists from Bulgaria to the east
He's calculating nuclei that are bound the least.

For such nuclei loosely bound
They think the THO is sound.

It cuts down the number of needed shells
And simplifies using diffuse nuclear wells.

He's worked throughout with many good girls and guys
Who help him a lot 'cause they're so wise.

Pedro, Jorge, Mario, Pepe, Jacek, Bruce, Petr, Sevdalina, Jonathon, Aleeeee,
Roelof, Witek, Dave, Piet, and others
So many really smart minds, working together like sisters and brothers.

Stu really loves to strut every new finding
From quarks to structure to nuclear binding.

His talks are great; have no fear
His audiences even rise up to cheer.

What's more, his talks are clear as a bell
Even though they're louder than hell.

Stu loves to travel hither and yon
And eat a mole or a wonton.

He travels the world from east to west
Counting countries is the competitive test.

Doing so took the brains of a physicist
To know which countries should make the list.

So he invented the rule of Bahrain
Can't just land there, got to get off the plane.

And going to Yugoslavia wasn't worth a dime
Since countries only count as they were at the time.

Anyway, no matter, I have more countries than he's got
And Susan has both of us beat by a lot.

Such trips yielded many adventures
About which he could give countless lectures.

In Viet Nam he shocked the VP
By telling how pretty the girls could be.

In India, his clothes never arrived, alas
So he refurbished at the Gent's Palace.

Moreover, he was heisted in Puri
Where to this day his camera may be.

He has been a fan for years of America Latina
With trips to Mexico and Argentina.

Mexico especially has provided much glee
Through all of his fun with dear Aleeee.

From police inspections of a scruffy Stu
To providing his nickname Dr. Tutu.

In Erice he listened to old Norbert the Thespian
Sitting next to he who this poem would pen.

Competing for chocolates Godiva (or was it rugalah)
He walked about so sprightly at the poopola cupola.

Saying to his friends with so much hoopla
They should scoopola the poopola off of the cupola.

Now there's a quadrupola rhyme
So it's off to other things at this time.

Now, here's something really weird
About our friend Stu, the guy with the beard.

He's so very proud of the strange fact that
He has such incredibly low total body fat.

You might not think this to look at him ·
But inside his clothes he says he's slim.

Is it eleven percent or more like seventeen?
Probably the latter – he's not that lean.

Stu has this influence on the ladies, what can I say,
Be it local ones or those far far away.

He first met Dixie whom he tried to woo
But later Vergados gave him Sue.

He loves to dance the night away
His face shows he's ready to play.

In Turkey, he showed how he could prance
When he got up to do the belly dance.

In Viet Nam we both met the cute girl from Hue
And that was a very memorable day.

Stu bragged about the information he was able to get,
When she told him that she wasn't married yet.

But I heard more interesting news
She still had her purity to lose.

Stu, we hope you have sixty years more
With much much physics still in store.

For nearly thirty years now a really great friend
For whom the peer review process we would gladly bend.

Your good points make you just so,
And one of the best friends that we know.

Now your brain may turn to jelly,
And we see a slightly bigger belly.

Your memory may no longer be clear as a bell
And maybe your sex life is going to hell.

But there's no need to cry or pout,
You have so much to be happy about.

As you grow old, don't fret or fuss
You have supportive friends like us.

So we praise and love you with all honesty tonight
Your past is great, may your future be bright.

Richard Casten
With help from Paula Fox
and advice from L.L. Riedinger
April, 2004

Opening remarks

Igal Talmi

Department of Particle Physics
Weizmann Institute of Science, Rehovot, Israel

It is a great pleasure to open the first session of the StuFiesta which is Spanish for StuFest. All of us are very happy to celebrate the 60eth birthday of our dear friend and colleague Stuart Pittel and to pay tribute to his work and achievements. He has been with us for many years, always young, enthusiastic, active, and productive. We are sure that he will go on being young, enthusiastic, active, and productive. So what is the meaning of this StuFiesta? It is simply a happy occasion to meet, celebrate and share ideas and problems with colleagues. We are grateful to Stuart for reaching this mature age and to the organizers for planning and making possible this conference in this beautiful location.

The distinguished audience gathered here is clear evidence for the wide interests and areas of research of Stuart. It demonstrates his standing among his peers and their warm feelings for him. We wish him to have many many happy years with Susan (up to 120!), and to be young, enthusiastic, active, and productive as before.

It is nice to see in the audience many old friends. It is even more impressive to see many young people. This strengthens the hope that nuclear structure physics has a bright future. There are several good reasons for this expectation. The first is the surge of activity in the experimental study of short lived radioactive nuclei. The facilities for "rare isotope beams" have already obtained important data on nuclei which could not be reached before, and more data will be obtained in the future. The venture into new regions of the periodic table promises to yield important information which is highly relevant for astrophysics. Help to other fields is certainly an important aspect of nuclear physics as it has always been. The new data, however, are very important for our understanding of more stable nuclei. They will furnish important tests of our current theories and models. We will learn their limitations and how to improve them.

One feature which was already established many years ago is related to shell closures. The order and spacings of single nucleon orbits in the shell model are not due to some schematic central potential well. They are determined by the mutual interaction between nucleons and primarily by the strong and attractive interaction

CP726, *Nuclear Physics, Large and Small: International Conference on*
Microscopic Studies of Collective Phenomena, edited by R. Bijker, R. F. Casten, and A. Frank
© 2004 American Institute of Physics 0-7354-0207-8/04/$22.00

between protons and neutrons. The order and spacings of neutron orbits are determined by the interaction of neutrons in those orbits with protons in closed shells and vice versa. In stable nuclei, bunching of single nucleon orbits gives rise to the well known magic numbers of major shell closures. In nuclei with more protons or more neutrons, the bunching of orbits and even their order, may change. Early examples can be seen in the ground state of ^{11}Be and in the disappearance of the magic number at $N=20$ in nuclei with large neutron excess. More such cases are expected to be found. Where protons and neutrons are outside closed shells, the situation is more complicated. Large scale shell model calculations which can deal with such cases are currently available.

This is due to a parallel important development of theoretical tools. The modern programs for diagonalizing giant matrices can deal with problems hitherto too difficult to tackle. Shell model calculations involving several orbits produce results in much better agreement with experiment. There is now hope for better predictions of properties of nuclei which become now available for experimental study. More fundamental calculations can now be carried out, promising better understanding of nuclear structure. This may eventually lead to establishing a solid foundation of the simple models which work very successfully in many cases.

New Generalizations of the Richardson-Gaudin Models

J. Dukelsky[*], V. G. Gueorguiev[1][*][†] and S. Pittel[**]

[*]*Instituto de Estructura de la Materia, CSIC, Serrano 123, 28006 Madrid, Spain*
[†]*Department of Physics and Astronomy, Louisiana State University, Baton Rouge, LA 70803, USA*
[**]*Bartol Research Institute, University of Delaware, Newark, Delaware 19716, USA*

Abstract. We discuss the extension of the Richardson-Gaudin models along two lines. We first describe a new family of exactly solvable atom-molecule Hamiltonians obtained by coupling a bosonic mode to the rational family. We then present the generalization of the Richardson-Gaudin models to any simple Lie algebra. As an example of relevance to nuclear structure, we show the exact solution of the T=1 pairing Hamiltonian with non-degenerate single-particle energies.

INTRODUCTION

Exactly-solvable models (ESMs) are powerful tools for studying the physics of strongly-correlated many-body quantum systems. This is especially true when the usual approximate many-body methods fail to adequately capture the physics of the system, as for the example in the case of one-dimensional (1D) models with short-range interactions, which can be solved exactly using the algebraic Bethe ansatz (ABA). Some examples in which this approach has proven fruitful include the Heisenberg model, which was first solved by Bethe in 1931, and the Hubbard model.

In the context of nuclear structure physics, ESMs based on dynamical symmetries have played a major role in providing an understanding of complex nuclear properties and in serving as testing grounds for approximate many-body theories. The simplest ESM based on a dynamical symmetry is the rank-1 $su(2)$ pairing model for one (degenerate) orbit, which is often used to introduce the chapter on superconductivity in quantum many-body textbooks. Among the rank-2 models, two that have been especially fruitful are the $su(3)$ Elliott model of the quadrupole-quadrupole interaction and the $so(5)$ model of generalized T=1 pairing, both in one degenerate shell.

All the above models are either constrained to 1D, as in the case of those solved by the algebraic Bethe ansatz, or involve dynamical symmetries that can only be realized in a single degenerate shell.

We will present here two new extensions of the so-called Richardson-Gaudin (RG) models, which have the virtue of being exactly solvable in any dimension and amenable to the inclusion of single-particle symmetry-breaking terms. The RG models were re-

---

[1] On leave from the Institute of Nuclear Research and Nuclear Energy, Bulgarian Academy of Sciences, Sofia 1784, Bulgaria.

CP726, *Nuclear Physics, Large and Small: International Conference on Microscopic Studies of Collective Phenomena*, edited by R. Bijker, R. F. Casten, and A. Frank
© 2004 American Institute of Physics 0-7354-0207-8/04/$22.00

cently proposed [1] as a generalization of Richardson's exact solution of the pairing model in non-degenerate orbits [2]. They also relate closely to Gaudin's model of a linear spin chain [3]. The RG models can describe rather general pairing Hamiltonians based on the $su(2)$ pair algebra for fermions systems or the $su(1,1)$ pair algebra for boson systems. In the first extension, we add a new bosonic degree of freedom to the boson or fermion pair algebras of the RG rational family [4]. As we will see, this gives rise to a new family of exactly-solvable atom-molecule Hamiltonians. The second extension, involves a generalization of the exactly-solvable RG models to any simple Lie algebra [5]. With this generalization, it is possible to include symmetry-breaking single-particle energies in several ESMs of interest in nuclear physics. As a specific example, we will show how the method can be used to provide exact solutions for the $so(5)$ T=1 pairing pairing model with non-degenerate single-particle energies.

THE RICHARDSON-GAUDIN MODELS FOR ATOM-MOLECULE HAMILTONIANS.

We now discuss how to extend the RG integrable models of the $su(2)$ fermion pair algebra or the $su(1,1)$ boson pair algebra to include a single bosonic degree of freedom.

We begin by introducing the generators of the $su(2)$ and $su(1,1)$ algebras, K_i^0, K_i^+ and $K_i^- = (K_i^+)^\dagger$, which satisfy the commutation relations

$$\left[K_i^0, K_j^+\right] = \delta_{ij} K_i^+ \; , \quad \left[K_i^0, K_j^-\right] = -\delta_{ij} K_i^- \; , \quad \left[K_i^+, K_j^-\right] = \mp 2\delta_{ij} K_i^0 \; . \tag{1}$$

The upper sign refers to the bosonic $su(1,1)$ algebra and the lower sign to the fermionic $su(2)$ algebra, as they will throughout the presentation.

In the quasi-spin or pair representation of these algebras, the generators are realized in terms of particle creation and annihilation operators as

$$K_j^0 = \frac{1}{2} \sum_m a_{jm}^\dagger a_{jm} \pm \frac{\Omega_j}{4} \; , \quad K_j^+ = \frac{1}{2} \sum_m a_{jm}^\dagger a_{j\bar{m}}^\dagger \; . \tag{2}$$

Here $a_{jm}^\dagger \left(a_{jm} \right)$ creates (annihilates) a boson or a fermion in the state $|jm\rangle$, $|j\bar{m}\rangle$ is the state obtained by acting with the time reversal operator on $|jm\rangle$, and Ω_j is the total degeneracy of single-particle level j.

Consider now the rational family of ESMs involving L degrees of freedom denoted by $j = 1, \cdots, L$. We now add to this family a new bosonic degree of freedom, denoted by $j = 0$ and represented by the creation (annihilation) operator $b^\dagger(b)$. The resulting integrals of motion of the new family of atom-molecule models are

$$R_j = K_j^0 + G \left[\sum_{i(\neq j)} \frac{1}{\left(\eta_i - \eta_j\right)} \{[K_i^+ K_j^- + K_i^- K_j^+] - 2K_i^0 K_j^0\} - [K_j^+ b + K_j^- b^\dagger] - \eta_j K_j^0 \right] \; . \tag{3}$$

$$R_0 = b^\dagger b + G \left[\sum_j \left(b^\dagger K_j^- + K_j^\dagger b \right) + \sum_j \eta_j K_j^0 \right], \qquad (4)$$

We claim that this set of $L+1$ operators still satisfies the conditions for an integrable model. They are hermitian, global, independent, and mutually commute with one another, thereby constituting a complete set of integrals of motion. An example of a Hamiltonian that can be expressed solely in terms of R_0 and thus can be solved exactly is

$$H = \omega R_0 \mp \frac{\omega G}{4} \sum_j \Omega_j \varepsilon_j = \omega b^\dagger b + \sum_{jm} \varepsilon_j \, a_{jm}^\dagger a_{jm} + V \sum_j \left(b^\dagger K_j^- + K_j^\dagger b \right), \qquad (5)$$

where $V = \omega G$ and $\varepsilon_j = V \eta_j / 2$. This Hamiltonian can be used to describe a system of bosonic or fermionic atoms coupled to diatomic molecules. Many other exactly-solvable models can also be constructed in this way by taking linear combinations of the complete set of integrals of motion and not just R_0.

It is also useful to remark here that the $su(2)$ algebra for fermions has another representation in terms of two-level atoms. When this representation is used, the Hamiltonian (5) provides a generalization of the Jaynes-Cummings model, much celebrated in quantum optics as a simple model of the interaction of atoms with a radiation field.

We now turn to the exact solutions for these ESMs. The common eigenstates of the complete set of integrals of motion (4-3) have the form·

$$|\Psi\rangle = \prod_{\alpha=1}^{M} \left(b^\dagger + \sum_l \frac{1}{x_\alpha - \eta_l} K_l^\dagger \right) |0\rangle , \qquad (6)$$

where the M parameters x_α are solutions of the M coupled Richardson equations

$$\frac{\omega}{2V} - \frac{1}{2} x_\alpha - \frac{1}{4} \sum_j \frac{\Omega_j}{\eta_j - x_\alpha} \mp \sum_{\beta(\neq\alpha)} \frac{1}{x_\beta - x_\alpha} = 0 . \qquad (7)$$

The eigenvalues of the Hamiltonian (5) can be obtained from the eigenvalues of the integral of motion R_0 and are given by $E = V \sum_\alpha x_\alpha$.

As noted earlier, the new exactly solvable atom-molecule Hamiltonians that can be derived in this way are well suited to treat problems involving a mixture of bosonic or fermionic atoms and molecules. We report here a numerical application to a system of bosonic atoms. More specifically, we considered a mixture of bosonic atoms confined to a 3D isotropic trap and a Feshbach resonance. This resonance is a molecular dimer or a composite boson and is represented by the b-boson operators. We have modeled the system by the Hamiltonian (5). The quantity ω is the molecular binding energy, which can be controlled with an external magnetic field. Also, V is the atom-molecule coupling strength, $\varepsilon j = j$ ($j = 0, 1, \ldots$) are the single-atom energies in a 3D isotropic trap, and $\Omega_j = (j+1)(j+2)/2$ are the associated level degeneracies.

The calculations were carried out for a system with $M = 500$ atom pairs plus molecules as a function of ω and the coupling strength V, both in units of the oscillator

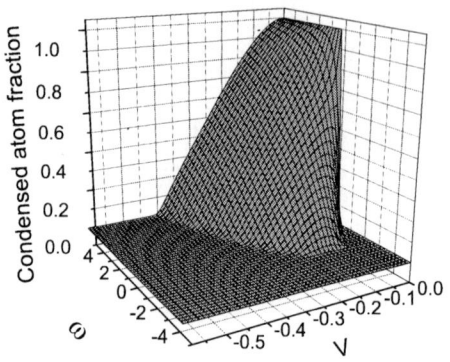

FIGURE 1. Condensed atom fraction as a function of ω and V for 1000 atoms in a $3D$ isotropic trap.

frequency. We only consider negative coupling strengths because the exact solutions do not depend on the sign of V, and include a total of $L = 51$ harmonic oscillator shells.

In Fig. 1, we show the occupation probability of the atom condensate (the atom fraction in the lowest oscillator state) as a function of V and ω. The occupation probabilities can be straightforwardly extracted from the exact solutions by making use of the Hellman-Feynman theorem. As can be seen from the figure, there is a line of second-order quantum phase transitions that take place at the points where the atomic fraction goes to zero. There have been speculations about the nature of this phase transition, which the exact solution seems well suited to investigate.

EXTENSION OF THE RG MODEL TO ANY SIMPLE SINGULAR LIE ALGEBRA.

Though most work on the RG models has concentrated on the rank-1 $su(2)$ and $su(1,1)$ algebras, there have been some efforts to extend these ESMs to larger groups. Ushveridze [6] showed that the Gaudin model could be extended to any simple singular Lie algebra $\mathscr{L}$ using purely algebraic methods. Similar results were recently reported in ref. [7] by establishing a relation between the Gaudin models and Chern-Simons theory. These were purely academic works, however, as their main interest was on the implications these models have for the theory of quantum integrability. Our interest is quite different. We would like to extend these models to include those ingredients that are necessary, *e.g.* one-body symmetry breaking terms to incorporate non-degenerate single-particle energies, to make them of practical use in applications to quantum many-body systems [5].

Because of lack of space, we present here the main results only. RG models of simple singular Lie algebras of rank r have r sets of spectral parameters $e_{a,\alpha}$ $(a = 1, \cdots, r)$ that are solutions of the generalized Richardson equations

4

TABLE 1. Semi-simple Lie algebras up to rank 4 and dynamical symmetry nuclear models.

rank n	$su(n+1)$	$so(2n+1)$	$sp(2n)$	$so(2n)$
1	$su(2)$ pairing	$so(3) \sim su(2)$	$sp(2) \sim su(2)$	$so(2) \sim u(1)$
2	$su(3)$ Elliott	$so(5)$ T=1 pairing	$sp(4) \sim so(5)$	$so(4) \sim su(2) \oplus su(2)$
3	$su(4)$ Wigner	$so(7) \subset so(8)$ FDSM *	$sp(6)$ FDSM *	$so(6) \sim su(4)$
4	$su(5)$	$so(9)$	$sp(8)$	$so(8)$ T=0,1 pairing, FDSM *

* Fermion Dynamical Symmetry Model

$$\sum_{b=1}^{r} \sum_{\beta=1}^{M^b} {}' \frac{\left(\pi^b, \pi^a\right)}{e_{b.\beta} - e_{a.\alpha}} - \sum_{i=1}^{L} \frac{\left(\Lambda_i, \pi^a\right)}{2\varepsilon_i - e_{a.\alpha}} = (\xi, \pi^a) \ , \tag{8}$$

where L is the number of copies of the algebra $\mathscr{L}$, and the symbols in parenthesis are related to the Cartan decomposition of $\mathscr{L}$ (see [7] for details). The equations (8) reduce to the well known Richardson equations for $r = 1$.

Using the Cartan classification of the simple Lie algebras, we display in Table 1 different algebras up to rank 4 that can be treated within this formalism and the corresponding fermionic nuclear models that can now be generalized by these new ESMs to include one-body symmetry-breaking terms.

As an example we will give here the set of equations that emerge for the $so(5)$ model of T=1 pairing in non-degenerate shells. As a reminder, Richardson proposed an exact solution for this problem back in the sixties [8], but it was recently [9] shown that his solution was wrong for more than two pairs due to a failure in the treatment of permutational symmetry. Being that $so(5)$ is a rank 2 algebra, the two sets of Richardson equations that emerge from (8) are

$$\frac{1}{g} = \sum_{i=1}^{L} \frac{\Omega_i}{2\varepsilon_i - e_\alpha} + \sum_{\beta \neq \alpha}^{M} \frac{2}{e_\alpha - e_\beta} + \sum_{\gamma=1}^{M-T} \frac{1}{w_\gamma - e_\alpha} \tag{9}$$

$$0 = \sum_{\alpha=1}^{M} \frac{1}{e_\alpha - w_\gamma} + \sum_{\delta \neq \gamma}^{M-T} \frac{1}{w_\gamma - w_\delta} \ , \tag{10}$$

where L is number of single-particle states with energies ε_i, M is the number of pairs, T is the isospin of the state, e_α are the pair energies associated with the $so(5)$ pairing algebra and w_γ are the parameters associated with the $su(2)$ subalgebra of isospin. The eigenvalues of the p-n pairing Hamiltonian are $E = \sum_\alpha e_\alpha$. We have checked numerically that these equations, which were also found recently by Links *et al.* [10], indeed yield the exact solution of the $so(5)$ problem.

CLOSING REMARKS

We have described two recent extensions of the RG models, which should significantly expand the range of applicability of these models to quite diverse quantum many-body systems. We first discussed the addition of a new bosonic degree of freedom to the rational family of RG models, and showed that this gives rise to a new family of exactly-solvable atom-molecule Hamiltonians of relevance to the physics of degenerate gases coupled to a Feshbach resonance. We then described a numerical application of one such model to a problem involving bosonic atoms coupled to a molecular dimer. By varying the energy of the molecular dimer ω and the atom-molecule interaction V, we can access several different quantum regimes, of relevance to the production of ultracold molecular BECs from bosonic atoms. The study of fermionic systems could be quite important in the description of ultracold fermionic gases, in which by dialing the energy of the Feshbach resonance we can control the interactions between the atoms and study, by means of the exact solution, the transition from a molecular BEC to a BCS Cooper pair condensate.

We then discussed the generalization of the RG models to any simple Lie algebra, pointing out some of the dynamical symmetry models in nuclear structure that can be extended in this way to include one-body symmetry breaking. Similar applications could also be envisioned for condensed matter systems with exotic pairing correlations. Finally, we presented the exact solution of the $so(5)$ rank-2 Lie algebra representing T=1 proton-neutron pairing in a set of non-degenerate single-particle orbits.

We believe that these new extensions of the RG models have enormous potential for providing unique insight into the properties of quantum many-body systems of interest in atomic and molecular physics, nuclear physics, condensed matter and quantum optics.

ACKNOWLEDGMENTS

This work was supported in part by the Spanish DGI under grant #s BFM2003-05316-C02-02, by the US National Science Foundation under grant # PHY-0140036, and by the NATO scientific program. Valuable contributions by P. Van Isacker, G.G. Dussel, C. Esebbag and G. Ortiz to the work reported herein are gratefully acknowledged.

REFERENCES

1. J. Dukelsky, C. Esebbag, and P. Schuck, *Phys. Rev. Lett.* **87**, 066403 (2001).
2. R.W. Richardson, *Phys. Lett.* **3**, 277 (1963).
3. M. Gaudin, J. Phys. (Paris) **37**, 1087 (1976).
4. J. Dukelsky, C. Esebbag, G.G. Dussel, and S. Pittel, submitted to *Phys. Rev. Lett.*
5. J. Dukelsky, V.G. Gueorguiev, and P. Van Isacker, in preparation.
6. A.G. Ushveridze, *Quasi Exactly Solvable Models in Quatum Mechanics* (IOP Publishing Ltd. 1994).
7. M. Asorey, F. Falceto, and G. Sierra, *Nucl. Phys.* B **622**, 593 (2002).
8. R.W. Richardson, *Phys. Rev.* **144**, 874 (1996).
9. F. Pan, and J. Draayer, *Phys. Rev. C* **66**, 044314 (2002).
10. J. Links, H.Q. Zhou, M.D. Gould, and R.H. McKenzie, *J. Phys.* A **35**, 6459 (2002).

Description of weakly bound or unbound nuclear states

A.T. Kruppa[*], N. Michel[†] and W. Nazarewicz[†**]

[*]Institute of Nuclear Research, Bem tér 18/c, 4026 Debrecen, Hungary
[†]Department of Physics and Astronomy, University of Tennessee, Knoxville, Tennessee 37996
Physics Division, Oak Ridge National Laboratory, P.O. Box 2008, Oak Ridge, Tennessee 37831
[**]Institute of Theoretical Physics, Warsaw University, ul. Hoża 69, PL-00681 Warsaw, Poland

Abstract. A major theoretical challenge when dealing with weakly bound nuclei is to obtain a consistent microscopic description of bound states, resonances, and the non-resonant continuum. In this talk, resonances in deformed nuclei are described within the coupled-channel approach employing the Gamow state formalism. The coupled-channel method is compared with the expansion schemes employing the harmonic oscillator basis and the Berggren ensemble.

INTRODUCTION

The goal of nuclear structure theory is to build a unified, comprehensive microscopic framework in which bulk nuclear properties, nuclear excitations, and nuclear reactions can all be described. Exotic radioactive nuclei are the critical new focus in this quest. The extreme isospin of these nuclei and their weak binding bring new phenomena to the fore and isolate and amplify important features of the nuclear many-body problem. Recent theoretical and experimental achievements are focusing new attention on a host of unsolved issues in nuclear structure and offer unprecedented opportunities for the next decade and beyond [1].

In this talk, theoretical models applied to the description of single-particle resonances in heavy deformed nuclei are discussed. All these models describe the radioactive parent nucleus in terms of a single nucleon interacting with a core (i.e., the daughter nucleus). Usually, the core is represented by a phenomenological collective model. Depending on the structure of the daughter nucleus, rotational or vibrational couplings are assumed. Generally speaking, theoretical approaches can be divided into two groups (see Ref. [2] and Refs. quoted therein). The models belonging to the first group, referred to as weak-coupling models or coupled-channel models, employ the coupled-channel formalism of reaction theory, which has been developed in the context of elastic or inelastic scattering. Models belonging to the second group, referred to as resonance Nilsson-orbit (or adiabatic) models, employ the framework of the deformed shell model. In general, the relation between adiabatic and non-adiabatic descriptions is not simple. For example, the resonance Nilsson-orbit model with a triaxial potential [3] (i.e., nonzero γ deformation) cannot be trivially related to a weak-coupling model extended to triaxial degrees of freedom [4].

CP726, Nuclear Physics, Large and Small: International Conference on
Microscopic Studies of Collective Phenomena, edited by R. Bijker, R. F. Casten, and A. Frank
© 2004 American Institute of Physics 0-7354-0207-8/04/$22.00

THE GAMOW STATES

An elegant way to describe particle decay is by assuming outgoing boundary conditions. This immediately leads to the notion of the Gamow or resonant states (the generalized eigenstates of the time-independent Schrödinger equation) which are regular at the origin and satisfy purely outgoing boundary conditions. Together with non-resonant scattering states, Gamow states form a complete set, the so-called Berggren ensemble [5], which can be used in a variety of applications [6], including the recently developed Gamow shell model [7, 8, 9, 10]. The Berggren ensemble contains resonant states and the complex non-resonant continuum.

The radial wave function of a Gamow state asymptotically behaves as an outgoing Coulomb wave:

$$u_c(r) \xrightarrow{\text{large } r} O_l(\eta, rk) = G_l(\eta, rk) + iF_l(\eta, rk), \tag{1}$$

where $k^2 = \frac{2m}{\hbar^2}\mathscr{E}$ and $\eta k = \frac{m}{\hbar^2}Ze^2$. Such boundary conditions are only satisfied for a discrete set of complex wave numbers k_c which define the generalized eigenvalues of the single-particle Schrödinger equation. The corresponding solutions are either bound states with negative real energies $\mathscr{E} < 0$ and pure imaginary wave numbers $k = i\gamma$ ($\gamma > 0$), or resonant states, $\mathscr{E} = E_{\text{res}} - i\frac{\Gamma_{\text{res}}}{2}$, with nonzero imaginary parts $\Gamma_{\text{res}} \neq 0$. The asymptotic behavior of the radial wave functions are determined by the wave number k. For Gamow states, these functions show oscillating behavior at large values of r so one must define a new normalization scheme. This is usually accomplished by means of a regularization procedure, e.g., that proposed by Berggren [5].

WEAK COUPLING DESCRIPTION

In the weak coupling approach, the particle-unstable parent nucleus is described here in terms of a single nucleon coupled to a collective core. The model Hamiltonian can be written as

$$H_{\text{tot}} = H_d - \frac{\hbar^2}{2m}\triangle_r + V_{\text{def}}(\mathbf{r}, \omega), \tag{2}$$

where H_d is the (collective) Hamiltonian of the daughter nucleus, the second term represents the relative proton-daughter kinetic energy, and V_{def} is the proton-core interaction, which depends on the position of the proton $\mathbf{r}$ and the orientation ω of the core.

It is straightforward to define V_{def} in the body-fixed frame. In our work, the intrinsic deformed field is given by a sum of the Woods-Saxon (WS), spin-orbit, and Coulomb potentials. By expanding the nuclear radius in multipoles (in our work we assume quadrupole deformations only) and performing multipole decomposition of the potential, one obtains the daughter-nucleon interaction in the laboratory system [11, 2]:

$$V_{\text{def}}(\mathbf{r}, \omega) = V_{\text{def}}^{(1)}(\mathbf{r}, \omega) + V_{\text{def}}^{(2)}(\mathbf{r}, \omega)$$

$$= \Sigma_{\lambda\mu} V_\lambda^{(1)}(r) D_{\mu 0}^\lambda Y_{\lambda,\mu}(\hat{r}) + \Sigma_{\lambda\mu} V_\lambda^{(2)}(r) \left(D_{\mu 2}^\lambda + D_{\mu -2}^\lambda \right) Y_{\lambda,\mu}(\hat{r}). \tag{3}$$

The Coupled Channel Equations

The states of the daughter nucleus are eigenvectors of H_d. In this work, the wave functions of the core, $\phi_{I\mu K}$, are approximated by those of the rigid rotor while the wave function of the parent nucleus can be written in the weak-coupling form

$$\Psi^{JM} = r^{-1} \sum_{IKlj} u^J_{IKlj}(r)\Phi_{JMIKlj}, \tag{4}$$

where $\Phi_{JMIKlj} = \sum_{\Omega\mu} \langle j\Omega I\mu | JM \rangle \mathscr{Y}_{lj\Omega}\phi_{I\mu K}$ is the channel function and $\mathscr{Y}_{lj\Omega}$ arises from the coupling of the proton spin with the orbital angular momentum. Due to the non-axial symmetric form of the nucleon-daughter interaction (3), the ground state $K = 0$ and the γ-vibrational $K=2$ band both contribute.

The radial functions $u^J_{IKlj}(r)$ are solutions of the set of coupled-channel equations:

$$\frac{\hbar^2}{2m}\left(-\frac{d^2}{dr^2} + \frac{l(l+1)}{r^2}\right) u^J_{IKlj} + \sum_{\lambda I'l'j'} A_\lambda(Ilj, I'l'j', J) B_\lambda(II'K) V^{(1)}_\lambda u^J_{I'Kl'j'} + \tag{5}$$

$$\sum_{\lambda I'K'l'j'} A_\lambda(Ilj, I'l'j', J) C_\lambda(IKI'K', a_2) V^{(2)}_\lambda u^J_{I'K'l'j'} = (E - E_{IK})u^J_{IKlj},$$

where E_{IK} is the energy of the daughter state described by the wave function $\phi_{I\mu K}$. The explicit expressions for the geometric coefficients A_λ and the reduced matrix elements B_λ and C_λ are given, e.g., in Ref. [11].

The Width of a Resonance

Once the resonance energy and radial wave function have been determined, there are different methods to calculate the width of the state. The simplest method is to take twice the imaginary part of the energy of the resonance as obtained from a direct solution of the Schrödinger equation. However, for narrow resonances, the accurate numerical calculation of Γ_{res} is difficult. Therefore, other methods are often used. One possibility is to calculate the partial width for each channel from the so-called current expression [12]

$$\Gamma_c(r) = i\frac{\hbar^2}{2\mu}\frac{u'^*_c(r)u_c(r) - u'_c(r)u^*_c(r)}{\sum_{c'}\int_0^r |u_{c'}(r')|^2 dr'}, \tag{6}$$

where the sum of the partial widths

$$\Gamma_{res} = \sum_c \Gamma_c(r) \tag{7}$$

gives the total decay width. Although values of $\Gamma_c(r)$ depend on r in the region where the coupling potential terms are not negligible, the total width (7) is independent of r, which reflects flux conservation.

If one neglects the imaginary part of k (which is a good approximation for very narrow resonances), the expression for the partial decay width can be written in a simple form:

$$\Gamma_c(r_{as}) \approx \frac{\hbar^2 k}{\mu}\frac{|u_c(r_{as})|^2}{|O_l(\eta, kr_{as})|^2 \sum_{c'}\int_0^{r_{as}} |u_{c'}(r')|^2 dr'}, \tag{8}$$

9

where r_{as} is the channel radius (the off-diagonal couplings are negligible for $r > r_{as}$).

The RMHO Technique

Unfortunately, the number of coupled equations rapidly increases with the number of excited states of the daughter nucleus taken into account. In addition, the solution of the eigenvalue problem of a very large set of coupled equations becomes numerically unstable at some point. A possible way out is to consider the R-matrix theory. However, even in this case, one has to deal with large sets of coupled differential equations. Recently, in Ref. [2] we proposed a simple method, based on the R-matrix formalism, to estimate the parameters of a resonance. The method is based on the expansion of the radial functions $u_{IKlj}(r)$ in the single-particle basis $\phi_{nl}^{HO}(r)$ of the spherical harmonic oscillator. In this basis, the total wave function (4) can be written in the form:

$$\Psi^{JM} = r^{-1} \sum_{IKlj} \sum_n C_{IKnlj}^J \phi_{nl}^{HO}(r) \Phi_{JMIKlj}. \tag{9}$$

The coefficients C_{IKnlj}^J can be obtained from the matrix eigenvalue equation:

$$\begin{aligned}
&\sum_{n'} \langle \phi_{nl}^{HO} | \tfrac{\hbar^2}{2m} \left(-\tfrac{d^2}{dr^2} + \tfrac{l(l+1)}{r^2} \right) | \phi_{n'l}^{HO} \rangle C_{IKn'lj}^J - (E_\lambda^{HO} - E_{IK}) C_{IKnlj}^J \\
&+ \sum_{\lambda I'n'l'j'} A_\lambda(Ilj, I'l'j', J) B_\lambda(II'K) \langle \phi_{nl}^{HO} | V_\lambda^{(1)} | \phi_{n'l'}^{HO} \rangle C_{I'Kn'l'j'}^J \\
&+ \sum_{\lambda I'K'n'l'j'} A_\lambda(Ilj, I'l'j', J) C_\lambda(IKI'K', a_2) \langle \phi_{nl}^{HO} | V_\lambda^{(2)} | \phi_{n'l'}^{HO} \rangle C_{I'K'n'l'j'}^J = 0.
\end{aligned} \tag{10}$$

The resulting radial functions define the *boundary condition function* at point r and the R-matrix formalism can be applied. This algorithm is called the *R-matrix method based on harmonic oscillator expansion* (RMHO). In RMHO, the energy and width of the resonance explicitly depend on r. However, for sufficiently large values of r, this dependence is extremely weak. As demonstrated in Ref. [2], RMHO is an easy and convenient method when one needs to significantly increase the number of states in the daughter nucleus to guarantee the convergence of the solution. In particular, RMHO gives an excellent reproduction of results obtained in the coupled-channel formalism.

GAMOW EXPANSION METHOD FOR THE DEFORMED POTENTIAL

Encouraged by the success of the RMHO expansion method, we applied the basis expansion method involving the s.p. Berggren ensemble to the Schrödinger equation with an axially deformed single-particle potential. In our calculations, we used the WS+Coulomb potential of Ref. [2] optimized for the proton emitter ^{141}Ho. The quadrupole deformation $\beta_2 = 0.244$ was assumed. The Berggren basis has been generated by a *spherical* WS+Coulomb potential. In the axially deformed case, the angular momentum projection Ω and parity π are good quantum numbers, and the Gamow completeness relation reads:

$$\sum_{l,j}\left[\sum_n |\phi_n^{l,j,\Omega}\rangle\langle\phi_n^{l,j,\Omega}| + \int_{L_+^{l,j,\Omega}} |\phi^{l,j,\Omega}(k)\rangle\langle\phi^{l,j,\Omega}(k)|\,dk\right] = 1, \qquad (11)$$

where the sum over n runs over all bound and resonant states lying between the real axis and the complex path $L_+^{l,j,\Omega}$, and $\phi^{l,j,\Omega}(k)$ are scattering states along this path [8]. In practice, the complex contour $L_+^{l,j,\Omega}$ has been discretized with $N_{scat}=60$ points. As far as the choice of $L_+^{l,j,\Omega}$ is concerned, we took the real path $(0, 2.0$ fm$^{-1})$ if there was no resonant state in a given channel. If there was a resonant state at $k=k_{res}$, a triangular contour was chosen corresponding to the two straight segments in the complex k-plane, joining the points: $k_0=0.0-i0.0$, k_1, $k_2=2.0-i0.0$ (all in fm^{-1}), where $\Re(k_1)=\Re(k_{res})$ and $\Im(k_1)=\max(5\cdot\Im(k_{res}),-0.05)$. Finally, we took $l_{max}=11$ in Eq. (11). For narrow resonances ($\Gamma<50$ keV) the imaginary part of the energy is a rather poor approximation to the width because of discretization effects and the fact that $\Gamma/\mathcal{E}$ is of the order of numerical precision. Consequently, for such states, we used the approximate current formula (8) at $r_{as}=13$ fm.

The results of calculations are displayed in Table 1 for $\Omega^\pi=\frac{1}{2}^+$ and $\frac{7}{2}^-$ deformed resonances in ^{141}Ho. The Gamow diagonalization technique has been compared with the expansion method in the spherical harmonic oscillator basis and with coupled-channel calculations. While for very narrow widths, all methods give similar results; for broader

TABLE 1. $\Omega^\pi=\frac{1}{2}^+$ (top) and $\frac{7}{2}^-$ resonances in the deformed WS proton potential for ^{141}Ho calculated in the coupled-channel method (CC), harmonic oscillator expansion method (HO diag.), and the Gamow expansion method (Gamow diag.). The first number is the energy $\mathcal{E}$ in MeV while the second number in parentheses is the width Γ (in keV).

Ω^π	CC	HO diag.	Gamow diag.
$\frac{1}{2}^+$	0.756 ($1.98\cdot10^{-20}$)	0.756 ($2.018\cdot10^{-20}$)	0.758 ($2.191\cdot10^{-20}$)
	3.968 (0.053)	3.968 (0.050)	3.970 (0.051)
	5.454 (0.035)	5.454 (0.037)	5.465 (0.032)
	10.214 (833)	-	10.534 (237)
	11.686 (729)	-	11.692 (651)
	21.777 (527)	-	21.809 (544)
$\frac{7}{2}^-$	1.190 ($3.24\cdot10^{-16}$)	1.190 ($3.29\cdot10^{-16}$)	1.194 ($2.66\cdot10^{-16}$)
	8.789 (17.51)	-	8.790 (17.53)
	9.933 (178)	-	9.934 (178)
	15.360 (104)	-	15.375 (104)

resonances, the harmonic oscillator cannot be used since the current expression (8) is not stable as a function of r_{as}. Here the Gamow diagonalization technique gives satisfactory results. The discrepancy for the width of the $\frac{1}{2}^+$ resonance at ~10.5 MeV is probably related to the fact that there are two broad states which overlap in the same energy region.

SUMMARY

Tying nuclear structure directly to nuclear reactions within a coherent framework is an important goal. On the nuclear structure side, the continuum shell model and modern mean-field theories allow for the consistent treatment of open channels, thus allowing the description of bound and unbound nuclear states and direct reactions on the same footing. On the reaction side, there has been an emphasis on better treatment of nuclear structure aspects. The battleground in this task is the territory of weakly bound nuclei where the structure and reaction aspects are intimately interwoven.

In this short presentation, the main emphasis was on the description of decaying states in deformed nuclei where coupling to collective excitations is important. While for very narrow resonances (e.g., on the proton-rich side of the nuclear landscape), the techniques based on harmonic oscillator expansions seem to work very well indeed; the methods based on resonant expansions are tools of choice for neutron-rich nuclei.

ACKNOWLEDGMENTS

This research was supported by the U.S. Department of Energy under Contract Nos. DE-FG02-96ER40963 (University of Tennessee), DE-AC05-00OR22725 with UT-Battelle, LLC (Oak Ridge National Laboratory), and DE-FG05-87ER40361 (Joint Institute for Heavy Ion Research), by the University Radioactive Ion Beam Consortium (UNIRIB), and by Hungarian OTKA Grant Nos. T37991 and T046791.

REFERENCES

1. J. Carlson, B. Holstein, X.D. Ji, G. McLaughlin, B. Müller, W. Nazarewicz, K. Rajagopal, W. Roberts, and X.-N. Wang, *A Vision for Nuclear Theory: Report to NSAC*, 2003, arXiv:nucl-th/0311056.
2. A.T. Kruppa and W. Nazarewicz, arXiv:nucl-th/040206; Phys. Rev C, in press.
3. C.N. Davids and H. Esbensen, Phys. Rev. C **69**, 034314 (2004).
4. A.T. Kruppa and W. Nazarewicz, *Proton-Emitting Nuclei*, Second International Symposium PRO-CON 2003, Eds. E. Maglione and F. Soramel, AIP Conference Proceedings, Volume **681**, p. 61, Melville, New York, 2003.
5. T. Berggren, Nucl. Phys. **A109**, 265 (1968).
6. T. Vertse, P. Curutchet, and R.J. Liotta, Lecture Notes in Physics **325** (Springer Verlag, Berlin 1987), p. 179.
7. N. Michel, W. Nazarewicz, M. Płoszajczak, and K. Bennaceur, Phys. Rev. Lett. **89**, 042502 (2002).
8. N. Michel, W. Nazarewicz, M. Płoszajczak, and J. Okołowicz, Phys. Rev. C **67**, 054311 (2003).
9. N. Michel, W. Nazarewicz, M. Płoszajczak, and J. Rotureau, Proc. XXVII Symposium on Nuclear Physics, Taxco, Guerrero, Mexico; arXiv:nucl-th/0401036.
10. R. Id. Betan, R.J. Liotta, N. Sandulescu, and T. Vertse, Phys. Rev. Lett. **89**, 042501 (2002).
11. T. Tamura, Rev. Mod. Phys. **37**, 679 (1965); Nucl. Phys. **73**, 241 (1965).
12. J. Humblet and L. Rosenfeld, Nucl. Phys. **26**, 529 (1961).

Time-Reversal Violation in Nuclei and Atomic Electric Dipole Moments

Jonathan Engel

Department of Physics and Astronomy, CB3255, University of North Caorolina, Chapel Hill, NC 27599-3255, USA

Abstract. The standard model of particle physics violates time-reversal invariance, but apparently not strongly enough to account for the excess of matter over antimatter in the universe. One of the best ways to search for other sources of time-reversal violation is to measure electric dipole moments of atoms. I argue that the best atoms to use are those with octupole-deformed nuclei, and describe a self-consistent calculation of the collective enhancement of time-reversal violation in one such nucleus, ^{225}Ra, an experiment on which is in preparation.

INTRODUCTION

Time-reversal invariance (T) is violated at a low level. Experiments with kaons and B-mesons conclusively demonstrate that. The results of these experiments can be explained by a small phase in the Cabibo-Kobayashi-Maskawa (CKM) matrix of the Standard Model. But the absence of antimatter in our universe is evidence that T invariance (or more precisely, CP invariance) was badly violated long ago. The CKM phase is unable to account for so large an effect, and so theorists believe there must be another source of T violation, this one from outside the Standard Model.

As we shall see, an atom in its ground state cannot have an electric dipole moment (EDM) without violating T. A number of experiments have searched for atomic EDMs, and the limits are very tight. But because CKM T violation shows up in first order only in flavor-changing processes, it should appear in atomic experiments only after the limits are improved by 5 or 6 orders of magnitude. The same constraint does not apply, however, to T violation in extensions to the Standard-Model. The most popular extension, supersymmetry, has many flavor-conserving phases, making EDM experiments ideal for testing it. Already these experiments are putting serious pressure on theory.

Here, after some preliminaries, I argue that experiments on atoms with octupole-deformed nuclei will be more sensitive to T-violation within the nucleus than the current best experiments. The enhancement of T violation in these nuclei is connected with the collective violation of intrinsic parity. This paper discusses work published by me together with Jim Friar and Anna Hayes [1] and Michael Bender, Jacek Dobacewski, Joao de Jesus, and Piotr Olbratowski [2].

CP726, Nuclear Physics, Large and Small: International Conference on
Microscopic Studies of Collective Phenomena, edited by R. Bijker, R. F. Casten, and A. Frank
© 2004 American Institute of Physics 0-7354-0207-8/04/$22.00

T VIOLATION AND EDMS

Why do EDMs require T violation?

It's obvious that for the negative-parity (P) dipole operator to have a non-zero expectation value in a non-degenerate state, P must be violated. But because states with good J, M are not eigenstates of the T operator, the usual argument from "good quantum numbers" doesn't work for T. Why must it be violated as well?

Consider a state $|g;J,M\rangle$ (g stands for "ground") with no degeneracy besides the $2J + 1$ spin multiplicity. Symmetry under rotations by π around the y axis implies that for a vector operator like $\vec{d} \equiv \Sigma_i e_i \vec{r}_i$,

$$\langle g : J,M|\vec{d}|g : J,M\rangle = -\langle g : J,-M|\vec{d}|g : J,-M\rangle . \tag{1}$$

The time reversal operator $\hat{T}$ takes $|g : J,M\rangle$ to a real phase times $|g : J,-M\rangle$ (under the usual phase conventions), just like rotations by π. But $\vec{d}$ doesn't change under $\hat{T}$, while the rotation flips its sign. So if the system is invariant under T, we also have

$$\langle g : J,M|\vec{d}|g : J,M\rangle = +\langle g : J,-M|\vec{d}|g : J,-M\rangle . \tag{2}$$

These two equations together imply that $\langle \vec{d} \rangle$ must vanish. If T is *violated*, the argument fails because $\hat{T}$ will take $|g : J,-M\rangle$ to a state with $J,-M$ that is not identical to the corresponding member of the ground-state multiplet. In that case, Eq. 2 doesn't hold.

How objects get EDMs and why atomic EDMs are suppressed

T-violation can work its way up from the most fundamental particles through to atoms. If the symmetry is violated by, e.g. supersymmetry, quarks will develop T-violating couplings, leading to effective T-violating πNN couplings. These in turn lead through pion exchange to an effective two-nucleon interaction of the form

$$\begin{aligned}
\hat{H}_T &= -\frac{g m_\pi^2}{8\pi m_N} \Bigg\{ (\vec{\sigma}_1 - \vec{\sigma}_2) \cdot (\vec{r}_1 - \vec{r}_2) \Bigg[\bar{g}_0 \, \tau_1 \cdot \tau_2 - \frac{\bar{g}_1}{2} (\tau_{1z} + \tau_{2z}) + \bar{g}_2 (3\tau_{1z}\tau_{2z} - \tau_1 \cdot \tau_2) \Bigg] \\
&\quad - \frac{\bar{g}_1}{2} (\vec{\sigma}_1 + \vec{\sigma}_2) \cdot (\vec{r}_1 - \vec{r}_2)(\tau_{1z} - \tau_{2z}) \Bigg\} \frac{\exp(-m_\pi|\vec{r}_1 - \vec{r}_2|)}{m_\pi|\vec{r}_1 - \vec{r}_2|^2} \Bigg[1 + \frac{1}{m_\pi|\vec{r}_1 - \vec{r}_2|} \Bigg],
\end{aligned} \tag{3}$$

where g is the normal strong πNN coupling constant and the three $\bar{g}$'s are dimensionless isoscalar, isovector, and isotensor T-violating πNN couplings. This interaction can cause the nucleus to develop an EDM, which in turn causes an atomic EDM. The goal of the atomic experiments (and this work) is to extract limits on the $\bar{g}$'s from experimental limits on an atomic EDM, or to determine them if an EDM is observed.

Unfortunately, atomic EDMs are suppressed. Any nuclear EDM induced by the interaction in Eq. 3 is shielded by the electrons, which rearrange themselves to create an electronic EDM in the opposite direction. Schiff proved [3] that the cancellation is exact in the limit of a point-like nucleus and nonrelativistic electrons.

Luckily, electrons are relativistic and the nucleus has a finite radius, so the shielding is not complete. It turns out, however, that after its effects are accounted for, the nuclear quantity that induces an EDM in the electrons is not the dipole moment $\vec{D}$, but rather a kind of weighted dipole moment (with a correction term) called the "Schiff moment":

$$\vec{S} = \frac{1}{10} \sum_p e_p \left[r_p^2 - \frac{5}{3} R^2 \right] \vec{r}_p \, , \tag{4}$$

where R^2 is the root-mean-squared nuclear charge-density.radius. If, as one would expect, $\langle \vec{S} \rangle \approx 0.1 R^2 \langle \vec{D} \rangle$, then the atomic EDM $\vec{d}$ is down from $\langle \vec{D} \rangle$ by $O(R^2 / 10 R_A^2) \approx 10^{-9}$. ($R_A$ is the atomic radius.) But the behavior of relativistic Coulomb wave functions near the origin partly offsets this terrible suppression via a factor $10 Z^2 \approx 10^5$ in heavy nuclei, so that the overall suppression of the atomic EDM from shielding is only about 10^{-4}. This number begins to approach the factor by which EDM limits on the neutron are worse than atomic limits, and the nuclear Schiff moment is more sensitive to neutral pion exchange than the neutron EDM, so atomic experiments are currently competitive in the search for some kinds of T violation.

ENHANCEMENT BY OCTUPOLE DEFORMATION

We can make atomic experiments even more attractive by finding the right atom, because some atoms are better places to look for an EDM than others. One reason is that octupole deformation of atomic nuclei enhances the nuclear Schiff moment dramatically.

Since the T-violating interaction $\hat{H}_T$ is very weak, it can be treated peturbatively, and the Schiff moment can be written as

$$\langle \vec{S} \rangle = \sum_m \frac{\langle 0 | \vec{S} | m \rangle \langle m | H_T | 0 \rangle}{E_0 - E_m} + c.c. \tag{5}$$

Two collective effects associated with octupole deformation make this expression large. The first is the existence of parity doubling. The intrinsic state has a shape that breaks parity symmetry. It contains both positive- and negative-parity components, and when projected onto good parity yields two states of opposite parity very close to one another in energy. In ^{225}Ra, for example, the $J^\pi = 1/2^+$ ground state ($|0\rangle$ in our notation) has a $J^\pi = 1/2^-$ partner $|\bar{0}\rangle$ just 55 keV higher. Since V_T is pseudoscalar, it connects $|0\rangle$ and $|\bar{0}\rangle$, and the single term with $|\bar{0}\rangle$ as the intermediate state dominates the sum in Eq. 5. Just like for quadrupole transitions, the transition matrix element $\langle 0 | \vec{S} | \bar{0} \rangle$ is proportional to the intrinsic Schiff moment, so that Eq. 5 becomes, to good approximation,

$$\langle \vec{S} \rangle \approx -2/3 \frac{\langle \vec{S}^{\text{intr.}} \rangle \langle H_T \rangle}{E_0 - E_{\bar{0}}} \, . \tag{6}$$

The second collective enhancement comes from robust intrinsic Schiff moments that often are much larger than R^2 times the intrinsic dipole moment. Although the intrinsic dipole moments in octupole-deformed nuclei are collective, they are often quite small.

The reason is that they depend on the distribution of charge *with respect to the center of mass*, and vanish in the limit that the neutron and proton densities coincide. Intrinsic Schiff moments are not subject to this kind of cancellation. As a result of this and the parity-doubling, the laboratory Schiff moment in a nucleus like ^{225}Ra is enhanced, according to collective model estimates[4, 5] by two or three orders of magnitude over that of ^{199}Hg, the atom the best experimental limit on its EDM.

Why the uncertainty of an order of magnitude, a factor large enough to deter experimentalists? The matrix element of H_T depends on the nuclear spin distribution, a delicate quantity. In simple collective models (such as the particle-rotor model) a single valence nucleon carries all the spin. In reality, however, the valence nucleon polarizes the core, an effect that can alter the spin distribution substantially. Even without core polarization, the matrix element of H_T depends sensitively on the wave function of the valence nucleon. To reduce the uncertainty in the Schiff moment to a reasonable level, we need a state-of-the art calculation.

CALCULATION OF $\langle S \rangle$ IN ^{225}RA

In ref. [2] we used the program HFODD [6] to do a completely self-consistent Skyrme-mean-field calculation of the intrinsic ground state of ^{225}Ra. The code allows the simultaneous breaking of rotational, invariance, parity symmetry, and T. The first two are needed to obtain octupole deformation, the last to polarize the core (i.e. break Kramers degeneracy). HFODD cannot yet treat pairing when it allows T to be broken, but pairing in T-odd channels is poorly understood. No existing codes can do more than HFODD in odd-A octupole-deformed nuclei.

We used the Skyrme interactions SIII, SkM*, SLy4, and SkO′, the last our preferred interaction because it was specially tuned in Ref. [7] to treat spin degrees of freedom (in particular isovector spin excitations). Figure 1 shows the shapes produced by SkO′ in the even Ra isotopes. The nucleus ^{225}Ra, with $N = 137$, will clearly be well deformed in both the quadrupole and octupole coordinates.

Figure 2 shows three parity-violating intrinsic quantities. In the top panel is the ground-state octupole deformation as a function of neutron number. The trend mirrors that in the density profiles shown earlier. The second panel shows the absolute values of intrinsic dipole moments $D_{\text{intr.}}$, along with experimental data extracted from $E1$ transition probabilities [8]. Both the experimental and calculated values change sign between $N = 134$ and $N = 138$, illustrating the delicacy of this quantity. None of the forces precisely reproduces the trend through all the isotopes, but the comparison has to be taken with a grain of salt because "data" derive from transitions between excited rotational. The intrinsic Schiff moment $\langle S_z \rangle$, as noted above, is more collective and under better control, as the bottom panel of the figure shows.

Finally, what about H_T? In the limit that the pion is very heavy, Eq. 3 reduces to an effective one-body potential

$$\hat{U}_T(\vec{r}) = -\frac{g}{2m_\pi^2 m_N} \sum_{i=1}^{A} \vec{\sigma}_i \tau_{z,i} \cdot \left[(\bar{g}_0 + 2\bar{g}_2)\vec{\nabla}\rho_1(\vec{r}) - \bar{g}_1\vec{\nabla}\rho_0(\vec{r}) \right] + \text{exchange} , \quad (7)$$

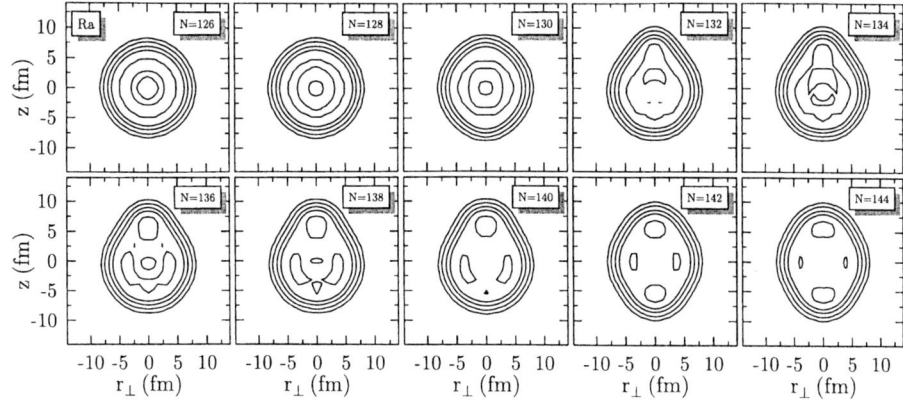

FIGURE 1. Contours of constant density for a series of even-N Radium isotopes. Contour lines are drawn for densities $\rho=0.01, 0.03, 0.07, 0.11$, and 0.15 fm^{-3}.

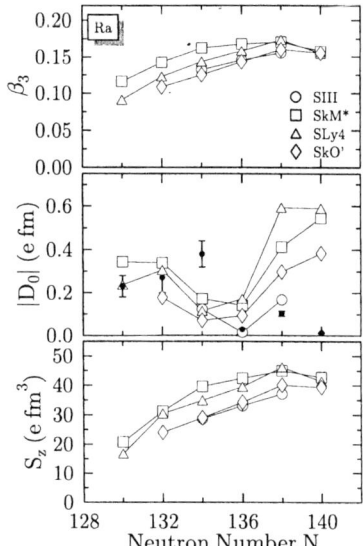

FIGURE 2. The predicted first-order octupole deformations (top), intrinsic dipole moments (middle) and intrinsic Schiff moments (bottom) for four Skyrme interactions in a series of even-N Radium isotopes. The experimental intrinsic dipole moments are also shown.

where ρ_0 and ρ_1 are the isoscalar and isovector densities and the exchange terms are very likely negligible. Table 1 shows matrix elements of the the most important operators in U_T (in the neutron-proton scheme). The zeros following the interaction names mean that the core-polarizing parts of the interactions, which were never fit for the older Skyrme forces, have been turned off. The differences between the lines labeled SkO′(0) and SkO′

TABLE 1. Intrinsic-state expectation values of important operators in U_T, in 10^{-3} fm^{-4}.

	$\langle \vec{\sigma}_n \cdot \vec{\nabla} \rho_n \rangle$	$\langle \vec{\sigma}_n \cdot \vec{\nabla} \rho_p \rangle$
SIII(0)	-0.577	-0.491
SkM*(0)	-0.619	-0.120
SLy4(0)	-0.628	-0.050
SkO'(—)	-0.331	-0.013
SkO'	-0.320	-0.114
particle-rotor [5]	-1.2	-0.8

indicate the effects of core polarization. Our final result for the Schiff moment obtained from the full finite-range force (though not yet with exchange terms) is

$$\langle S_z \rangle_{\mathrm{Ra}} = -1.90\, g\bar{g}_0 + 6.31\, g\bar{g}_1 - 3.80\, g\bar{g}_2\ [e\,\mathrm{fm}^3].\tag{8}$$

A recent calculation for ^{199}Hg [9] gave

$$\langle S_z \rangle_{\mathrm{Hg}} = .0004\, g\bar{g}_0 + .055\, g\bar{g}_1 + .009\, g\bar{g}_2\ [e\,\mathrm{fm}^3].\tag{9}$$

Our Schiff moment, though smaller than particle-rotor estimates, is still over 100 times larger (and significantly more than that if $\bar{g}_1$ is much smaller than the other two couplings) than that of ^{199}Hg. Combined with an additional factor-of-three enhancement from atomic physics [10], this result bodes well for upcoming experiments [11] to measure the EDM of ^{225}Ra.

ACKNOWLEDGMENTS

I wish to thank my collaborators on the two papers summarized here: M. Bender, J. Dobaczewski, J. Friar, A.C. Hayes, J. de Jesus, and P. Olbratowski. This work was supported in part by the U.S. Department of Energy under Grant DE-FG02-97ER41019.

REFERENCES

1. Engel, J., Friar, J. L., and Hayes, A. C., *Phys. Rev. C*, **61**, 035502 (1999).
2. Engel, J., Bender, M., Dobaczewski, J., de Jesus, J., and Olbratowski, P., *Phys. Rev. C*, **68**, 025501 (2003).
3. Schiff, L. I., *Phys. Rev.*, **132**, 2194 (1963).
4. Auerbach, N., Flambaum, V. V., and Spevak, V., *Phys. Rev. Lett.*, **76**, 4316 (1996).
5. Spevak, V., Auerbach, N., and Flambaum, V. V., *Phys. Rev. C*, **56**, 1357 (1997).
6. Dobaczewski, J., and Olbratowski, P., *Comp. Phys. Comm.*, **158**, 158 (2004), and references therein.
7. Bender, M., Dobaczewski, J., Engel, J., and Nazarewicz, W., *Phys. Rev. C*, **65**, 054322 (2002).
8. Butler, P. A., and Nazarewicz, W., *Rev. Mod. Phys.*, **68**, 349 (1996).
9. Dmitriev, V. F., and Sen'kov, R. A., *Phys. Atom. Nucl.*, **66**, 1940 (2003).
10. Dzuba, V. A., Flambaum, V. V., Ginges, J. S. M., and Kozlov, M. G., *Phys. Rev. A.*, **66**, 012111 (2002).
11. Holt, R., http://mocha.phys.washington.edu/~int_talk /WorkShops/int_02_3/People/Holt_R/ (2002)

18

Derivation of the generators of the pseudo SU(3) symmetry group of heavy nuclei in the shell model picture

M. Moshinsky[1]

Instituto de Física
Universidad Nacional Autónoma de México
Apartado Postal 20-364, 01000 México D.F., México
moshi@fisica.unam.mx

Abstract. We derive the explicit form of the unitary operator connecting the normal parity harmonic oscillator eigenstates with the full set of eigenstates of a pseudo oscillator. This unitary operator allow us to analyze the shell model of medium and heavy nuclei with the help of a pseudo SU(3) group.

My recent interest have centered on "Relativistic many body problems" and "Dissipation in quantum mechanics" but this conference is in honor of the 60th birthday of Stuart Pittel whose contributions in nuclear physics, and his collaboration with physicists in Mexico, mainly through the Oaxtepec meetings, is well known.

Due to the above fact I wanted to speak in this conference on subjects of interest to Stuart and one of them is the title of this talk which I will now introduce.

One of the fields of interest to Stuart is the energy level structure of medium and heavy nuclei, particularly as they can be understood with the help of symmetries associated with pseudo SU(3) group.

How does this group appear when we study these nuclei with the help of the shell model? This is an old problem that Stuart has considered and which I myself was interested some years ago in collaboration with Castaños, Quesne, Balentekin, del Sol Mesa and others[1, 2, 3]. What I will present today is not new but I think it is not very well known and besides I will try to discuss it in a somewhat different manner.

To begin with I will consider the shell model in which all the nucleons move in an harmonic oscillator potential to which is added first a strong spin orbit coupling interaction to reproduce the right magic number for closed shells of medium and heavy nuclei and second, for these nuclei, the nucleons with higher orbital angular momentum l feel a deeper potential.

In order to take the effects into account Nilsson proposed[4, 5] a single body H Hamiltonian of the form.

$$(H/\hbar\omega) = H_0 - 2k\mathbf{L} \cdot \mathbf{S} - k\mu L^2 \tag{1}$$

---

[1] Member of El Colegio Nacional and Sistema Nacional de Investigadores

CP726, *Nuclear Physics, Large and Small: International Conference on*
Microscopic Studies of Collective Phenomena, edited by R. Bijker, R. F. Casten, and A. Frank
© 2004 American Institute of Physics 0-7354-0207-8/04/$22.00

where H_0 is the number of quanta operator

$$H_0 = \eta \cdot \xi = \tfrac{1}{2}(p^2 + r^2) - \tfrac{3}{2} \tag{2}$$

with

$$\eta = \frac{1}{\sqrt{2}}(\mathbf{r} + i\mathbf{p}), \ \xi = \frac{1}{\sqrt{2}}(\mathbf{r} - i\mathbf{p}) \tag{3}$$

being the creation and annhilations operators of an oscillator of unit frequency. The $\mathbf{L}$ is the orbital angular momentum

$$\mathbf{L} = \mathbf{r} \times \mathbf{p} = i(\eta \times \xi) \tag{4}$$

and $\mathbf{S}$ the spin given in terms of the Pauli matrix vector σ as

$$\mathbf{S} = \tfrac{1}{2}\sigma \tag{5}$$

If the parameter $k = 0$ we just have the number operator $\eta \cdot \xi$ which obviously commutes with the generators of an U(3) group[5] given by

$$C_{ij} \equiv \eta_i \xi_j, \qquad i,j = 1,2,3 \tag{6}$$

and satisfies the commutation relations

$$[C_{ij}, C_{i'j'}] = C_{ij'}\delta_{ji'} - C_{ji'}\delta_{ij'} \tag{7}$$

For light nuclei the single particle shell model Hamiltonian (1) with $k = 0$ is a good approximation and thus it has an SU(3) symmetry which was used by Elliot[6] in the analysis of spectra in the s-d shell nuclei.

For medium an heavy nuclei taking $k = 0$ does not give even the right magic numbers and describes poorly their energy level spectra. Thus[1] since the work of Hecht, Draayer and others[2] one can speak only of a pseudo SU(3) symmetry for these nuclei as indicated in the next paragraph.

The answer lies in the fact that it is possible to apply a unitary transformation to the Hamiltonian H of Eq. (1) so as to diminish the contribution of the $\mathbf{L} \cdot \mathbf{S}$ and L^2 terms and thus get a new Hamiltonian H' in which only the number operator predominates. By applying the same unitary transformation to C_{ij} of Eq. (6), we get the generators of a group, known in the literature as pseudo SU(3), that becomes a symmetry group of the medium and heavy nuclei in the shell model picture.

We proceed to determine the operator form of this unitary transformation as a function of η, ξ and $\mathbf{S}$.

Let us consider a function of the observables

$$F(\eta, \xi, \mathbf{S}) \tag{8}$$

As we want it to have a quantum mechanical meaning as an operator we may restrict ourselves to polynomial functions of the observables in a prescribed order.

The F as an operator is not, in general, unitary but it may be made so with the help of the Hermitian operator

$$(\eta,\xi,\mathbf{S}) \equiv F^\dagger(\eta,\xi,\mathrm{s})F(\eta,\xi,\mathbf{S}) \tag{9}$$

and writing

$$U = F(\eta,\xi,\mathbf{S})\left[(\eta,\xi,\mathbf{S})\right]^{-1/2} \tag{10}$$

implying

$$U^\dagger = \left[(\eta,\xi,\mathbf{S})\right]^{-1/2} F^\dagger(\eta,\xi,\mathbf{S}) \tag{11}$$

Clearly then U is a unitary operator as

$$U^\dagger U =$$
$$\left[(\eta,\xi,\mathbf{S})\right]^{-1/2}\left[(\eta,\xi,\mathbf{S})\right]\left[(\eta,\xi,\mathbf{S})\right]^{-1/2} = I \tag{12}$$

with I being the unit operator.

For heavy nuclei the single particle energy of the orbitals $j = l + 1/2$ and $j = (l+2) - 1/2$ are close so that the difference of their energies

$$(\Delta\varepsilon/\hbar\omega) = -k(2\mu - 1)(2l + 3) \tag{13}$$

is small compared to $\hbar\omega$. This can be achieved if the value of μ is taken as close to $1/2$ and, in particular, for $\mu = 1/2$ this pair of orbitals is exactly degenerate for all the values of l compatible with N.

>From now on we shall take $F(\eta,\xi,\mathbf{S}) = (\xi \cdot \mathbf{S})$ and if we apply the unitary transformation U to the terms appearing in the Nilsson Hamiltonian we get the following results.

$$
\begin{array}{rcl}
UH_0U^\dagger & = & H_0' \doteq H_0 + I \tag{14}\\
US^2U^\dagger = S'^2 & = & S^2 \tag{15}\\
UL^2U^\dagger = L'^2 & = & L^2 + 2 \tag{16}\\
U[\mathbf{L}\cdot\mathbf{S}]U^\dagger & = & \mathbf{L}'\cdot\mathbf{S}' = \mathbf{L}\cdot\mathbf{S} - \tag{17}
\end{array}
$$

where Q is the operator

$$\equiv (\eta\cdot\xi + 3 + \mathbf{L}\cdot\mathbf{S})^{-1}(L^2 + \eta\cdot\xi + 3) \tag{18}$$

The total angular momentum $\mathbf{J} = \mathbf{L} + \mathbf{S}$ is an integral of motion of the Nilsson Hamiltonian of Eq. (1) as it is a scalar under rotations of the vectors $\mathbf{L},\mathbf{S},\eta,\xi$.

>From the fact that

$$J^2 = (\mathbf{L}+\mathbf{S})^2 = L^2 + S^2 + 2\mathbf{L}\cdot\mathbf{S}, \tag{19}$$

we get

$$\mathbf{L}\cdot\mathbf{S} = \tfrac{1}{2}(J^2 - L^2 - S^2) \tag{20}$$

so that under unitary transformation U of Eq. (10) we have

$$U J^2 U^\dagger = J'^2 = J^2 \tag{21}$$

which is corroborated by Eqs. (14-17).

The states of the Hamiltonian Eq. (1) can be obtained in terms of those of a particle with spin $1/2$ in an harmonic oscillator potential by the ket

$$|N(l,\tfrac{1}{2})jm> = R_{Nl}(r)\sum_{\mu,\sigma} <l\mu,\tfrac{1}{2}\sigma|jm> Y_{l\mu}(\theta,\varphi)\chi_{\frac{1}{2}\sigma} \tag{22}$$

where N is the total number of quanta, l the orbital and j total angular momentum.

For a given N, l can take the values

$$l = N, N-2, \ldots 1 \quad \text{or} \quad 0 \tag{23}$$

depending on wether N is even or odd. Besides $l = j \pm 1/2$ as $\mathbf{L} = \mathbf{J} - \mathbf{S}$.

As we mentioned we replace $F(\eta,\xi,\mathbf{S})$ by $(\xi\cdot\mathbf{S})$ and this last operator acting on the states in Eq. (22) gives us a state of $N-1$ quanta and different orbital angular momentum which we denote as l'. Thus, using the short hand notation

$$|N(j\pm\tfrac{1}{2},\tfrac{1}{2})jm> \equiv |\pm)\ ,\ |N-1(j\pm\tfrac{1}{2},\tfrac{1}{2})jm> \equiv |\pm> \tag{24}$$

we easily obtain from elementary Racah analysis

$$< +|\xi\cdot\mathbf{S}|+) = < -|\xi\cdot\mathbf{S}|-) = 0 \tag{25}$$

$$< -|\xi\cdot\mathbf{S}|+) = -(\tfrac{1}{2})(N+j+\tfrac{3}{2})^{\frac{1}{2}} \tag{26}$$

$$< +|\xi\cdot\mathbf{S}|-) = -(\tfrac{1}{2})(N-j+\tfrac{1}{2})^{\frac{1}{2}} \tag{27}$$

The application of $\xi\cdot\mathbf{S}$ to the states of N quanta, where $l = j \pm 1/2$, transforms them into states of $N-1$ quanta, where $l' = j \mp 1/2$.

Writing the results in terms of l rather than $j = l \pm \tfrac{1}{2}$ we see from Eq. (25)-(27) that

$$(\xi\cdot\mathbf{S})|N(l,\tfrac{1}{2})l+\tfrac{1}{2},m> =$$
$$(\tfrac{1}{2})(N-l)^{\frac{1}{2}}|N-1,\ (l+1,\tfrac{1}{2})l+\tfrac{1}{2},m> \tag{28}$$
$$(\xi\cdot\mathbf{S})|N(l,\tfrac{1}{2})l-\tfrac{1}{2},m> =$$
$$-(\tfrac{1}{2})(N+l+1)^{\frac{1}{2}}|N-1,\ (l-1,\tfrac{1}{2})l-\tfrac{1}{2},m> \tag{29}$$

For $j = l + 1/2$ and $l = N$ the right hand side of Eq. (28) vanishes and thus the corresponding states of $N-1$ quanta does not appear. The rest of them are precisely those used in pseudo SU(3) model.

Take, for example $N = 6$, where in the spectroscopic notation we have the degenerate states $s_{1/2}, d_{3/2}, d_{5/2}, g_{7/2}, g_{9/2}, i_{11/2}, i_{13/2}$. We disregard $i_{13/2}$ as indicated in the previous paragraph as the corresponding state of $N - 1$ quanta does not appear and are left with states with 5 quanta $p_{1/2}, p_{3/2}, f_{5/2}, f_{7/2}, h_{9/2} h_{11/2}$ as one should expect for $N - 1 = 5$.

It is clear that the operator $(\xi \cdot S)$ is a good choice to arrive from states of N quanta of the standard oscillator with U(3) symmetry to those of $N - 1$ quanta that have the same symmetry which could be denoted by pseudo SU(3)[7, 8].

As our unitary transformation U of Eq. (10) does not change the commutator relations we have still a SU(3) symmetry for the heavy nuclei but it is not of the same type as that for light nuclei, thus the name of pseudo SU(3).

CONCLUSION

The generators of the pseudo SU(3) symmetry group of the shell model Hamiltonian for medium and heavy nuclei are then given by

$$U^{\dagger} C_{ij} U \tag{30}$$

when U is the operator in Eq. (10) when $F(\eta, \xi, S) = (\xi \cdot S)$ and the C_{ij} defined in Eq. (19). As our main interest would concern the matrix element of these operator with respect to the states of Eq. (24) with N quanta, they could also be given by the effect C_{ij} on $N - 1$ quanta and thus are already determined in the case of an harmonic oscillator interaction.

REFERENCES

1. O. Castaños, M. Moshinsky, C. Quesne "Seminar in Honor of K. T. Hecht on Symmetries in Nuclear Physics (World Scientific 1992), Eds. J. P. Draayer and J. Janecke
2. O. Castaños, M. Moshinsky, C. Quesne, *Phys. Lett.* **B277**, 238 (1992)
3. A. B. Batantekin, O. Castaños, M. Moshinsky , *Phys. Lett.* **B284**, 1 (1992)
4. S. G. Nilsson, *Mat. Fys. Med. K. Dan. Vidensl. Selsk.* **29** (16) (1955);
5. A. Bohr, J. Hamamoto, B.R. Mottelson, *Phys. Scrp.* **26**, 267 (1982)
6. J. P. Elliot *Proc. Roy. Soc.* **A245** 562 (1958)
7. K. T. Hecht and A. Adler *Nucl. Phys.* **A137**, 129 (1969); R. D. Ratna Raju, T. P. Draayer and K. T. Hecht *Nucl. Phys.* **A202**, 433, (1973).
8. J. P. Draayer and K. J. Weeks *Ann. Phys.* **156**, 4, (1984).

Supersymmetry (SUSY), Pseudospin Symmetry and Identical Bands.

Peter von Brentano

IKP Universität zu Köln, Germany

Abstract. Supersymmetry applied to identical bands is discussed. The role of a nearly decoupled pseudospin is investigated. A schematic model of proton neutron supersymmetry is discussed and compared to data on ^{195}Pt. Keywords: U(6/12) supersymmetry, pseudospin, identical bands, mixed symmetry states

Introduction. Supersymmetry is a symmetry which connects a bosonic system with a fermionic system. Supersymmetry has been successfully transfered from particle physics to nuclear physics by Franco Iachello and coworkers [1, 2, 3, 4, 5, 6, 7, 8]. Jolie and Graw were successful in describing four neighboring nuclei by the same SUSY Hamiltonian [9].

Identical Bands (IB) [10, 11, 12] were discovered by Byrski et. al.(Strasbourg-Liverpool). They found an agreement better than 0.3% in the gamma energies of two superdeformed (SD) bands in ^{151}Tb and ^{152}Dy. The data on these SD bands are still incomplete [12].Complete data on IB are available at normal deformation [12] [16], e.g., in the pair ^{174}Hf and ^{173}Hf shown in Fig.1. We discuss SUSY models for IB [13] [14] using Hamiltonians without dynamical supersymmetries following Jolos et al.[15] [16] and Frank et al. [2].

Pseudo-Spin -Symmetry (PSS). Pseudo-Spin Symmetry (PSS) (Arima [17], Hecht [18]) is an approximate symmetry, based on a relativistic symmetry (Ginocchio [19]).

In the PSS scheme the $p_{1/2}$, $f_{5/2}$, $p_{3/2}$, $h_{9/2}$, $f_{7/2}$ s.p. levels are relabeled with the pseudo-spin quantum numbers ($\tilde{l}$, $\tilde{s}$), as : $\tilde{s}_{1/2}$, $\tilde{d}_{5/2}$, $\tilde{d}_{3/2}$, $\tilde{g}_{9/2}$, and $\tilde{g}_{7/2}$. Angular momenta of particles are l = 1, 3, 5. The corresponding pseudo angular momenta are $\tilde{l}$ = 0, 2, 4. The decoupled pseudo-spin case is particularly simple: both the pseudospin $\tilde{s}$ and the total pseudo angular momentum $\tilde{L}$ are good quantum numbers in the odd nucleus. The proof is easy: $[H,\tilde{s}] = 0$ and $J = \tilde{L} + \tilde{s}$ imply $[H,\tilde{L}] = [H,J] - [H,\tilde{s}] = 0$. For a decoupled pseudospin all states in the odd nucleus are either a singlet $J = 1/2$ or a doublet of states with spins $J = \tilde{L} + 1/2$ and $J = |\tilde{L} - 1/2|$.

A beautiful example of nearly decoupled pseudo-spin doublets is found in data for ^{195}Pt by Jolie and Graw [20]. As is shown in Fig. 3 each level in ^{195}Pt (below 0.7 MeV) comes either as a doublet or as a singlet. The origin of the very small splitting of the doublets in ^{195}Pt with a minimum splitting of about 30 KeV needs discussion.

Of course this very small splitting can be due to near degeneracies in the s.p. spectrum as shown in Fig.2 (left row) or there is only one doublet because of an accidental location

CP726, Nuclear Physics, Large and Small: International Conference on
Microscopic Studies of Collective Phenomena, edited by R. Bijker, R. F. Casten, and A. Frank
© 2004 American Institute of Physics 0-7354-0207-8/04/$22.00

EXP — EXP — CALC

Figure 1 level energies

^{174}Hf (EXP):

J^π	E
(16^+)	3209
(14^+)	2598
(12^+)	2020
(10^+)	1486
8^+	1010
6^+	608
4^+	297
2^+	91
0^+	0

^{173}Hf (EXP):

J	E
33/2	3001
31/2	2969
29/2	2392
27/2	2358
25/2	1832
23/2	1796
21/2	1330
19/2	1294
17/2	894
15/2	862
13/2	536
11/2	508
9/2	262
7/2	242
5/2	81
3/2	69
1/2	0

^{173}Hf (CALC):

J	E
33/2	2946
31/2	2893
29/2	2387
27/2	2340
25/2	1857
23/2	1817
21/2	1368
19/2	1335
17/2	932
15/2	905
13/2	563
11/2	542
9/2	
7/2	277
5/2	262
3/2	86
1/2	78
	0

FIGURE 1. Fig. 1 : Experimental level energies in ^{174}Hf and in ^{173}Hf [16].

Pseudospin (PS)

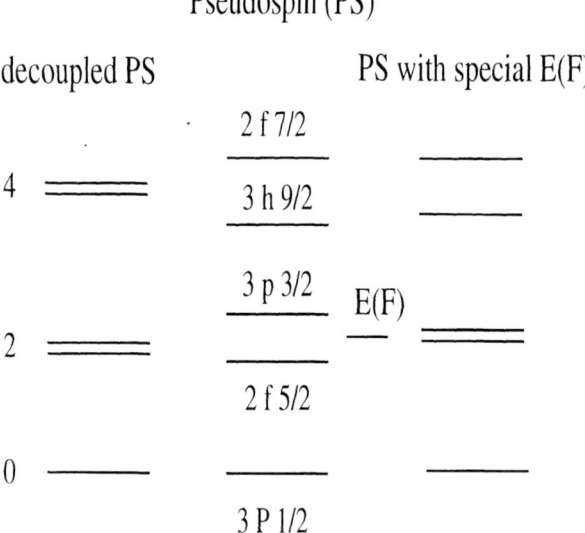

decoupled PS

PS with special E(F)

4	2 f 7/2
	3 h 9/2
	3 p 3/2 E(F)
2	2 f 5/2
0	3 P 1/2

FIGURE 2. Fig. 2 : The figure shows schematic neutron quasiparticle energies in the Pb region (middle). The energies of a fully decoupled PS (left) and a partially decoupled PS (right) are also shown.

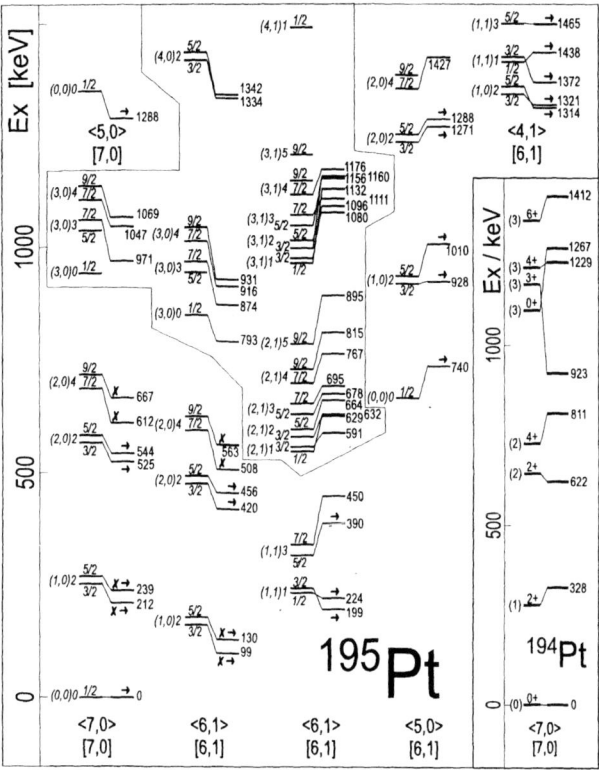

FIGURE 3. Fig 3 : Excitation energies in ^{195}Pt(taken from [20]).

of the Fermi energy E(F) between the $2f_{5/2}$ and the $3p_{3/2}$ levels (right row). The latter seems more likely , because there is no reason for the observed very small PS splittings in ^{195}Pt. Another example of small PS splittings are the Identical Bands in rare earth nuclei [12], e.g., ^{174}Hf and ^{173}Hf shown in Fig. 1 [16].

A schematic model for proton neutron supersymmetry (pn-SUSY SM) for neutrons in $3p_{1/2}$, $2f_{5/2}$, $3p_{3/2}$ *shells.* SUSY works particularly well for neutrons in these shells [3? , 20]. The pseudo angular momenta $\tilde{l}$ of the odd particle have integer values and thus are bosonic excitations. They have the values $\tilde{l} = 0,2$ corresponding to the spins $L = 0,2$ of s and d bosons in the IBA. Thus, following T. Otsuka [21], it is natural to describe them by quasi bosons with the pseudo spins $\tilde{l} = 0 : s^f$ and $\tilde{l} = 2 : d^f$. Three kinds of bosons are used (neutron bosons b^n , proton bosons b^p, and quasi bosons b^f). Furthermore pn-SUSY SM assumes :

Equal energies $\varepsilon_d^n = \varepsilon_d^p = \varepsilon_d^f$ of the bosons : b^n , b^p, b^f .

Equal interactions V_{ik} among the bosons b^n , b^p, and b^f as well as a fully decoupled pseudo-spin.

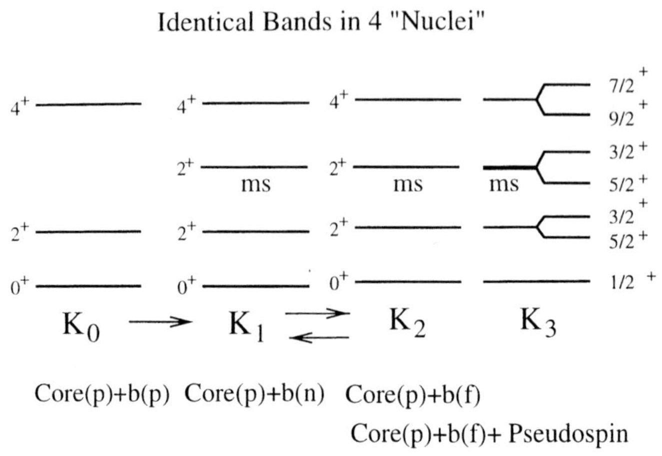

Identical Bands in 4 "Nuclei"

FIGURE 4. Fig.4 : The figure shows identical structures in 4 nuclei. Core(p) is a nucleus with only proton bosons.

The symmetries of the pn-SUSY SM are shown schematically in Fig.4 . In particular one finds identical structures in three bosonic nuclei (K_0, K_1, K_2). $K_0 = Core(p) + b^p$, $K_1 = Core(p) + b^n$, $K_2 = Core(p) + b_n^f$, where $Core(p)$ is a nucleus with only proton bosons . One finds, that every state in K_0 corresponds to a state in K_1 with the same energy and spin L. There are more states in K_1, however. These are the mixed symmetry states. They occur because the last boson in K_1 is a neutron boson and is different from the proton bosons in $Core(p)$. There is a one-to-one correspondence of the states in K_1 and in K_2 with the same energy and spin. There is also a one-to-one correspondence between states in K_2 and multiplets in the odd nucleus K_3 with pseudospin. This implies that to each pseudo-spin multiplet of the odd nucleus K_3 there belongs a unique state in both nuclei $K_2 = Core(p) + b_n^f$ and $K_1 = Core(p) + b^n$ with the same energy and angular momentum $\tilde{L} = L$. This result means, that there are as many states in the even nucleus K_1 as there are pseudo-spin multiplets in the odd nucleus.

"Extra" low lying states (pseudospin multiplets) in ^{195}Pt as mixed symmetry states. Schematic pn-SUSY model vs data.   A) To each level of even ^{194}Pt with an energy E below about 0.7 MeV there corresponds a doublet or singlet in ^{195}Pt with similar centroid energy E_{cent} and the same pseudo angular momentum $\tilde{L} = L$.

B) There are "extra" low lying multiplets in ^{195}Pt below 0.7 MeV with pseudo angular momenta $\tilde{L} = 2$ and $\tilde{L} = 0, 1, 2, 3, 4$ which do not correspond to low lying states in ^{194}Pt

C) In the pn-SUSY SM these states in ^{195}Pt are mixed symmetry (ms) states . They are the 2_{ms}^+ one phonon ms state and the $2_s^+ \otimes 2_{ms}^+$ two phonon ms multiplet.

D) The corresponding ms multi phonon states states in ^{194}Pt lie at much higher energy. Thus pn-SUSY SM is broken in the energies of the mixed symmetry states. IBA-2 calculations indicate that the breaking affects mostly the energies and much less the wavefunctions, however.

ACKNOWLEDGMENTS

Acknowledgments: I thank R. F. Casten, A. Dewald, G. Graw, A. Gelberg , J. Jolie, R. V. Jolos, N. Pietralla, T. Otsuka and V. Werner for discussions and K. Jessen for reading the manuscript carefully. Supported by DFG under contract Br 799/12-1.

REFERENCES

1. F. Iachello and P. Van Isacker, The Interacting Boson–Fermion Model, (Cambridge University Press, Cambridge, 1991).
2. A. Frank and P. Van Isacker, Algebraic Methods in Molecular and Nuclear Physics, (John Wiley & Sons, New York, 1994).
3. P. Van Isacker, Rep. Prog. Phys. 62, 1661 (1999).
4. F. Iachello, Phys. Rev. Lett. 44, 772 (1980).
5. F. Iachello and S. Kuyucak, Ann. Phys. (N.Y.) 136, 19 (1981).
6. A. B. Balantekin, I. Bars and F. Iachello, Nucl. Phys. A370, 284 (1981).
7. J. Jolie, P. Van Isacker, K. Heyde, and A. Frank, Phys. Rev. Lett. 55, 653 (1985).
8. R. Bijker and F. Iachello, Ann. Phys. (N.Y.) 161, 360 (1985).
9. A. Metz, J. Jolie, G. Graw, R. Hertenberger, R. Gröger, J. Guenther, C. Warr and Y. Eisermann, Phys. Rev. Lett. 83, 1542 (1999).
10. T. Byrski, F. A. Beck, D. Curien, C. Schuck et al., Phys. Rev. Lett. 64, 1650 (1990).
11. W. Nazarewicz, P. J. Twin, P. Fallon, and J. D. Garrett, Phys. Rev. Lett. 64, 1654 (1990).
12. C. Baktash, B. Haas, W. Nazarewicz, Ann. Rev. Nucl. Part. Sci. 45, 485 (1995).
13. A. Gelberg, P. von Brentano, R. F. Casten, J. Phys. G 16, L143 (1990).
14. R. D. Amado, R. Bijker, F. Cannata, and J. P. Dedonder, Phys. Rev. Lett. 67, 2777 (1991).
15. R. V. Jolos and P. von Brentano, Phys. Rev. C 60, 064318 (1999).
16. R. V. Jolos and P. von Brentano, Phys. Rev. C 63, 024304 (2001) and references given therein.
17. A. Arima, M. Harvey, and K. Shimizu, Phys. Lett. 30 B, 517 (1969).
18. K. T. Hecht and A. Adler, Nucl. Phys. A 137, 129 (1969).
19. J. N. Ginocchio, Phys. Rev. Lett. 78, 436 (1997).
20. A. Metz, Y. Eisermann, A. Gollwitzer, R. Hertenberger, G. Graw and J. Jolie, Phys. Rev. C 61 064313 (2000).
21. T. Otsuka , private communication.

New correlations induced by nuclear supersymmetry

J. Barea*, R. Bijker* and A. Frank*

*ICN-UNAM, A.P. 70-543, 04510 México, D.F., México

Abstract. We show that the nuclear supersymmetry model (n-susy) in its extended version, predicts correlations in the nuclear structure matrix elements which characterize transfer reactions between nuclei that belong to the same supermultiplet. These correlations are related to the fermionic generators of the superalgebra and if verified experimentally can provide a direct test of the model.

The extended n-susy model [1] correlates spectroscopic properties of adjacent nuclei. It has been particularly sucessful in describing the quartet formed by the isotopes ^{194}Pt, ^{195}Pt, ^{195}Au and ^{196}Au using the $Spin(6)$ limit of the dynamical supersymmetry $U_\nu(6/12) \otimes U_\pi(6/4)$ [1, 2] in which the odd-proton is allowed to occupy the $\pi d_{3/2}$ orbit, the odd-neutron occupies the $\nu p_{1/2}$, $\nu p_{3/2}$ and $\nu f_{5/2}$ orbits, even-even nuclei are described by the $SO(6)$ limit of the interacting boson model.

In·this framework, all states belong to the same irreducible representation (irrep) of the initial product of supergroups and can be labeled by different irreps of the subgroups present in the chain of groups. The states in the quartet are said to form a supermultiplet and the excitation spectra is obtained using the same hamiltonian. In this way the excitation energies are correlated. Transitions probabilities and moments are also correlated through the use of the same operators.

Both the hamiltonian and the electromagnetic transition operators are based on the bosonic generators of the superalgebra, which transform bosons into bosons and fermions into fermions, inducing transitions inside each nucleus. In this work we explore the fermionic generators, which transform bosons into fermions and viceversa. They are thus associated to transitions among different nuclei, which we can associate to transfer reactions.

The purpose of this contribution is to show that n-susy establishes correlations between different transfer reactions. We start with one-nucleon transfer reactions and then we present the first results, to our knowledge, of a susy analysis of the two-nucleon transfer reaction ^{198}Hg$(\vec{d}, \alpha)^{196}$Au, in which the wave function correlations of proton-neutron clusters in the target can be tested.

CORRELATIONS IN ONE-NUCLEON TRANSFER REACTIONS

The nuclear structure information which can be extracted from one-nucleon transfer reactions is contained in the so called spectroscopic intensity, which is the modulus

CP726, Nuclear Physics, Large and Small: International Conference on
Microscopic Studies of Collective Phenomena, edited by R. Bijker, R. F. Casten, and A. Frank
© 2004 American Institute of Physics 0-7354-0207-8/04/$22.00

square of the reduced matrix element of the transfer operator $T^{(J)}$ between the ground state of the target nucleus in the reaction, and the final state of the residual nucleus

$$I = |\langle \alpha_f J_f || T^{(J)} || \alpha_i J_i \rangle|^2. \tag{1}$$

It is possible to use a transfer operator deduced from microscopic asumptions to calculate these quantities, but here we use an alternative method based on symmetry considerations. In this case, the transfer operator can be written as a tensor operator under the subgroups that appear in the group chain of the dynamical supersymmetry and has the advantage of giving rise to selection rules and closed expressions for the spectroscopic intensities [3, 4, 5].

For the one-proton transfer reactions, the transfer tensor operators read [5]

$$T_{1,\pi}^{\langle \frac{1}{2},\frac{1}{2},-\frac{1}{2} \rangle \langle \frac{1}{2},\frac{1}{2} \rangle \frac{3}{2}} = -\sqrt{\frac{1}{6}} \left(\tilde{s}_\pi \times a_{\pi,\frac{3}{2}}^\dagger \right)^{\left(\frac{3}{2}\right)} + \sqrt{\frac{5}{6}} \left(\tilde{d}_\pi \times a_{\pi,\frac{3}{2}}^\dagger \right)^{\left(\frac{3}{2}\right)} \tag{2}$$

$$T_{2,\pi}^{\langle \frac{3}{2},\frac{1}{2},\frac{1}{2} \rangle \langle \frac{1}{2},\frac{1}{2} \rangle \frac{3}{2}} = \sqrt{\frac{5}{6}} \left(\tilde{s}_\pi \times a_{\pi,\frac{3}{2}}^\dagger \right)^{\left(\frac{3}{2}\right)} + \sqrt{\frac{1}{6}} \left(\tilde{d}_\pi \times a_{\pi,\frac{3}{2}}^\dagger \right)^{\left(\frac{3}{2}\right)}, \tag{3}$$

where the upper indices specify the tensorial properties under $Spin(6)$, $Spin(5)$ and $Spin(3)$. The tensorial character under $Spin(6)$ of $T_{1,\pi}$ implies that it only excites the ground state of the odd-even nucleus from the ground state of the even-even nucleus. However, $T_{2,\pi}$ allows the transfer to an excited state in the odd-even nucleus. The ratio of the intensities for each transfer operator is given by [5]

$$R_1(\text{ee} \rightarrow \text{oe}) = \frac{I_{gs \rightarrow exc}}{I_{gs \rightarrow gs}} = 0, \tag{4}$$

$$R_2(\text{ee} \rightarrow \text{oe}) = \frac{I_{gs \rightarrow exc}}{I_{gs \rightarrow gs}} = \frac{9(N+1)(N+5)}{4(N+6)^2}, \tag{5}$$

where N is taken as the number of bosons in the odd-odd nucleus of the quartet and ee and oe refer to even-even and odd-even respectively. In the case of the one-proton transfer ^{194}Pt $\rightarrow ^{195}$Au, the second ratio is $R_2 = 1.12$ ($N = 5$), but the relatively small strength to excited $J = \frac{3}{2}$ states suggests that the operator $T_{1,\pi}$ can be used to describe the data.

Due to the F-spin symmetry structure of the wave functions it is possible to establish the following correlations with the reaction which involves the even-odd (eo) and odd-odd(oo) nuclei:

$$R_1(\text{ee} \rightarrow \text{oe}) = R_1(\text{eo} \rightarrow \text{oo}) = 0, \tag{6}$$

$$R_2(\text{ee} \rightarrow \text{oe}) = R_2(\text{eo} \rightarrow \text{oo}) = \frac{9(N+1)(N+5)}{4(N+6)^{2\cdot}}. \tag{7}$$

A different way to understand this result is through the use of a tensor operator which transforms as a scalar in the pseudo-l degree of freedom [6] (upper and lower indices specify the tensorial properties under $Spin(6)$ and $Spin(5)$, $Spin(3)$ and $SU(2)$ respectively)

$$P^{\langle 0,0,0\rangle}_{(0,0)0,\frac{1}{2}} = \left(\tilde{s}_v \times a^{\dagger}_{v,\frac{1}{2}}\right)^{\left(\frac{1}{2}\right)} - \sqrt{2}\left(\tilde{d}_v \times a^{\dagger}_{v,\frac{3}{2}}\right)^{\left(\frac{1}{2}\right)} + \sqrt{3}\left(\tilde{d}_v \times a^{\dagger}_{v,\frac{5}{2}}\right)^{\left(\frac{1}{2}\right)}. \tag{8}$$

This operator links the wave functions of the even-odd and odd-odd nuclei in the quartet to the wave functions of the even-even and odd-even nuclei, respectively. The use of this property and the fact that P conmutes with the one-proton transfer operators $T_{1,\pi}$ and $T_{2,\pi}$, allows to write the equations (6) and (7) [7, 8].

We thus find a direct correlation between two different one-proton reactions, which for the case of $^{195}\text{Pt} \rightarrow ^{196}\text{Au}$ can be tested experimentally. Very recently this reaction has been measured [9] and the experimental data confirms that $T_{1,\pi}$ is capable of describing it, given that the strength to the ground state of ^{196}Au is relatively strong and the excited state which $T_{2,\pi}$ can excite is not seen populated in the measurements. This fact confirms also that the correlation seems to apply.

For the one neutron transfer reaction there is a similar correlation. Let us consider the following transfer tensor operator (labels like in (8)) [6]

$$T^{(2,0,0)}_{(1,0)2,j,v} = \sqrt{\frac{1}{2}}\left(\tilde{s}_v \times a^{\dagger}_{v,j}\right)^{(j)} - \sqrt{\frac{1}{2}}\left(\tilde{d}_v \times a^{\dagger}_{v,\frac{1}{2}}\right)^{(j)} \qquad j = \frac{3}{2}, \frac{5}{2}. \tag{9}$$

The same argument about the link between the wave functions of the even-even and the odd-even nuclei permits us to correlate the one neutron transfer reaction even-even $\rightarrow$ odd-even with the inverse reaction odd-even $\rightarrow$ even-even [8]. Taking the following ratios

$$R(\text{ee} \rightarrow \text{oe}) = \frac{\left|\left\langle \text{oe};[N_1,N_2]\langle\sigma_1,\sigma_2,\sigma_3\rangle(1,0)2;J \left\| T^{(2,0,0)}_{(1,0)2,j,v} \right\| \text{ee}; \text{g.s.}\right\rangle\right|^2}{\left|\left\langle \text{oe};[N+1,1]\langle N+1,1,0\rangle(1,0)2;J \left\| T^{(2,0,0)}_{(1,0)2,j,v} \right\| \text{ee}; \text{g.s.}\right\rangle\right|^2}, \tag{10}$$

$$R(\text{oe} \rightarrow \text{ee}) = \frac{\left|\left\langle \text{ee};[N_1,N_2]\langle\sigma_1,\sigma_2,\sigma_3\rangle(1,0)2 \left\| \tilde{T}^{(2,0,0)}_{(1,0)2,j,v} \right\| \text{oe}; \text{g.s.}\right\rangle\right|^2}{\left|\left\langle \text{ee};[N+1,1]\langle N+1,1,0\rangle(1,0)2 \left\| \tilde{T}^{(2,0,0)}_{(1,0)2,j,v} \right\| \text{oe}; \text{g.s.}\right\rangle\right|^2}, \tag{11}$$

we find the following relations

$$R(\text{oe} \rightarrow \text{ee}) = R(\text{ee} \rightarrow \text{oe}) \text{ for } N_2 = 1,$$

$$R(\text{oe} \rightarrow \text{ee}) = R(\text{ee} \rightarrow \text{oe})\frac{(N+1)(N_v+1)}{(N_\pi+1)} \text{ for } N_2 = 0,$$

In table 1 we show these ratios and quote the experimental values [10] and the calculated ones for the isotopes ^{194}Pt and ^{195}Pt. We can observe a consistence between the experimental and calculated values, but clearly more experimental work is needed to confirm if these correlations hold.

Table 1. Intensity ratios for one-neutron transfer reactions.

$[N_1,N_2]\langle\sigma_1,\sigma_2,\sigma_3\rangle$	$R_{ee\to oe}$	^{194}Pt $\to$ ^{195}Pt			^{195}Pt $\to$ ^{194}Pt	
		calc.	exp.		calc.	exp.
			$j=\frac{3}{2}$	$j=\frac{5}{2}$		
$[N+2]\langle N+2,0,0\rangle$	$\frac{2(N+4)}{(N+1)(N+3)(N+6)}$	0.034	0.264	0.052	0.511	——
$[N+2]\langle N,0,0\rangle$	$\frac{N(N+4)(N+5)}{2(N+2)(N+3)(N+6)^2}$	0.033	——	——	0.498	——
$[N+1,1]\langle N,0,0\rangle$	$\frac{N^2(N+5)}{2(N+2)(N+6)^2}$	0.148	0.087	——	0.148	——

CORRELATIONS IN TWO-NUCLEON TRANSFER REACTIONS

In contrast to the one-nucleon transfer reactions in which the single particle structure of nuclear states is examined, two-nucleon transfer reactions are very sensitive to the correlation between the transfered nucleons. As a consequence, this kind of transfer reactions supply a stringent test for the nuclear wave functions.

Two factors determine the strength of a two nucleon transfer reaction, which are related to the two-nucleon fractional parentage coefficients. On the one hand it depends on how similar the state of $A+2$ nucleons is to the state of A plus two additional nucleons, and on the other, if the correlation of these two nucleons in the $A+2$ state is similar to the correlation in the state of the light nucleus. The nuclear structure information that can be extracted from a model is related to both factors. We shall follow the formalism developed by Glendenning [11] and briefly sketch the main ingredients as applied to the reaction ^{198}Hg$(\vec{d},\alpha)^{196}$Au, which has been measured recently to study the odd-odd nucleus ^{196}Au [9].

The nuclear structure information is contained in the spectroscopic strengths, G_{LJ}. This quantities are written as

$$G_{LJ} = \left| \sum_{Nj_\nu j_\pi} \beta_{j_\nu j_\pi} G_{NLJ}^{j_\nu j_\pi} \right|^2 , \tag{12}$$

where

$$\beta_{j_\nu j_\pi} = \left\langle {}^{196}\text{Au};J \left\| T_{j_\nu j_\pi}^{(J)} \right\| {}^{198}\text{Hg};0_{gs} \right\rangle \tag{13}$$

$$G_{NLJ}^{j_\nu j_\pi} = \sqrt{3}\hat{L}\hat{j}_\nu\hat{j}_\pi \begin{Bmatrix} l_\nu & \frac{1}{2} & j_\nu \\ l_\pi & \frac{1}{2} & j_\pi \\ L & 1 & J \end{Bmatrix} \langle n0NL|n_\nu l_\nu n_\pi l_\pi ;L\rangle . \tag{14}$$

For the two nucleon transfer operator $T_{j_\nu j_\pi}^{(J)}$ we have chosen the simplest possible form:

$$T_{j_\nu j_\pi}^{(J)} = \alpha_{j_\nu} \left(a_{j_\nu}^\dagger \times a_{j_\pi}^\dagger \right)^{(J)} . \tag{15}$$

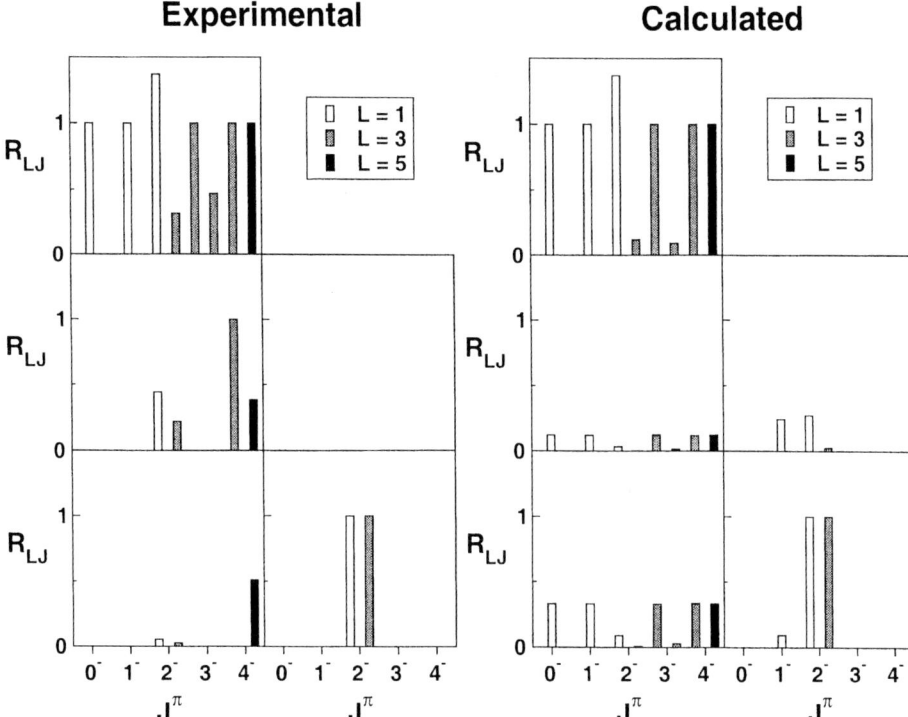

Figure 1. Ratios of spectroscopic strengths. The first column in each frame correspond to states with $\left(\frac{3}{2}, \frac{1}{2}\right)$ $Spin(5)$ labels and the second with $\left(\frac{1}{2}, \frac{1}{2}\right)$ labels, respectively. Each row from the bottom to the top corresponds to states with labels $[6,0]\langle 6,0\rangle(13/2,1/2,1/2)$, $[5,1]\langle 5,1\rangle(11/2,1/2,1/2)$ and $[5,1]\langle 5,1\rangle(11/2,3/2,1/2)$ for the groups $U_{\nu\pi}^{BF}(6)$, $SO_{\nu\pi}^{BF}(6)$ and $Spin(6)$.

The parameters α_{j_ν} are determined by using a least square fit of the spectroscopic strengths to the experimental values. To compare the experimental and calculated spectroscopic strengths we have choosen seven states of reference, each one corresponding to each of the seven possible LJ transfers that the n-susy model allows, and we have calculated for each LJ the transfer ratio

$$R_{LJ} = \frac{G_{LJ}}{G_{LJ}^{ref}}, \qquad (16)$$

where G_{LJ}^{ref} is the spectroscopic strength for the reference state for a particular LJ tranfer. In figure 1 we show the experimental and calculated ratios R_{LJ}. We observe from this figure that the comparison is quite reasonable if we take into account the simple form we adopted for the two nucleon transfer operator. In concluding this section we can say that the n-susy model can reproduce remarkably well the main features of the neutron-proton correlations which are present in the nuclear states involved.

35

SUMMARY

We have found new correlations predicted by extended n-susy. These correlations relate spectroscopic strengths of different one-nucleon transfer reactions between the nuclei which are described by the model. We focused on the supermultiplet formed by ^{194}Pt, ^{195}Pt, ^{195}Au and ^{196}Au whose spectroscopic properties have been previously described in terms of this model. We have shown that these correlations are partially fulfilled with the availaible experimental data, but clearly more experimental work is necessary to test their full validity.

We have calculated the spectroscopic strengths associated to the two nucleon transfer reaction ^{198}Hg$(\vec{d}, \alpha)^{196}$Au, which was recently measured. The comparison between the calculated and experimental data shows good agreement considering the simple form adopted for the two nucleon transfer operator.

We plan to search for other experimental examples to which extended n-susy and its correlations can be applied, eventually relaxing the constraints set by dynamical symmetry [12, 13]. We wish to emfasize that nuclear susy may be a model whose range of applicability is wider than was previously realized and which may lay the foundations of a new and unifying point of view in nuclear structure.

ACKNOWLEDGMENTS

This work was supported by CONACyT. We are grateful to G. Graw for sharing the new experimental data on the transfer reactions ^{195}Pt$(^{3}$He,d$)^{196}$Au and ^{198}Hg$(\vec{d}, \alpha)^{196}$Au prior to publication. Many enlighting discussions with J. Gómez-Camacho and P. Van Isacker are acknowledged.

REFERENCES

1. P. van Isacker, J. Jolie, K. Heyde and A. Frank, *Phys. Rev. Lett.* **54**, 653 (1985).
2. A. Metz, J. Jolie, G. Graw, R. Hertenberger, J. Gröger, C. Günther, N. Warr and Y. Eisermann, *Phys. Rev. Lett.* **83**, 1542 (1999).
3. F. Iachello and S. Kuyucak, *Ann. Phys. (N.Y.)* **136**, 19 (1981).
4. R. Bijker and F. Iachello, *Ann. Phys. (N.Y.)* **161**, 360 (1985).
5. J. Barea, R. Bijker, A. Frank and G. Loyola, *Phys. Rev. C* **64**, 064313 (2001).
6. R. Bijker, Ph. D. Thesis (1984).
7. R. Bijker, J. Barea and A. Frank, in *Proceedings of the International Conference 'Symmetries in Nuclear Structure'* (Erice, 2003).
8. J. Barea, R. Bijker and A. Frank, to be published.
9. G. Graw, private communication.
10. Y. Yamazaki and R. K. Sheline, *Phys. Rev. C* **24**, 531 (1976). G. Berrier-Ronsin *et al.*, *Phys. Rev. C* **23**, 2425 (1981).
11. N. K. Glenndening, *'Direct nuclear reactions'*, Academic Press (1983).
12. A. Frank, J. Barea and R. Bijker, in *Proceedings of the International Conference 'Symmetries in Nuclear Structure'* (Erice, 2003).
13. A. Frank, P. Van Isacker and D. D. Warner, *Phys. Lett.* B **197**, 474 (1987).

The *Ab Initio* Large-Basis No-Core Shell Model

B. R. Barrett*, P. Navrátil†, A. Nogga**, W. E. Ormand†, I. Stetcu*, J. P. Vary‡ and H. Zhan*

*Department of Physics, P.O. Box 210081, University of Arizona, Tucson, AZ 85721
†University of California, Lawrence Livermore National Laboratory, Livermore, CA 94551
**Institute for Nuclear Theory, University of Washington, Box 351550,Seattle, WA 98195
‡Department of Physics and Astronomy, Iowa State University, Ames, IA 50011

Abstract. We describe the development and application of the *ab initio* No-Core shell Model, in which the effective Hamiltonians are derived microscopically from realistic, high-quality nucleon-nucleon (NN) potentials plus various realistic three-nucleon (NNN) potentials, as a function of the finite harmonic-oscillator (HO) basis space. For presently feasible no-core model spaces, we evaluate the effective Hamiltonians in a cluster approach, which is guaranteed to provide exact results for sufficiently large model spaces and/or sufficiently large clusters. A number of recent applications of the NCSM are given.

1. INTRODUCTION

The major outstanding problem in nuclear-structure physics is to calculate the properties of finite nuclei starting from the basic interactions among the nucleons. Such calculations have been performed so far only for light nuclei up to $A = 10$ [1]. We have developed a new *ab initio* technique for accurately computing nuclear properties, in which all A nucleons are taken to be active, interacting by realistic NN plus theoretical NNN interactions. We call this approach the No-Core Shell Model (NCSM) [2, 3, 4, 5, 6, 7, 8, 9, 10, 11].

2. NO-CORE SHELL-MODEL APPROACH

The NCSM is based on an effective Hamiltonian derived from realistic "bare" interactions and acting within a finite Hilbert space. All A-nucleons are treated on an equal footing. The approach is both computationally tractable and demonstrably convergent to the exact result of the full (infinite) Hilbert space.

Initial investigations used two-body interactions [2] based on a G-matrix approach. Later, we implemented the Lee-Suzuki procedure [12] to derive two-body [3] and three-body [5] effective interactions, based on realistic NN and NNN interactions.

For pedagogical purposes, we outline the NCSM approach only with NN interactions and refer the reader to the literature on how to include NNN interactions. However, some results with NNN interactions will be given. We begin with the purely intrinsic

CP726, *Nuclear Physics, Large and Small: International Conference on*
Microscopic Studies of Collective Phenomena, edited by R. Bijker, R. F. Casten, and A. Frank
© 2004 American Institute of Physics 0-7354-0207-8/04/$22.00

Hamiltonian for the A-nucleon system, i.e.,

$$H_A = T_{\text{rel}} + \mathcal{V} = \frac{1}{A} \sum_{i<j}^{A} \frac{(\vec{p}_i - \vec{p}_j)^2}{2m} + \sum_{i<j=1}^{A} V_N^{ij}, \tag{1}$$

where m is the nucleon mass and V_N^{ij}, the NN interaction in pair (ij), with both strong and electromagnetic components. Note the absence of a phenomenological single-particle potential. We may use either local potentials in coordinate-space, such as the Argonne potentials [1] or non-local ones, such as the CD-Bonn [13].

Next, we add the center-of-mass (CM) HO Hamiltonian, $H_{\text{CM}} = T_{\text{CM}} + U_{\text{CM}}$, where $U_{\text{CM}} = \frac{1}{2} A m \Omega^2 \vec{R}^2$, $\vec{R} = \frac{1}{A} \sum_{i=1}^{A} \vec{r}_i$, to H_A. In the full Hilbert space the added H_{CM} term has no influence on the intrinsic properties. However, when we introduce our cluster approximation below, the added H_{CM} term will be very important for the determination of the effective interactions. The modified Hamiltonian, with a pseudo-dependence on the HO frequency Ω, can be cast into the form

$$H_A^\Omega = H_A + H_{\text{CM}} = \sum_{i=1}^{A} \left[\frac{\vec{p}_i^2}{2m} + \frac{1}{2} m \Omega^2 \vec{r}_i^2 \right] + \sum_{i<j=1}^{A} \left[V_N^{ij} - \frac{m\Omega^2}{2A} (\vec{r}_i - \vec{r}_j)^2 \right]. \tag{2}$$

In the spirit of Da Providencia and Shakin [14] and Lee, Suzuki and Okamoto [12], we introduce a unitary transformation, which is able to accommodate the short-range two-body correlations in a nucleus, by choosing an antihermitian operator S, acting only on intrinsic coordinates, such that

$$\mathcal{H} = e^{-S} H_A^\Omega e^S. \tag{3}$$

In our approach, S is determined by the requirements that $\mathcal{H}$ and H_A^Ω have the same symmetries and eigenspectra over the subspace $\mathcal{K}$ of the full Hilbert space. In general, both S and the transformed Hamiltonian are A-body operators. Our simplest, non-trivial approximation to $\mathcal{H}$ is to develop a two-body ($a = 2$) effective Hamiltonian, where the upper bound of the summations "A" is replaced by "a", but the coefficients remain unchanged. A three-body effective Hamiltonian, ($a = 3$), is obtained in a similar manner.

If the full Hilbert space is divided into a finite model space ("P-space") and a complementary infinite space ("Q-space"), using the projectors P and Q with $P + Q = 1$, it is possible to determine the transformation operator S_a from the decoupling condition

$$Q_a e^{-S^{(a)}} H_a^\Omega e^{S^{(a)}} P_a = 0, \tag{4}$$

and the simultaneous restrictions $P_a S^{(a)} P_a = Q_a S^{(a)} Q_a = 0$. Note that a-nucleon-state projectors (P_a, Q_a) appear in Eq. (4). P_a is chosen to project onto the set of all a body states, which are included in P. The unitary transformation and decoupling condition, introduced by Suzuki and Okamoto and referred to as the unitary-model-operator approach (UMOA) [15], has a solution that can be expressed in the following form

$$S^{(a)} = \text{arctanh}(\omega - \omega^\dagger), \tag{5}$$

with the operator ω satisfying $\omega = Q_a \omega P_a$, and solving its own decoupling equation,

$$Q_a e^{-\omega} H_a^{\Omega} e^{\omega} P_a = 0 . \tag{6}$$

Given the eigensolutions, $H_a^{\Omega}|k\rangle = E_k|k\rangle$, then the operator ω can be determined from

$$\langle \alpha_Q | \omega | \alpha_P \rangle = \sum_{k \in \mathscr{K}} \langle \alpha_Q | k \rangle \langle \tilde{k} | \alpha_P \rangle , \tag{7}$$

where we denote by tilde the inverted matrix of $\langle \alpha_P | k \rangle$, i.e., $\sum_{\alpha_P} \langle \tilde{k} | \alpha_P \rangle \langle \alpha_P | k' \rangle = \delta_{k,k'}$ and $\sum_k \langle \alpha_P' | \tilde{k} \rangle \langle k | \alpha_P \rangle = \delta_{\alpha_P', \alpha_P}$, for $k, k' \in \mathscr{K}$. In the relation (7), $|\alpha_P\rangle$ and $|\alpha_Q\rangle$ are the model-space and the Q-space basis states, respectively, and $\mathscr{K}$ denotes a set of d_P eigenstates, whose properties are reproduced in the model space. Necessarily, d_P is equal to the dimension of the model space.

With the help of the solution for ω (7) we obtain a simple expression for the matrix elements of the hermitian effective Hamiltonian

$$\langle \alpha_P | \bar{H}_{a-\text{eff}} | \alpha_P' \rangle = \sum_{k \in \mathscr{K}} \sum_{\alpha_P''} \sum_{\alpha_P'''} \langle \alpha_P | (P_a + \omega^{\dagger} \omega)^{-1/2} | \alpha_P'' \rangle \langle \alpha_P'' | \tilde{k} \rangle E_k \langle \tilde{k} | \alpha_P''' \rangle$$
$$\times \langle \alpha_P''' | (P_a + \omega^{\dagger} \omega)^{-1/2} | \alpha_P' \rangle . \tag{8}$$

We note that in the limit $a \to A$, we obtain the exact solutions for d_P states of the full problem for any finite basis space.

On account of our cluster approximation, a dependence of our results on the model-space size and on the HO frequency Ω arises. For a fixed cluster size, the smaller the basis space, the larger the dependence on Ω.

In order to construct the operator ω (7) we need to select the set of eigenvectors $\mathscr{K}$. Because of the added CM Hamiltonian, the a-body clusters are confined, which ensures that all eigenvectors are bound states. We keep the lowest states obtained in each two-body channel. It turns out that these states also have the largest overlap with the model space for the range of $\hbar\Omega$ and the P-spaces we have investigated.

We input the effective Hamiltonian, now consisting of a relative 2-body operator and the pure H_{CM} term introduced earlier, into an m-scheme Lanczos diagonalization process to obtain the P-space eigenvalues and eigenvectors. At this stage we also add the term H_{CM} again with a large positive coefficient to separate the physically interesting states with $0s$ CM motion from those with excited CM motion. We retain only the states with pure $0s$ CM motion when evaluating observables. All observables that are expressible as functions of relative coordinates, such as the rms radius and radial densities, are then evaluated free of CM motion effects.

We close our presentation on the theoretical framework with the observation that all observables require the same transformation as implemented on the Hamiltonian. To date, we have found rather small effects on the rms radius operator when we transformed it to a P-space effective rms operator at the a=2 cluster level [6]. On the other hand, substantial renormalization was observed for the kinetic energy operator when using the a=2 transformation to evaluate its expectation value [16].

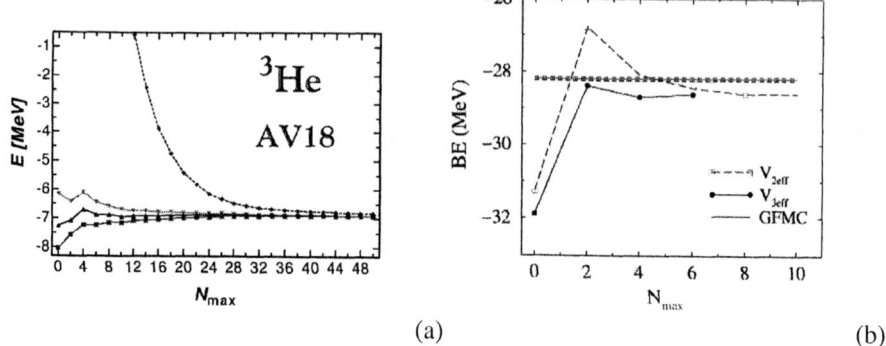

FIGURE 1. (a) Ground-state energy dependence on the model-space size for ^{3}He interacting by the AV18 NN potential The dashed line shows the result based on the bare interaction. The solid lines with down-pointing triangles, up-pointing triangles and squares are results based on the effective interaction for $\hbar\Omega = 32$, 28 and 24 MeV, respectively. (b) Same for ^{6}Li using the AV8' NN potential with Coulomb. The plotted energies occur at the HO frequency minima for the given value of N_{max} The results using both the two-body and the three-body effective interaction are compared with the GFMC results from Ref. [1]. The figure is from Ref. [8].

3. RESULTS

Our $A = 3$ results indicate the feasibility of our approach, showing that accurate values of physics properties can be obtained in sufficiently large model spaces, which we define by the maximal HO excitation above the minimal A-body configuration N_{max} included in the basis. Even for the complicated AV18 NN potential [1], the $N_{max} = 50$ model space is sufficient for obtaining a converged result with an error less than 10 keV, as shown in Fig. 1(a). It is also seen that the utilization of the effective NN interaction speeds up the convergence significantly compared with the bare interaction.

A bigger challenge for the NCSM is the p-shell, where model spaces increase rapidly in size with increasing N_{max}. Consequently, model spaces larger than $N_{max} = 8$ are not presently feasible for most p-shell nuclei. However, besides increasing N_{max} to improve convergence, one can also increase the cluster size of the effective interaction. This has been investigated by Navrátil and Ormand [8] for several p-shell nuclei. E.g. for ^{6}Li, it was demonstrated that three-body effective interactions accelerate convergence. This is is shown in Fig. 1(b).

Our ability to calculate the effective Hamiltonian at the three-body cluster level as well as for two-body cluster makes it possible for us to investigate the nature of different NNN interaction models. The spectra of the light nuclei are well suited for analyzing NNN forces, because they are especially sensitive to their spin/isospin structure. In addition, the NCSM results for the spectra typically converge faster than the binding energies (see, *e.g.*, [6]) and are generally more accurately predicted than the binding energies. To exemplify the sensitivity of the spectra to the three-nucleon interaction, we show in Fig. 2(a) our results for the excitation energy of the 3^+ state in ^{6}Li. NN forces alone overpredict this observable. Note that the combinations AV8' with the

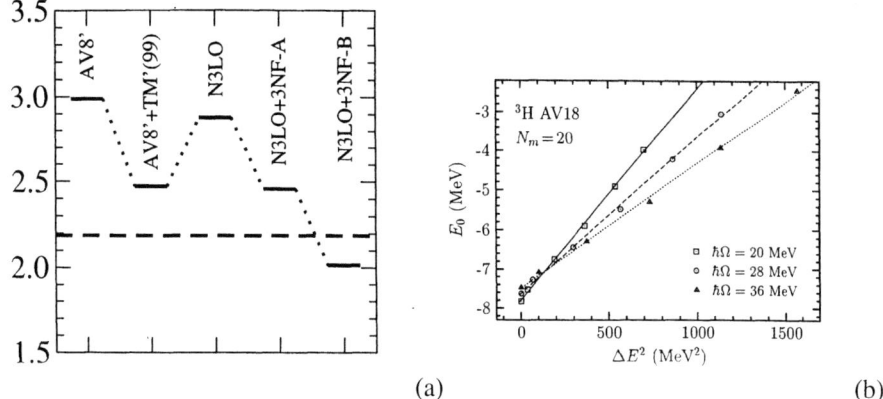

(a)   (b)

FIGURE 2. (a) Excitation energy of the 3^+ state of ^{6}Li for NN and NNN interactions. The dashed line marks experiment. All results are for $N_{max} = 6$. (b) The linear relation between E_0 and ΔE^2, for ^{3}H with the AV18 potential. Symbols from right to left correspond to $\tilde{N}_m = 10, 12, , ..., 20$ for each model. The lines fit the results from $\tilde{N}_m = 10$ to 18.

2π exchange Tucson-Melbourne (TM'(99)) NNN force [17] are already close to the experimental number. This becomes even more pronouced for ^{10}B, where TM'(99) corrects the wrong ordering of ground and excited state predicted by NN interactions only or in combination with the Urbana-IX 3N force [1]. We also studied the chiral NN interaction Idaho-N3LO [18] in combination with the consistently defined leading chiral 3N interaction [19]. Here we identified two sets of parameters, which describe the ^{3}H and ^{4}He binding energies equally well. The excitation energy is different for both sets of parameters, clearly showing the sensitivity of this quantity to the NNN force structure. Calculations for other nuclei are in progress.

To increase the accuracy of our predictions, it is desirable to further increase the model space sizes of these calculations. A recently developed extrapolation method based on second-order perturbation theory will be very important for estimating the binding energies of NCSM calculations without diagonalizing the complete Hamiltonian in the extremely large basis space[20]. We obtained the result that the converged binding energy scales with the energy variance, ΔE^2, as shown in Fig. 2(b). We are presently extending this procedure to heavier mass nuclei.

4. CONCLUSIONS

In this contribution we described the *ab initio* NCSM approach and demonstrated its usefulness by applications to the $A = 3$ system, for which we obtain well-converged results. For the accurate description of p-shell nuclei not only NN, but also NNN interactions are important: In order to include the latter ones, we need three-body effective interactions. These also improve the rate of convergence in smaller model spaces, although more work remains to be done regarding the role and properties of NNN and perhaps NNNN effective interactions. We have shown how the NCSM approach at the NNN cluster level can be used to analyze the nature of different realistic models for the NNN forces. It should be noted that our calculations contain no adjustable parameters. The favorable comparison with available data that we obtain is a consequence of the underlying NN and NNN interaction. To extend our calculations to even larger model spaces, a new extrapolation method will be a very useful tool.

ACKNOWLEDGMENTS

B.R.B., I.S. and H.Z. acknowledge partial support by NSF grants PHY0070858 and PHY0244389. J.P.V. acknowledges partial support by USDOE grant No. DE-FG-02-87ER-40371. The work was performed in part under the auspices of the U. S. Department of Energy by the University of California, Lawrence Livermore National Laboratory under contract No. W-7405-Eng-48. P.N. and W.E.O. received support from LDRD contract 00-ERD-028. A.N. acknowledges support by USDOE grants DE-F02-01ER41187 and DE-FG02-00ER41132.

REFERENCES

1. R. B. Wiringa, V. G. J. Stoks and R. Schiavilla, Phys. Rev. C **51**, 38 (1995); B. S. Pudliner, V. R. Pandharipande, J. Carlson, S. C. Pieper and R. B. Wiringa, Phys. Rev. C **56** 1720, (1997); R. B. Wiringa, Nucl. Phys. **A 631**, 70c (1998); S. Pieper and R. B. Wiringa, Annu. Rev. Nucl.Part.Sci. 51, 53 (2001); S. Pieper, K. Varga and R. B. Wiringa, Phys. Rev. C 66, 044310 (2002).
2. D. C. Zheng, B. R. Barrett, L. Jaqua, J. P. Vary, and R. L. McCarthy, Phys. Rev. C **48**, 1083 (1993); D. C. Zheng, J. P. Vary, and B. R. Barrett, Phys. Rev. C **50**, 2841 (1994); D. C. Zheng, B. R. Barrett, J. P. Vary, W. C. Haxton, and C. L. Song, Phys. Rev. C **52**, 2488 (1995).
3. P. Navrátil and B. R. Barrett, Phys. Rev. C **54**, 2986 (1996); Phys. Rev. C **57**, 3119 (1998).
4. P. Navrátil and B. R. Barrett, Phys. Rev. C **57**, 562 (1998), Phys. Rev. C **59**, 1906 (1999).
5. P. Navrátil, G. P. Kamuntavičius and B. R. Barrett, Phys. Rev. C **61**, 044001 (2000). E-print archive No. nucl-th/9907054.
6. P. Navrátil, J. P. Vary and B. R. Barrett, Phys. Rev. Lett. **84**, 5728 (2000); Phys. Rev. C **62**, 054311 (2000).
7. P. Navrátil, J. P. Vary, W. E. Ormand, and B. R. Barrett, Phys. Rev. Lett. **87**, 172502 (2001).
8. P. Navrátil and W. E. Ormand, Phys. Rev. Lett. **88**, 152502 (2002).
9. C.P. Viazminsky and J.P. Vary, J. Math. Phys., **42**, 2055(2001).
10. P. Navrátil and W. E. Ormand, Phys. Rev. C 68, 034305 (2003).
11. P. Navrátil and E. Caurier, Phys. Rev. C 69, 014311 (2004).
12. K. Suzuki and S.Y. Lee, Prog. Theor. Phys. **64**, 2091 (1980); K. Suzuki, Prog. Theor. Phys. **68**, 246 (1982). K. Suzuki and R. Okamoto, Prog. Theor. Phys. **70**, 439 (1983).
13. R. Machleidt, F. Sammarruca and Y. Song, Phys. Rev. C **53**, 1483 (1996); R. Machleidt, Phys. Rev. C **63**, 024001 (2001).
14. J. Da Providencia and C. M. Shakin, Ann. of Phys. **30**, 95 (1964).
15. K. Suzuki, Prog. Theor. Phys. **68**, 1999 (1982); K. Suzuki and R. Okamoto, Prog. Theor. Phys. **92**, 1045 (1994).
16. H. Kamada, et. al, Phys. Rev. C **64**, 044001. (2001).
17. S. A. Coon and H. K. Han, Few Body Systems **30**, 131 (2001).
18. D. R. Entem and R. Machleidt, Phys. Rev. C 68, 041001(R) (2003).
19. U. van Kolck, Phys. Rev. C 49, 2932(1994); E. Epelbaum et. al., Phys. Rev. C 66, 064001 (2002).
20. H. Zhan, A. Nogga, B. R. Barrett, J. P. Vary and P. Navrátil, Phys. Rev. C **69**, 034302 (2004).

Chaos and Symmetry

Takaharu Otsuka*† and Noritaka Shimizu**†

*Department of Physics and Center for Nuclear Study, University of Tokyo, Hongo, Bunkyo-ku, Tokyo, 113-0033, Japan
†RIKEN, Hirosawa, Wako-shi, Saitama, 351-0198, Japan
**Department of Physics, University of Tokyo, Hongo, Bunkyo-ku, Tokyo, 113-0033, Japan

Abstract. The mechanism of the dominance (preponderance) of the 0^+ ground state for random interactions is proposed to be the chaotic realization of the highest rotational symmetry. This is a consequence of a general principle on the chaos and symmetry that the highest symmetry is given to the ground state if sufficient mixing occurs in a chaotic way by a random interaction. Under this symmetry-realization mechanism, the ground-state parity and isospin can be predicted so that the positive parity is favored over the negative parity and the isospin $T = 0$ state is favored over higher isospin. It is further suggested how one can enhance the realization of highest symmetries within random interactions. Thus, chaos and symmetry are shown to be linked deeply.

INTRODUCTION

The properties of the ground state with a random interaction have been attracting much interest since Johnson, Bertsch and Dean reported that the angular momentum of such a ground state is $J = 0$ predominantly [1]. In other words, in a shell model calculation in the sd shell with 6 neutrons, the ground state spin becomes $J = 0$ in the probability of 76 % with randomly generated two-body matrix elements, whereas only 9.8 % of the Hilbert space of 6 neutrons in the sd shell is $J = 0$ [1].

Since this striking report [1], many works have been carried out [2, 3, 4, 5, 6, 7, 8, 9, 10, 11]. The major points seem to be summarized as

1. A $J = 0$ ground state appears dominantly even if the pairing interaction is switched off in a random interaction.
2. No collectivity can be seen.
3. No general connection to symmetries has been established.

In particular, the basic mechanism for the preponderance of the $J = 0$ ground state has not been known, except for special cases with some algebraic structures [6, 7]. We shall discuss, in this talk, a possible mechanism and its generality.

RANDOM INTERACTION

Like previous works, the following shell model Hamiltonian is diagonalized:

$$H = \Sigma_{ijklL} v_{ijkl}^{(L)}([a_i^\dagger a_j^\dagger]^{(L)}[a_l a_k]^{(L)}), \tag{1}$$

CP726, Nuclear Physics, Large and Small: International Conference on Microscopic Studies of Collective Phenomena, edited by R. Bijker, R. F. Casten, and A. Frank
© 2004 American Institute of Physics 0-7354-0207-8/04/$22.00

where some mathematical details are omitted for brevity. Here, i, j, k and l denote single-particle orbits, L means their coupled angular momenta, and v stands for two-body matrix element (TBME). In the discussions below, the v's are generated as random numbers, unless otherwise specified. These random numbers are generated actually so as to obey the Gaussian distribution, and will be called TBRE. In the Hamiltonian above, single-particle energies are assumed to be zero because they are not essential.

CHAOTIC REALIZATION OF SYMMETRY

In considering properties of the ground state of a random interaction discussed just above, we propose the following Ansatz.

- A certain type of random interaction mixes various states strongly in a chaotic way without preferences, in eigenstates of a many-body system.
- In this situation, the maximum binding can be gained by being invariant with respect to transformations of coordinates, because all orientations should be equal.
- The highest symmetries are then given to the ground state ; a distortion of wave function results in loss of binding energy.

At this point, we mention that a randomly generated interaction does not necessarily fulfill the above feature of mixing. Namely, such an interaction can have certain dynamics *accidentally*. If such an accidental dynamics is absent, a random two-body interaction may mix various states with no preferred orientation in any sense. The invariance with respect to orientation occurs in the level of many-body wave function. Thus, the quantum chaotic mixing can produce the maximum binding energy, while the wave function acquires highest symmetries for relevant transformations. This will be called the chaotic realization of (highest) symmetries.

The consequence of the above mechanism for the rotation is that the ground state for an even number of particles should be of $J = 0$. This is the case as well known since [1].

The same mechanism is applied to a single-particle space comprised of orbits of both parities. Namely, one should expect the preponderance of positive-parity ground state for random interactions, because the positive-parity states are invariant with respect to the inversion. Figure 1 shows the probability distribution of the appearance of the ground state with given angular momentum and parity, J^π. The single-particle orbits are $1f_{5/2}$, $2p_{3/2,1/2}$ and $1g_{9/2}$, and 10 neutrons are put. The ground state is predominantly of positive parity, and the $J^\pi = 0^+$ ground state has the largest probability [12].

The same mechanism can be applied to a more abstract symmetry such as isospin. The states of $T = 0$ are invariant with respect to the rotation in the isospace. Therefore, one should expect the preponderance of the $T = 0$ ground state, if there are an equal number of protons and neutrons in the same single-particle space. Figure 2 shows the probability distribution of the appearance of the ground state with given angular momentum, J, and isospin, T. The single-particle space is the sd shell, and four protons and four neutrons are assumed. The ground state is really dominated by $T = 0$ as predicted. This phenomenon was shown in [8].

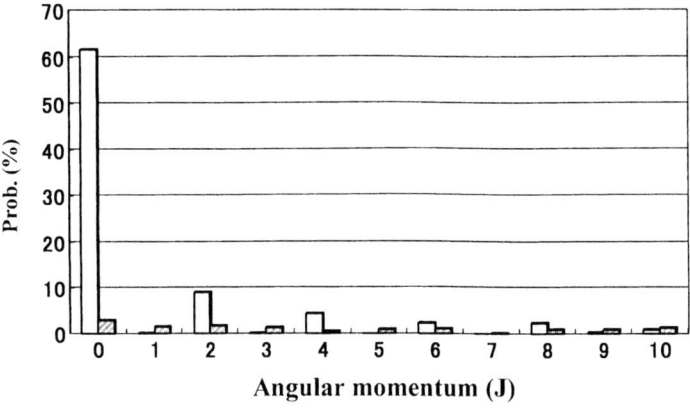

FIGURE 1. Probability distribution of the appearance of the ground state with given angular momentum and parity. The positive parity states are indicated by white histograms, while the negative parity states by gray ones.

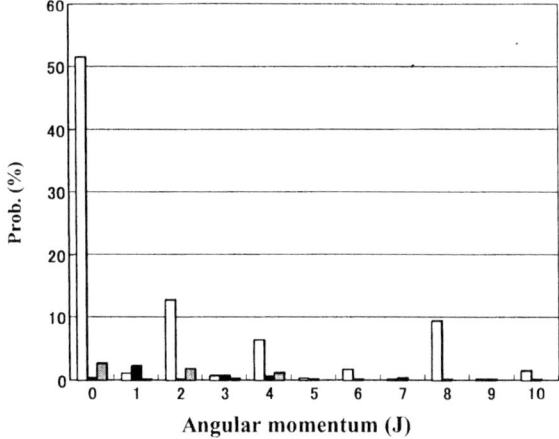

FIGURE 2. Probability distribution of the appearance of the ground state with given angular momentum, J, and isospin, T. The white, black and gray histograms indicate $T=0$, 1 and 2, respectively.

The cases with an odd number of particles can be understood along the same line. If there are N particles with N being an odd integer, $(N-1)$ particles tend to form a state with the highest symmetry, while the last (odd) particle is coupled to this *core*. Therefore, states with the same quantum numbers as those of the single-particle orbits are likely favored. Figure 3 shows the probability distribution of the appearance of the ground state with a given angular momentum, J. There are five neutrons in the sd shell. The ground state is really dominated by $J=1/2$, 3/2 and 5/2 as predicted.

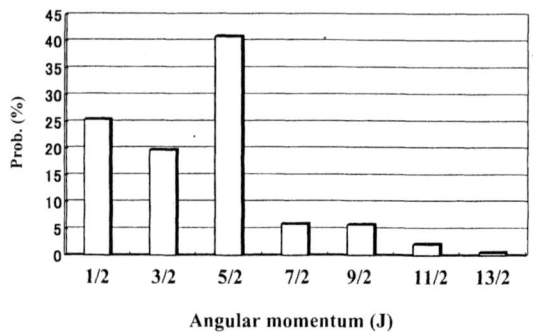

FIGURE 3. Probability distribution of the appearance of the ground state with a given angular momentum, J for the 5-body system in the *sd* shell.

QUANTUM CHAOTIC MIXING

We now discuss what type of random interactions generate the chaotic realization of (highest) symmetries. For this purpose, we introduce energy eigenvalues of 2-body system, E_{2-body}, because TBME's are subject to the choice of the single-particle bases but the eigenvalues are independent. We classify E_{2-body} according to the angular momentum L of the 2-body system. Figure 4 (a) and (b) indicate two typical cases. In (a), E_{2-body}'s are distributed, but the lowest one is far away from the rest. In (b), all E_{2-body}'s are close to each other. In fact, in (a), the lowest eigenvalue can produce a certain dynamical effect, resulting in a $J \neq 0$ ground state, if it happens with $L \neq 0$. On the other hand, in (b), various states are well mixed without much dynamics or preferences, and the ground state likely possesses the highest symmetry. For (a), one can shift the lowest eigenvalue to the middle of the rest as shown in Fig. 4 (c). There are many cases that the ground state is not $J = 0$ originally, but the $J = 0$ ground state appears after this manipulation.

This illustration suggests that random E_{2-body}'s in a compact energy region seem to be the key factor for the chaotic realization of symmetries.

Figure 5 indicates 2-body spectra for different L values. The TBRE can produce situations like the left hand side of Fig. 5. In this case, exceedingly low eigenvalues of $L=2$ may produce some special effects, leading to a $J \neq 0$ ground state. However, if all E_{2-body}'s get closer over all L values like the right hand side of Fig. 5, the quantum chaotic mixing determines the ground state, and a $J=0$ ground state should arise. This can be verified in Fig. 6 where the probability distribution is presented for two kinds of random interactions. One of them is the original TBRE. In the other one, all E_{2-body}'s of a given L are equally separated from neighbors, with their mean value being zero and their variance being a common fixed value. Of course, the wave functions of 2-body systems are random. Such confined random interactions produce the probability distribution shown by white histograms in Fig. 6, favoring $J=0$ ground states by $\sim 20\%$ more than the original TBRE does as shown by the hatched histograms.

(a)

$E_{2\text{-}body}$

(b)

$E_{2\text{-}body}$

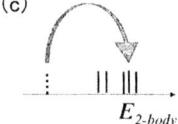

(c)

$E_{2\text{-}body}$

FIGURE 4. Schematic illustration of 2-body spectra. Vertical bars indicate energy eigenvalues.

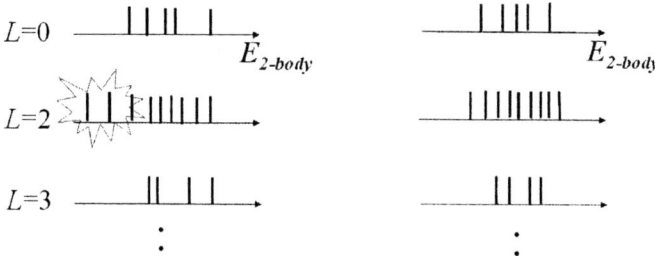

FIGURE 5. Schematic illustration of 2-body spectra of various L states. See the caption of Fig. 4.

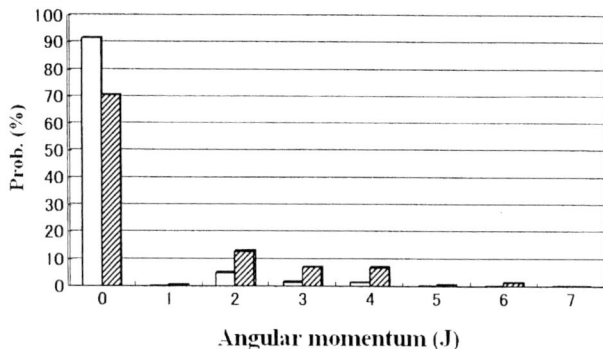

FIGURE 6. Probability distribution of the appearance of the ground state with a given angular momentum, J. White histograms are obtained from equally separated E_{2-body}'s (a model of confined random interaction), while hatched ones from the usual TBRE.

CONCLUDING REMARKS

We have proposed the mechanism of the chaotic realization of highest symmetries, and applied it to the angular momentum, parity and isospin. For the angular momentum, the third kind of the 0^+ state arises after the first kind from the pairing (or BCS) and the second kind from the rotational band (Nambu-Goldstone mechanism). The present mechanism can be generalized to other quantum many-body systems. A search for experimental examples is very intriguing, too. The chaotic realization of symmetries can be viewed also in the following way: if there is only a random interaction, it cannot gain much binding energy by localization, deformation, etc., and only the possible way is to exploit the phase space. To exploit the phase space at maximum, the state must be isotropic for relevant variable, which means the highest symmetry.

The present mechanism can be applied not only to the ground state but also to the state highest in energy, as proved by changing the sign of the interaction. Concerning bosons, the quantum chaotic mixing may be suppressed because possible boson condensation into the lowest 2-body eigenstate. This is consistent with what has been reported [2, 4]. Thus, the Pauli principle may play a certain role. Finally, the entropy should be clearly smaller in ground states with the chaotic realization of symmetries [13], although they gain the maximum binding from random interactions.

ACKNOWLEDGMENTS

One of the authors (TO) is grateful to the ECT* for its workshop "Recent Advances in the Nuclear Shell Model" where the initial idea was conceived. The authors acknowledge Professors C.W. Johnson and P. Van Isacker for useful information. This work was supported in part by Grant-in-Aid for Specially Promoted Research (13002001) from the Ministry of Education, Science, Sport and Culture.

REFERENCES

1. C.W. Johnson, and G.F. Bertsch and D.J. Dean, Phys. Rev. Lett. **80**, 2749 (1998).
2. R. Bijker and A. Frank, Phys. Rev. Lett. **84**, 420 (2000); Phys. Rev. C **62**, 014303 (2000).
3. D. Mulhall, A. Volya, and V. Zelevinsky, Phys. Rev. Lett. **85**, 4016-4019 (2000).
4. D. Kusnezov, N. V. Zamfir, and R. F. Casten, Phys. Rev. Lett. **85**, 1396 (2000).
5. L. Kaplan, T. Papenbrock, and C. W. Johnson, Phys. Rev. C **63**, 014307 (2001).
6. P. Chau Huu-Tai, A. Frank, N. A. Smirnova, and P. Van Isacker, Phys. Rev. C **66**, 061302 (2002).
7. Y. M. Zhao, A. Arima, and N. Yoshinaga, Phys. Rev. C **66**, 064322 (2002); 064323 (2002).
8. M. Horoi, A. Volya, and V. Zelevinsky, Phys. Rev. C **66**, 024319 (2002).
9. Y. M. Zhao, S. Pittel, R. Bijker, A. Frank, and A. Arima, Phys. Rev. C **66**, 041301 (2002).
10. V. Velazquez and A. P. Zuker, Phys. Rev. Lett. **88**, 072502 (2002).
11. V. Zelevinsky and A. Volya, Phys. Rep. **391**, 311 (2004).
12. The preponderance of positive parity ground states is pointed out also in the talk by Arima.
13. V. Zelevinsky, B.A. Brown, N. Frazier and M. Horoi, Phys. Rep. **276**, 85 (1996).

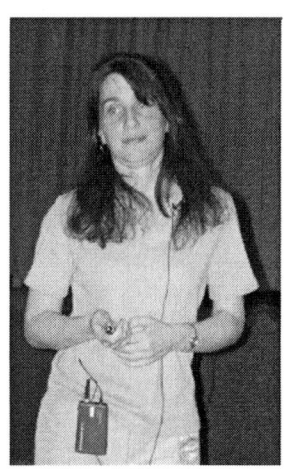

Skyrme-HFB deformed nuclear mass table

J. Dobaczewski[*†], M.V. Stoitsov[†**] and W. Nazarewicz[†*]

*Institute of Theoretical Physics, Warsaw University ul. Hoża 69, PL-00681 Warsaw, Poland
†Department of Physics and Astronomy, The University of Tennessee Knoxville, TN 37996, USA,
Joint Institute for Heavy Ion Research, Oak Ridge, TN 37831, USA
Physics Division, Oak Ridge National Laboratory, P.O. Box 2008, Oak Ridge, TN 37831, USA
**Institute of Nuclear Research and Nuclear Energy, Bulgarian Academy of Sciences
Sofia-1784, Bulgaria

Abstract. The Skyrme-Hartree-Fock-Bogoliubov code HFBTHO using the axial (2D) Transformed Harmonic Oscillator basis is tested against the HFODD (3D Cartesian HO basis) and HFBRAD (radial coordinate) codes. Results of large-scale ground-state calculations are presented for the SLy4 and SkP interactions.

1. INTRODUCTION

The code HFBTHO [1] solves the self-consistent HFB equations by using the axial (2D)·Transformed Harmonic Oscillator (THO) basis [2], which allows for a correct treatment of the single-quasiparticle wave function asymptotics. As discussed recently [3], the THO technique is a method of choice for performing massive nuclear structure calculations including weakly bound systems. In order to fully test the formalism, in the present study we present results obtained with the axial (2D) HFBTHO (v1.64) code compared to those obtained with two other codes: HFODD (v2.08i) [4], which uses a Cartesian (3D) Harmonic Oscillator (HO) basis and spherical (1D) HFBRAD [5], which uses a lattice of points in the radial coordinate.

In Ref. [3] we have published the first complete mass table of even-even nuclei obtained by using the THO method for the SLy4 Skyrme force [6]. Here we discuss one specific improvement of the method, and also present new results obtained with the SkP Skyrme force [7]. More details, including downloadable tables of ground-state properties can be found at http://www.fuw.edu.pl/~dobaczew/thodri/thodri.html.

2. TESTS

In this section, we discuss results of two numerical tests. First, by switching off the Local Scaling Transformation (LST) of THO, we run HFBTHO in the axial HO basis and test it against HFODD. For a given Skyrme interaction and zero-range, density-dependent pairing force, both codes should give exactly the same results. Since technical details of the inner structure of both codes are completely different, such calculations constitute

CP726, Nuclear Physics, Large and Small: International Conference on
Microscopic Studies of Collective Phenomena, edited by R. Bijker, R. F. Casten, and A. Frank
© 2004 American Institute of Physics 0-7354-0207-8/04/$22.00

an extremely stringent test of both codes.

Second, by switching the LST on, we could test the code HFBTHO against the spherical code HFBRAD [5]. Here, results of both codes cannot be *exactly* identical, because the phase spaces in which the solutions are obtained are significantly different.

Table 1 displays the results of test calculations performed for the SLy4 Skyrme interaction [6] and for the mixed zero-range pairing force [8]: $V(\vec{r})=V_0(1-\rho(\vec{r})/\rho_0)$ for $\rho_0=0.32$ fm^{-3}. The cutoff energy of $\varepsilon_{cut}=60$ MeV was used for summing up contributions of the HFB quasiparticle states to density matrices [2]. For a given phase space, the strength of the pairing force V_0 was adjusted so as to reproduce the experimental neutron pairing gap in ^{120}Sn. The resulting values are $V_0=-285.88$, -284.10, and -284.36 MeV fm^3 for the HO (THO) bases of 680 and 3276 states, and for the radial box of $R_{box}=30$ fm, respectively. The radial HFBRAD calculations were performed with 300 points (i.e., the $\Delta r=0.1$ fm grid spacing), and the wave functions were included up to $j_{max}=39/2$. We checked that even with $j_{max}=33/2$, all energies were stable within 1 eV. The nucleon-mass and elementary-charge parameters were fixed at $\hbar^2/2m=20.73553$ MeV fm^2 and $e^2=1.439978$ MeV fm, respectively.

Table 1 displays the following quantities: N_0 is the maximum number of the HO oscillator quanta included in the basis (for the deformed basis we give the numbers of quanta in the perpendicular ($N_\perp$) and axial (N_z) directions); N_{st} is the number of the lowest deformed HO states included in the basis; N_n^{qp} and N_p^{qp} are the numbers of (doubly degenerate) neutron and proton quasiparticle states with equivalent single-particle energies [2] below the cutoff energy ε_{cut}; $b_\perp$ and b_z are the oscillator constants in the perpendicular and axial directions; λ_n and λ_p are the neutron and proton Fermi energies, which, for vanishing pairing correlations, are taken as the s.p. energies of the last occupied states; Δ_n and Δ_p are the average pairing gaps [7]; R_n and R_p are the rms radii; Q_n and Q_p are the quadrupole moments $\langle 2z^2-x^2-y^2 \rangle$; ε_n^{gs} and ε_p^{gs} are the s.p. energies of the most bound neutron and proton states; Σ_n^ε and Σ_p^ε are sums of the canonical energies weighted by the corresponding occupation probabilities; E_n^{pair} and E_p^{pair} are the pairing energies; E_n^{kin} and E_p^{kin} are the kinetic energies; E_{cen} and E_{SO} are the energies corresponding to the central and spin-orbit parts of the Skyrme energy density functional; E_{dir} and E_{exc} are the direct and exchange parts of the Coulomb energy; and E_{stab} is the stability energy characterizing the level of self-consistency. In the code HFODD, E_{stab} is estimated from the sum of s.p. energies [9]; in the code HFBTHO E_{stab} is estimated from the maximum difference of all matrix elements of s.p. potentials calculated in two consecutive iterations; and in the code HFBRAD it is calculated as a variance of the total binding energy, E_{tot}, over the last five iterations.

Calculations for ^{208}Pb yield a spherical solution with vanishing pairing gaps. HFBTHO and HFODD give the total binding energies that differ by 627 eV, and this difference can be (primarily) traced back to the direct Coulomb energy. We have checked that without the Coulomb interaction, this difference decreases to 202 eV. The axial-basis HFBTHO calculation gives a very small total quadrupole moment of 39 μb. This suggests that the THO basis generates a slight deviation from the spherical symmetry due to a different numerical treatment of z- and $\perp$-direction. In this respect, HFODD calculations should be considered more accurate.

TABLE 1. (Color online) Benchmark results of the HFB calculations performed for the SLy4 interaction and mixed δ pairing. All energies are in MeV, lengths in fm, and quadrupole moments in barns. Boldface colored digits differ between the HFBTHO and HFODD/HFBRAD calculations. See text for details.

Nucleus:	^{208}Pb		^{168}Er		^{168}Er		^{120}Sn	
Code:	HFBTHO	HFODD	HFBTHO	HFODD	HFBTHO	HFODD	HFBTHO	HFBRAD
Basis:	2D-HO	3D-HO	2D-HO	3D-HO	2D-HO	3D-HO	2D-THO	Radial
N_0	14	14	14	14	$N_\perp=13, N_z=17$	$N_\perp=13, N_z=17$	25	n.a.
N_{st}	680	680	680	680	680	680	3276	n.a.
N_n^{qp}	532	532	489	489	497	497	924	4260
N_p^{qp}	481	481	448	448	451	451	855	4003
$b_\perp$	2.2348121	2.2348121	2.1566616	2.1566616	2.0581218	2.0581218	2.0390141	n.a.
b_z	2.2348121	2.2348121	2.1566616	2.1566616	2.3681210	2.3681210	2.0390141	n.a.
λ_n	−8.114095	−8.114020	−6.936061	−6.936058	−6.943872	−6.943858	−8.016795	−8.018081
λ_p	−8.810501	−8.810445	−7.156485	−7.156477	−7.152114	−7.152007	−11.107284	−11.107777
Δ_n	0	0	0.394570	0.394578	0.392326	0.392327	1.244750	1.244648
Δ_p	0	0	0.390601	0.390605	0.397728	0.397746	0	0
R_n	5.619758	5.619757	5.357578	5.357578	5.360037	5.360044	4.730466	4.730184
R_p	5.460080	5.460090	5.225538	5.225539	5.227218	5.227231	4.593884	4.593653
Q_n	−0.000022	6.6E-11	11.473921	11.473920	11.567875	11.567983	−0.001055	0
Q_p	−0.000017	4.7E-11	7.880228	7.880224	7.930128	7.930227	−0.000631	0
ε_n^{gs}	−58.001139	−58.001145	−56.014966	−56.014973	−55.996356	−55.996370	−55.756516	−55.755837
ε_p^{gs}	−44.042810	−44.042814	−44.422148	−44.422167	−44.486154	−44.486271	−46.629670	−46.631739
Σ_n^e	−3009.265452	−3009.264720	−2401.023343	−2401.023305	−2401.701865	−2401.698888	−1667.965633	−1668.063705
Σ_p^e	−1678.791400	−1678.790238	−1439.480739	−1439.480826	−1439.922261	−1439.913577	−1123.812244	−1123.857483
E_n^{pair}	0	0	−1.716956	−1.717024	−1.703028	−1.703045	−12.467146	−12.466964
E_p^{pair}	0	0	−1.528611	−1.528643	−1.584308	−1.584480	0	0
E_n^{kin}	2525.991268	2525.991925	1974.613878	1974.613824	1973.986024	1973.980663	1340.457995	1340.668648
E_p^{kin}	1334.854465	1334.854465	1118.313614	1118.313442	1118.487643	1118.487818	830.735396	830.848077
E_{cen}	−6194.978513	−6194.978930	−4944.027994	−4944.027545	−4943.869108	−4943.856093	−3475.705844	−3476.043789
E_{SO}	−96.374920	−96.375003	−80.186775	−80.186826	−80.216433	−80.214900	−49.167364	−49.196956
E_{dir}	827.607126	827.607885	602.810399	602.810352	602.694020	602.697867	366.472441	366.503834
E_{exc}	−31.248467	−31.248462	−25.935909	−25.935905	−25.935633	−25.935528	−19.102496	−19.103705
E_{stab}	8.1E-09	3.5E-11	1.0E-08	3.4E-06	9.6E-09	3.8E-06	9.9E-09	8.8E-08
E_{tot}	−1634.148747	−1634.148120	−1357.658354	−1357.658322	−1358.132823	−1358.127702	−1018.777019	−1018.790854

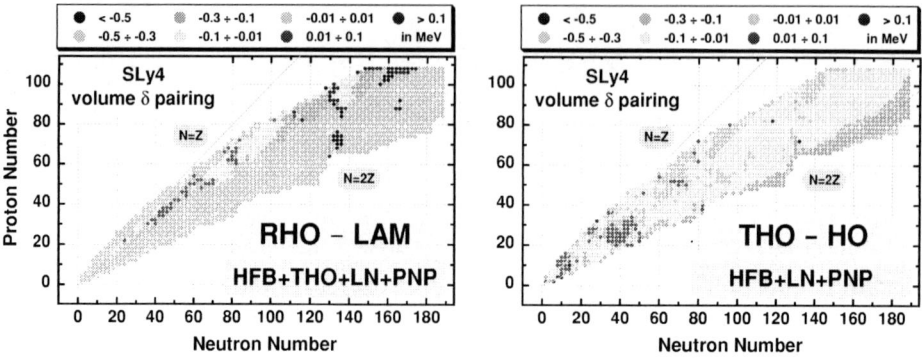

FIGURE 1. (Color online) Left: differences in E_{tot} obtained in HFBTHO by using LST based on RHO or LAM conditions. Right: differences between E_{tot} obtained in THO and HO bases. Calculations were performed using the SLy4 interaction with volume pairing and 20 oscillator shells. Lipkin-Nogami method followed by the exact particle-number projection was used to correct for the particle number nonconservation in HFB.

Calculations for ^{168}Er performed within a spherical HO basis, $b_\perp = b_z$, yield a well-deformed and weakly paired prolate ground state. Here, the total binding energies and quadrupole moments obtained within HFBTHO and HFODD differ only by 32 eV and 5 μb, respectively. When the same calculation is performed in a deformed HO basis, $b_\perp \neq b_z$, the differences grow to 5.1 keV and 207 μb, respectively. Again, without the Coulomb interaction, the difference in the total binding energy is only 96 eV. It is seen that by employing the deformed basis, the binding energy decreases, as expected.

Comparison with the coordinate-space code HFBRAD for ^{120}Sn shows that E_{tot} in HFBTHO is correct up to 14 keV for $N_0=25$. However, the kinetic energy still differs by as much as 221 keV, which is compensated by a similar difference in the interaction energy. Within the HO basis and $N_0=25$, the corresponding differences are larger: 41 and 337 keV. The analogous differences obtained for $N_0=20$ are 142 and 1103 keV (THO), and 152 and 964 keV (HO), respectively. Nevertheless, the above comparison shows that the $N_0=20$ calculations yield E_{tot} with a precision of a couple of hundred keV.

3. MASS TABLES

The LST employed in Ref. [3] was based on HO densities corrected in the asymptotic region by the contribution from the lowest-energy quasiparticle. Since a common LST has to be carried out for both neutrons and protons, for each nucleus one is forced to make a decision whether the LST is to be based on neutron or proton density. In Ref. [3] we used a prescription (referred to as LAM) that the neutron densities were used for $\lambda_n \geq \lambda_p$ and *vice versa*. In this work, we use the condition $\rho_n(R_{min}) \geq \rho_p(R_{min})$, where R_{min} is the point where the neutron or proton logarithmic density has a minimum as a function of r. In practice, the above condition, dubbed RHO, does not depend on whether the neutron or proton R_{min} is considered.

In Fig. 1 (left panel) we show the differences in E_{tot} obtained in HFBTHO by using the LST condition employing the Fermi energies (LAM) [3] or the densities (RHO). One can see that in the majority of neutron-rich nuclei both prescriptions lead to identical results. However, in many proton-rich nuclei the new prescription decreases binding up to about 500 keV, and for some medium-mass proton-rich nuclei the RHO method *decreases* binding by up to 100 keV. This latter effect is due to a better description of asymptotics in the pairing channel, which leads to extended pairing fields and reduced pairing energies [10]. The right panel of Fig. 1 shows differences in E_{tot} obtained in THO and HO bases. In most nuclei, by using the THO basis, one obtains a small energy gain of up to 10 keV. This grows to $\sim$500 keV for the very neutron-rich systems. Again, in lighter nuclei, a better asymptotics may lead to a reduced binding. In fact, our results show that improvements in density profiles at large distances cannot be treated variationally. First, E_{tot} is quite insensitive to the precise description of nucleonic densities in outer nuclear regions. Second, due to the pairing-space cutoff, the pairing energy is not reacting variationally on the improvement of the wave function.

Figures 2 and 3 present HFBTHO results obtained with the SLy4 and SkP Skyrme forces. It is obvious that without further improvements these traditional Skyrme forces describe nuclear masses rather poorly. The rms deviations between calculated and measured masses are as large as 3.14 MeV for SkP and 5.10 MeV for SLy4, respectively, as compared to about 0.70 MeV deviations obtained for forces fitted specifically to masses (see Ref. [11] for a review). Moreover, pronounced kinks obtained at magic numbers suggest that the quality of the description of (semi)magic and open-shell systems is not the same. This may point to a need to systematically include dynamical zero-point corrections [12]. Work in this direction is in progress.

ACKNOWLEDGMENTS

This work was supported in part by the Polish Committee for Scientific Research (KBN); by the Foundation for Polish Science (FNP); by the U.S. Department of Energy under Contract Nos. DE-FG02-96ER40963 (University of Tennessee), DE-AC05-00OR22725 with UT-Battelle, LLC (Oak Ridge National Laboratory), and DE-FG05-87ER40361 (Joint Institute for Heavy Ion Research); and by the National Nuclear Security Administration under the Stewardship Science Academic Alliances program through DOE Research Grant DE-FG03-03NA00083.

REFERENCES

1. M.V. Stoitsov *et al.*, to be published in Comput. Phys. Commun.
2. M.V. Stoitsov, W. Nazarewicz, and S. Pittel, Phys. Rev. **C58**, 2092 (1998); M.V. Stoitsov, J. Dobaczewski, P. Ring, and S. Pittel, Phys. Rev. C **61**, 034311 (2000).
3. M.V. Stoitsov *et al.*, Phys. Rev. C **68**, 054312 (2003).
4. J. Dobaczewski and P. Olbratowski, Comput. Phys. Commun. **158**, 158 (2004).
5. K. Bennaceur and J. Dobaczewski, to be published in Comput. Phys. Commun.
6. E. Chabanat, P. Bonche, P. Haensel, J. Meyer, and F. Schaeffer, Nucl. Phys. **A635**, 231 (1998).
7. J. Dobaczewski, H. Flocard, and J. Treiner, Nucl. Phys. **A422**, 103 (1984).

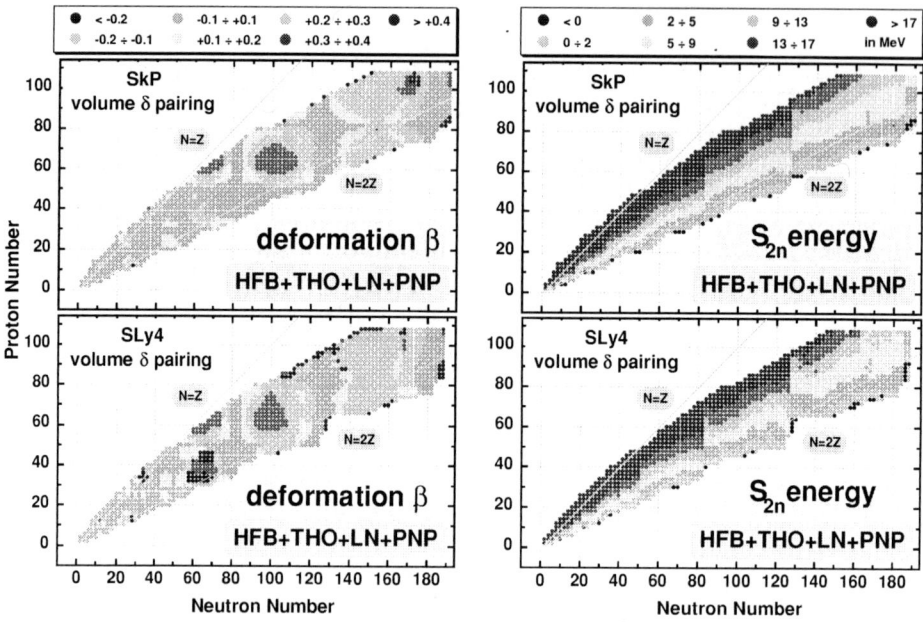

FIGURE 2. (Color online) Ground-state deformations β (left) and two-neutron separation energies S_{2n} (right) obtained within HFBTHO using SkP (top) and SLy4 (bottom) interactions.

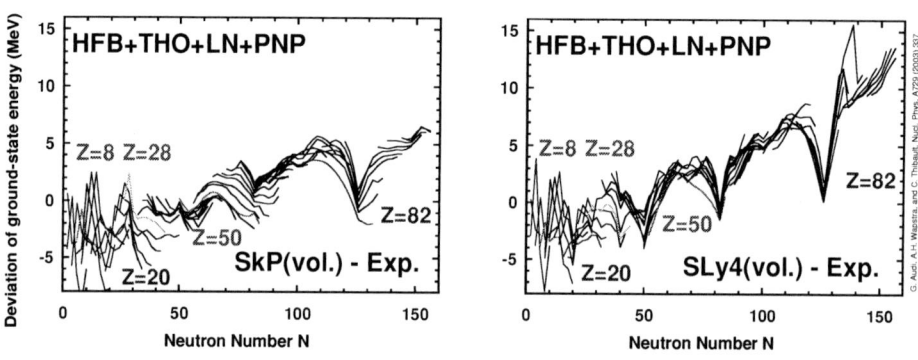

FIGURE 3. (Color online) Deviations of ground-state HFBTHO energies from experiment [13] for SkP (left) and SLy4 (right) interactions. Positive values correspond to underbound nuclei. No corrections beyond mean field were included.

8. J. Dobaczewski, W. Nazarewicz, and M.V. Stoitsov, in *The Nuclear Many-Body Problem 2001*, eds. W. Nazarewicz and D. Vretenar (Kluwer, Dordrecht, 2002), p. 181.
9. J. Dobaczewski and J. Dudek, Comput. Phys. Commun. **102**, 166 (1997); **102**, 183 (1997).
10. J. Dobaczewski *et al.*, Phys. Rev. **C53**, 2809 (1996).
11. D. Lunney, J.M. Pearson, and C. Thibault, Rev. Mod. Phys. **75**, 1021 (2003).
12. M. Bender, G. Bertsch, and P.-H. Heenen, Phys. Rev. **C69**, 034340 (2004).
13. G. Audi, A.H. Wapstra, and C. Thibault, Nucl. Phys. **A729**, 337 (2003).

Particle-Number-Projected HFB Method

M.V. Stoitsov*†, J. Dobaczewski**†, W. Nazarewicz†** and J. Terasaki‡

*Institute of Nuclear Research and Nuclear Energy, Bulgarian Academy of Sciences,
Sofia-1784, Bulgaria
†Department of Physics and Astronomy, University of Tennessee, Knoxville, TN 37996, USA,
Joint Institute for Heavy Ion Research, Oak Ridge, TN 37831, USA
Physics Division, Oak Ridge National Laboratory, P.O. Box 2008, Oak Ridge, TN 37831, USA
**Institute of Theoretical Physics, Warsaw University, ul. Hoża 69, 00-681 Warsaw, Poland
‡Department of Physics and Astronomy, University of North Carolina at Chapel Hill, Phillips Hall,
Chapel Hill, NC 27599-3255, USA

Abstract. Particle-number restoration before variation is implemented in the HFB method employing the Skyrme force and contact delta pairing. Results are compared with those obtained within the Lipkin-Nogami method, with or without the particle-number projection after variation.

1. INTRODUCTION

Pairing correlations play a central role in describing properties of atomic nuclei. In mean-field approaches, they are best treated in the Hartree-Fock-Bogoliubov (HFB) approximation [1]. The HFB product ansatz for the nuclear wave function, however, breaks the particle-number symmetry. The symmetry needs to be restored, in principle, especially if one looks at observables that strongly vary as functions of particle number.

Recently, it has been shown [2] that the total energy in the particle-number-projected (PNP) HFB approach can be expressed as a functional of the unprojected HFB density matrix and pairing tensor. Its variation leads to a set of HFB-like equations with modified Hartree-Fock fields and pairing potentials. The method has been illustrated within schematic models [3], and also implemented in HFB calculations with the finite-range Gogny force [4]. In the present paper we adopt it for the Skyrme functional and delta pairing, where the method must relay on the spatial locality of densities and mean fields. The HFB results using Lipkin-Nogami (LN) approximation followed by the particle-number projection *after variation* (PLN) are compared to the HFB results with projection *before variation* (PNP).

2. PARTICLE-NUMBER PROJECTED HFB METHOD

In the HFB method, the total energy is expressed as the energy functional of the one-body ρ and pairing κ density matrices

$$E[\rho,\kappa] = \frac{\langle\Phi|H|\Phi\rangle}{\langle\Phi|\Phi\rangle} = \mathrm{Tr}\left[(e+\tfrac{1}{2}\Gamma)\rho\right] - \tfrac{1}{2}\mathrm{Tr}[\Delta\kappa^*], \tag{1}$$

CP726, Nuclear Physics, Large and Small: International Conference on
Microscopic Studies of Collective Phenomena, edited by R. Bijker, R. F. Casten, and A. Frank
© 2004 American Institute of Physics 0-7354-0207-8/04/$22.00

where

$$\Gamma_{n_1 n_3} = \sum_{n_2 n_4} \bar{v}_{n_1 n_2 n_3 n_4} \rho_{n_4 n_2}, \quad \Delta_{n_1 n_2} = \frac{1}{2} \sum_{n_3 n_4} \bar{v}_{n_1 n_2 n_3 n_4} \kappa_{n_3 n_4}, \tag{2}$$

and e is a one-body (kinetic) term. Variation of Eq. (1) with respect to ρ and κ results in the HFB equations:

$$\begin{pmatrix} e + \Gamma - \lambda & \Delta \\ -\Delta^* & -(e+\Gamma)^* + \lambda \end{pmatrix} \begin{pmatrix} U \\ V \end{pmatrix} = E \begin{pmatrix} U \\ V \end{pmatrix}, \tag{3}$$

where the Lagrange multiplier λ is introduced to fix the correct *average* number of particles.

In Ref. [2] it has been pointed out that the PNP HFB energy,

$$E^N[\rho, \kappa] = \frac{\langle \Phi | H P^N | \Phi \rangle}{\langle \Phi | P^N | \Phi \rangle} = \frac{\int d\phi \langle \Phi | H e^{i\phi(\hat{N}-N)} | \Phi \rangle}{\int d\phi \langle \Phi | e^{i\phi(\hat{N}-N)} | \Phi \rangle}, \tag{4}$$

is again an energy functional of the unprojected densities ρ and κ. Variation of Eq. (4) with respect to ρ and κ results in the PNP HFB equations [2]:

$$\begin{pmatrix} \varepsilon^N + \Gamma^N + \Lambda^N & \Delta^N \\ -(\Delta^N)^* & -(\varepsilon^N + \Gamma^N + \Lambda^N)^* \end{pmatrix} \begin{pmatrix} U \\ V \end{pmatrix} = E^N \begin{pmatrix} U \\ V \end{pmatrix}. \tag{5}$$

Equation (5) has the same structure as Eq. (3) except that the new PNP fields read

$$\varepsilon^N = \frac{1}{2} \int d\phi\, y(\phi)\, (Y(\phi) \mathrm{Tr}[e\rho(\phi)] + [1 - 2ie^{-i\phi} \sin\phi\, \rho(\phi)] eC(\phi)) + \mathrm{h.c.},$$

$$\Gamma^N = \frac{1}{4} \int d\phi\, y(\phi)\, (Y(\phi) \mathrm{Tr}[\Gamma(\phi)\rho(\phi)] + 2[1 - 2ie^{-i\phi} \sin\phi\, \rho(\phi)] \Gamma(\phi) C(\phi)) + \mathrm{h.c.},$$

$$\Lambda^N = -\frac{1}{4} \int d\phi\, y(\phi)\, (Y(\phi) \mathrm{Tr}[\Delta(\phi)\bar{\kappa}^*(\phi)] - 4ie^{-i\phi} \sin\phi\, C(\phi)\Delta(\phi)\bar{\kappa}^*(\phi)) + \mathrm{h.c.},$$

$$\Delta^N = \frac{1}{2} \int d\phi\, y(\phi) e^{-2i\phi} C(\phi)\Delta(\phi) - \text{transposed},$$

$$\tag{6}$$

with

$$\Gamma_{n_1 n_3}(\phi) = \sum_{n_2 n_4} \bar{v}_{n_1 n_2 n_3 n_4} \rho_{n_4 n_2}(\phi),$$

$$\Delta_{n_1 n_2}(\phi) = \frac{1}{2} \sum_{n_3 n_4} \bar{v}_{n_1 n_2 n_3 n_4} \kappa_{n_3 n_4}(\phi), \quad \bar{\Delta}^*_{n_3 n_4}(\phi) = \frac{1}{2} \sum_{n_1 n_2} \bar{\kappa}^*_{n_1 n_2}(\phi) \bar{v}_{n_1 n_2 n_3 n_4}, \tag{7}$$

where I is the unit matrix and

$$\begin{aligned}
C(\phi) &= e^{2i\phi} \left(1 + \rho(e^{2i\phi} - 1)\right)^{-1}, & \rho(\phi) &= C(\phi)\rho, \\
\kappa(\phi) &= C(\phi)\kappa, & \bar{\kappa}(\phi) &= e^{2i\phi} C^\dagger(\phi)\kappa, \\
x(\phi) &= \frac{1}{2\pi} \frac{e^{-i\phi N} \det(e^{i\phi}I)}{\sqrt{\det C(\phi)}}, & y(\phi) &= \frac{x(\phi)}{\int d\phi' x(\phi')}, \\
Y(\phi) &= ie^{-i\phi} \sin\phi\, C(\phi) - i \int d\phi' y(\phi') e^{-i\phi'} \sin\phi'\, C(\phi').
\end{aligned} \tag{8}$$

58

3. PARTICLE-NUMBER-PROJECTED SKYRME-HFB METHOD

Due to the zero-range character of the Skyrme force, the Skyrme-HFB energy (1) becomes the energy functional,

$$E[\rho,\tilde{\rho}] = \frac{\langle\Phi|H|\Phi\rangle}{\langle\Phi|\Phi\rangle} = \int d\mathbf{r}\,\left[H(\mathbf{r}) + \tilde{H}(\mathbf{r})\right], \tag{9}$$

of *local* particle and pairing densities, where $H(\mathbf{r})$ and $\tilde{H}(\mathbf{r})$ are normal and pairing energy densities, respectively. Their explicit expressions [5] are given in terms of particle (pairing) local density $\rho(\mathbf{r})$ ($\tilde{\rho}(\mathbf{r})$), kinetic energy density $\tau(\mathbf{r})$ ($\tilde{\tau}(\mathbf{r})$) and spin-current density $\mathbf{J}_{ij}(\mathbf{r})$ ($\tilde{\mathbf{J}}_{ij}(\mathbf{r})$). All these local densities are completely determined by particle $\rho_{n'n}$ and pairing $\tilde{\rho}_{n'n}$ density-matrix elements in the configurational space, i.e.,

$$\rho(\mathbf{r}\sigma,\mathbf{r}'\sigma') = \sum_{nn'}\rho_{nn'}\,\psi_{n'}^*(\mathbf{r}',\sigma')\psi_n(\mathbf{r},\sigma),$$

$$\tilde{\rho}(\mathbf{r}\sigma,\mathbf{r}'\sigma') = \sum_{nn'}\tilde{\rho}_{nn'}\,\psi_{n'}^*(\mathbf{r}',\sigma')\psi_n(\mathbf{r},\sigma), \tag{10}$$

where $\psi_n(\mathbf{r},\sigma)$ constitute a complete and orthonormal set of single-particle wave functions that define the configurational space. The use of pairing density matrix $\tilde{\rho}(\mathbf{r}\sigma,\mathbf{r}'\sigma') = -2\sigma'\kappa(\mathbf{r},\sigma,\mathbf{r}',-\sigma')$ instead of the pairing tensor κ is convenient when the time-reversal symmetry is assumed [5].

Having in mind representation (10), one can see that the energy (9) is in fact a function $E(\rho,\tilde{\rho})$ of the matrix elements $\{\rho_{nn'}\}$ of ρ and $\{\tilde{\rho}_{nn'}\}$ of $\tilde{\rho}$. Its derivatives with respect to $\rho_{nn'}$ and $\tilde{\rho}_{nn'}$ define the particle-hole and particle-particle matrices

$$h_{nn'} = \frac{\partial E(\rho,\tilde{\rho})}{\partial\rho_{n'n}} = \sum_{\sigma\sigma'}\int d\mathbf{r}\,\psi_n^*(\mathbf{r},\sigma)h(\mathbf{r},\sigma,\sigma')\psi_{n'}(\mathbf{r},\sigma'),$$

$$\tilde{h}_{nn'} = \frac{\partial E(\rho,\tilde{\rho})}{\partial\tilde{\rho}_{n'n}} = \sum_{\sigma\sigma'}\int d\mathbf{r}\,\psi_{n'}^*(\mathbf{r},\sigma)\tilde{h}(\mathbf{r},\sigma,\sigma')\psi_n(\mathbf{r},\sigma'), \tag{11}$$

respectively, which enter the Skyrme-HFB equations

$$\begin{pmatrix} h-\lambda & \tilde{h} \\ \tilde{h} & -h+\lambda \end{pmatrix}\begin{pmatrix} U \\ V \end{pmatrix} = E\begin{pmatrix} U \\ V \end{pmatrix}. \tag{12}$$

In the case of the Skyrme force, the projected energy (4) reads

$$E^N[\rho,\tilde{\rho}] = \frac{\langle\Phi|HP^N|\Phi\rangle}{\langle\Phi|P^N|\Phi\rangle} = \int d\phi\,y(\phi)\int d\mathbf{r}\,(H(\mathbf{r},\phi)+\tilde{H}(\mathbf{r},\phi)), \tag{13}$$

where the gauge-angle dependent energy densities $H(\mathbf{r},\phi)$ and $\tilde{H}(\mathbf{r},\phi)$ are derived from the unprojected ones by simply replacing particle (pairing) local densities $\rho(\mathbf{r})$ ($\tilde{\rho}(\mathbf{r})$), $\tau(\mathbf{r})$ ($\tilde{\tau}(\mathbf{r})$), $\mathbf{J}_{ij}(\mathbf{r})$ ($\tilde{\mathbf{J}}_{ij}(\mathbf{r})$) by their gauge-angle dependent counterparts $\rho(\mathbf{r},\phi)$

$(\tilde{\rho}(\mathbf{r},\phi))$, $\tau(\mathbf{r},\phi)$ $(\tilde{\tau}(\mathbf{r},\phi))$, $\mathbf{J}_{ij}(\mathbf{r},\phi)$ $(\tilde{\mathbf{J}}_{ij}(\mathbf{r},\phi))$. The latter ones depend on the gauge-angle dependent density matrices, i.e.,

$$\rho(\mathbf{r}\sigma,\mathbf{r}'\sigma',\phi) = \sum_{nn'} \rho_{nn'}(\phi)\, \psi_{n'}^*(\mathbf{r}',\sigma')\psi_n(\mathbf{r},\sigma),$$

$$\tilde{\rho}(\mathbf{r}\sigma,\mathbf{r}'\sigma',\phi) = \sum_{nn'} \tilde{\rho}_{nn'}(\phi)\, \psi_{n'}^*(\mathbf{r}',\sigma')\psi_n(\mathbf{r},\sigma). \tag{14}$$

Since

$$\rho_{n'n}(\phi) = \frac{\langle\Phi|c_n^\dagger c_{n'} e^{i\phi\hat{N}}|\Phi\rangle}{\langle\Phi|e^{i\phi\hat{N}}|\Phi\rangle} = \sum_m C_{nm}(\phi)\rho_{mn'},$$

$$\tilde{\rho}_{n'n}(\phi) = e^{-i\phi}\sum_m C_{nm}(\phi)\tilde{\rho}_{mn'}, \tag{15}$$

the projected energy (13) is again a functional of the unprojected density matrices ρ and $\tilde{\rho}$ and a function $E^N(\rho,\tilde{\rho})$ of their matrix elements $\rho_{n'n}$, $\tilde{\rho}_{n'n}$. In order to take the derivatives of $E^N(\rho,\tilde{\rho})$ with respect to ρ and $\tilde{\rho}$, one should take first the derivatives of $E^N(\rho,\tilde{\rho})$ with respect to $\rho(\phi)$ and $\tilde{\rho}(\phi)$, and then the derivatives of $\rho(\phi)$ and $\tilde{\rho}(\phi)$ with respect to unprojected ρ and $\tilde{\rho}$. Of course, one should take also the derivative of $y(\phi)$ with respect to ρ. In this way, one obtains the PNP Skyrme-HFB equations in the form

$$\begin{pmatrix} h^N & \tilde{h}^N \\ \tilde{h}^N & -h^N \end{pmatrix} \begin{pmatrix} U \\ V \end{pmatrix} = E^N \begin{pmatrix} U \\ V \end{pmatrix}, \tag{16}$$

with particle-hole and particle-particle hamiltonians

$$h^N = \int d\phi y(\phi)Y(\phi)E(\phi) + \int d\phi y(\phi)e^{-2i\phi}\, C(\phi)h(\phi)C(\phi)$$

$$- \left[\int d\phi y(\phi)2ie^{-i\phi}\sin(\phi)\tilde{\rho}(\phi)\tilde{h}(\phi)C(\phi) + \text{h.c.}\right], \tag{17}$$

$$\tilde{h}^N = \int d\phi y(\phi)e^{-i\phi}\{\tilde{h}(\phi)C(\phi) + \text{transposed}\}$$

The gauge-angle dependent fields $h(\mathbf{r},\sigma,\sigma',\phi)$ and $\tilde{h}(\mathbf{r},\sigma,\sigma',\phi)$ entering the matrix elements,

$$h_{nn'}(\phi) = \sum_{\sigma\sigma'} \int d^3\mathbf{r}\, \psi_n^*(\mathbf{r}\sigma)h(\mathbf{r},\sigma,\sigma',\phi)\psi_{n'}(\mathbf{r}\sigma'),$$

$$\tilde{h}_{nn'}(\phi) = \sum_{\sigma\sigma'} \int d^3\mathbf{r}\, \psi_{n'}^*(\mathbf{r}\sigma)\tilde{h}(\mathbf{r},\sigma,\sigma',\phi)\psi_n(\mathbf{r}\sigma'), \tag{18}$$

are obtained by simply replacing in the unprojected fields (11) the particle (pairing) local densities $\rho(\mathbf{r})$ $(\tilde{\rho}(\mathbf{r}))$, $\tau(\mathbf{r})$ $(\tilde{\tau}(\mathbf{r}))$, and $\mathbf{J}_{ij}(\mathbf{r})$ $(\tilde{\mathbf{J}}_{ij}(\mathbf{r}))$ with their gauge-angle dependent counterparts $\rho(\mathbf{r},\phi)$ $(\tilde{\rho}(\mathbf{r},\phi))$, $\tau(\mathbf{r},\phi)$ $(\tilde{\tau}(\mathbf{r},\phi))$, and $\mathbf{J}_{ij}(\mathbf{r},\phi)$ $(\tilde{\mathbf{J}}_{ij}(\mathbf{r},\phi))$. When using delta pairing forces, one has to restrict the quasiparticle space in order to avoid the divergences associated with the zero range. Within the unprojected HFB calculations, a pairing cut-off is introduced by using the so-called equivalent single-particle spectrum

[5, 6]. After each iteration, performed with a given chemical potential λ, one calculates an equivalent spectrum $\bar{e}_n$ and pairing gaps Δ_n:

$$\bar{e}_n = (1 - 2P_n)E_n, \quad \bar{\Delta}_n = 2E_n\sqrt{P_n(1 - P_n)}, \tag{19}$$

where E_n are the quasiparticle energies and P_n denotes the norms of the lower HFB wave functions. Due to the similarity between $\bar{e}_n$ and the single-particle energies, one can take into account only those quasiparticle states for which $\bar{e}_n$ is less than the cut-off energy ε_{cut} (usually around 60 MeV).

One can see that this procedure cannot be directly applied to the PNP HFB calculations, because the Lagrange multiplier λ entering the unprojected HFB Eqs. (12) is no longer available in Eq. (16). This means that the local densities emerging after each HFB diagonalization (16) are not automatically normalized to the particle number N. As a result, all auxiliary quantities, as e.g. the analogues of the quasiparticle energies, E_n^N, and probabilities, P_n^N, do not have usual meaning. However, one can always reintroduce the Lagrange multiplier λ into Eq. (16), without changing results, and adjust it independently, so as to have a correct *average* particle number in the *unprojected* state. In practice, it is enough to calculate for the solutions of Eq. (16) the average values E_n of the unprojected HFB matrix and use them in Eq. (19) together with $P_n \equiv P_n^N$. This allows for defining the Lagrange multiplier and implementing the cut-off procedure. Such an approach is not too expensive, since the unprojected quantities have to be calculated anyhow at zero gauge-angle, and their use is straightforward.

4. SAMPLE RESULTS

Fig. 1 gives the PNP HFB results for the complete chain of the calcium isotopes (proton-neutron drip to drip line), calculated for the Sly4 force [7] and mixed delta pairing [8]. Comparison is also made with the HFB Lipkin-Nogami (LN) results and projected (after variation) Lipkin-Nogami results (PLN). One can conclude that the PLN approximation is good for open-shell nuclei, where the total energy differences between various variants are less than 250 keV. For closed-shell nuclei [9], however, the energy differences increase to more than 1 MeV. In such cases, one can improve the PLN results by applying the projection to the LN solutions obtained in neighboring nuclei [10], as is illustrated in the top panel of Fig. 1.

ACKNOWLEDGMENTS

This work was supported in part by the U.S. Department of Energy under Contract Nos. DE-FG02-96ER40963 (University of Tennessee), DE-AC05-00OR22725 with UT-Battelle, LLC (Oak Ridge National Laboratory), and DE-FG05-87ER40361 (Joint Institute for Heavy Ion Research); by the National Nuclear Security Administration under the Stewardship Science Academic Alliances program through DOE Research Grant DE-FG03-03NA00083; by the Polish Committee for Scientific Research (KBN); and by the Foundation for Polish Science (FNP).

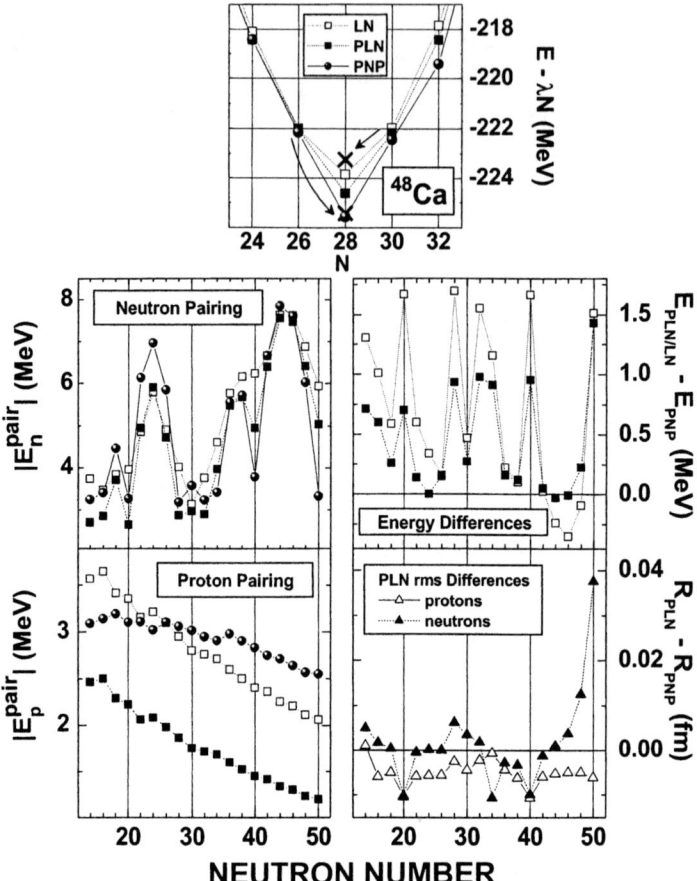

FIGURE 1. The LN and PLN (projection after variation), and PNP HFB (projection before variation) results obtained for the SLy4 force and mixed delta pairing. Arrows in the top panel indicate projection results from neighboring nuclei.

REFERENCES

1. P. Ring and P. Schuck, *The Nuclear Many-Body Problem* (Springer Verlag, New York, 1980).
2. J.A. Sheikh and P. Ring, Nucl. Phys. **A665** (2000) 71.
3. J.A. Sheikh, P. Ring, E. Lopes, and R. Rossignoli, Phys. Rev. **C66**, 044318 (2002).
4. M. Anguiano, J.L. Egido and L.M. Robledo, Nucl. Phys. **A696**, 476 (2001).
5. J. Dobaczewski, H. Flocard, and J. Treiner, Nucl. Phys. **A422**, 103 (1984).
6. M.V. Stoitsov, W. Nazarewicz, and S. Pittel, Phys. Rev. **C58**, 2092 (1998); M.V. Stoitsov, J. Dobaczewski, P. Ring, and S. Pittel, Phys. Rev. C **61**, 034311 (2000).
7. E. Chabanat, P. Bonche, P. Haensel, J. Meyer, and F. Schaeffer, Nucl. Phys. **A635**, 231 (1998).
8. J. Dobaczewski, W. Nazarewicz, and M.V. Stoitsov, in *The Nuclear Many-Body Problem 2001*, eds. W. Nazarewicz and D. Vretenar (Kluwer, Dordrecht, 2002), p. 181.
9. J. Dobaczewski and W. Nazarewicz, Phys. Rev. **C47**, 2418 (1993)
10. P. Magierski, S. Ćwiok, J. Dobaczewski, and W. Nazarewicz, Phys. Rev. **C48**, 1686 (1993)

Shifted-Contour Monte Carlo Method for Nuclear Structure

G.S. Stoitcheva and D.J. Dean

Physics Division, Oak Ridge National Laboratory, P.O. Box 2008, Oak Ridge, TN 37831, USA

Abstract. We propose a new approach for alleviating the 'sign' problem in the nuclear shell model Monte Carlo method. The approach relies on modifying the integration contour of the Hubbard-Stratonovich transformation to pass through an imaginary stationary point in the auxiliary-field associated with the Hartree-Fock density.

1. INTRODUCTION

The interacting nuclear shell model has been applied successfully to the microscopic description of nuclei in the p, sd, and pf-shells using two-body interactions that give a highly accurate description of nuclear properties. However, the application of standard diagonalization of the shell-model Hamiltonian matrix is limited by the combinatorial increase of the rank of the matrix with increasing model space or increasing numbers of valence particles within a given model space. One often resorts to truncation techniques in order to circumvent this combinatorial growth. An alternative approach requires recasting the standard diagonalization problem into a problem of multidimensional integration.

The Shell Model Monte Carlo (SMMC) method [1, 2] avoids an explicit enumeration of the many body-states and circumvents the difficulties related to the dimensions of the many-body space. Furthermore, it is capable of providing exact results for a wide range of observables. Nevertheless, the power of the SMMC technique has been limited by a sign problem associated with the Monte Carlo weight function when using realistic interactions. Different attempts were pursued that alleviate the sign difficulties. One of these employed an extrapolation approach that overcomes the sign problem and was previously validated [3]. However, the approach is not exact and does rely on extrapolations.

In this paper, we exploit a new method, the Shifted-Contour Shell Model Monte Carlo (SC-SMMC) method for nuclear structure. The shifted contour technique has been successfully applied for electron-structure calculations of ground and low-lying excited-states [4]. The foundation of this new method is based on shifting the Monte Carlo integration to a region in the complex plane where fluctuations around the Hartree-Fock solution are minimized.

We consider a generalized two-body Hamiltonian

$$\hat{H} = \sum_{\alpha\beta} \varepsilon_\alpha a_\alpha^\dagger a_\alpha + \frac{1}{2} \sum_{\alpha\beta\gamma\delta} V_{\alpha\beta\gamma\delta} a_\alpha^\dagger a_\beta^\dagger a_\delta a_\gamma, \tag{1}$$

CP726, *Nuclear Physics, Large and Small: International Conference on Microscopic Studies of Collective Phenomena*, edited by R. Bijker, R. F. Casten, and A. Frank
© 2004 American Institute of Physics 0-7354-0207-8/04/$22.00

where $a_\alpha^\dagger$ and a_α are anti-commuting fermion creation and annihilation operators associated with the single particle state α defined by the complete set of quantum numbers $(nlmjt_z)$, and ε_α and $V_{\alpha\beta\gamma\delta}$ are the single-particle energies and the two-body matrix elements of the residual interaction, respectively. The SMMC technique relies on the ability to employ the imaginary-time evolution operator, $\hat{U}=\exp(-\beta\hat{H})$, where β is the inverse temperature, to project out the ground state. To achieve a simplification of the propagator, $\hat{U}$, the Hamiltonian in Eq. (1) can be brought into quadratic form

$$\hat{H} = \varepsilon\hat{\Theta} + \frac{1}{2}\Lambda\hat{\Theta}\hat{\Theta}, \tag{2}$$

where $\hat{\Theta}$ is a density operator, Λ is the strength of the two-body interaction, and ε is a single-particle energy. Then, by using the Hubbard-Stratonovich (HS) transformation [6]

$$e^{\frac{1}{2}\Lambda\hat{\Theta}^2} = \sqrt{\frac{|\Lambda|}{2\pi}} \int d\sigma\, e^{-\frac{1}{2}\Lambda\sigma^2 + s\sigma\Lambda\hat{\Theta}}, \tag{3}$$

where the phase, $s = \pm 1$ if $\Lambda \geq 0$ or $\pm i$ if $\Lambda < 0$, the two-body term of the Hamiltonian is presented as a superposition of one-body operators

$$e^{-\beta\hat{H}} = \sqrt{\frac{\beta|\Lambda|}{2\pi}} \int d\sigma\, e^{-\frac{1}{2}\beta|\Lambda|\sigma^2} e^{-\beta\hat{h}}, h = (\varepsilon + s\Lambda\sigma)\hat{\Theta}. \tag{4}$$

The integral is a weighted sum over all possible densities σ and the weight, $e^{-\frac{1}{2}\beta|\Lambda|\sigma^2}$, is Gaussian. Furthermore, fluctuating auxiliary fields which are coupled to the protons and the neutrons are generated and the exponential is split into N_t time-slices with $\beta = N_t\Delta\beta$. In these fields the fermions are considered as non-interacting particles.

The associated expectation value of a given observable is expressed by

$$\langle\hat{\Omega}\rangle_A = \frac{\int D[\sigma]G(\sigma)\Phi(\sigma)\langle\Omega\rangle_\sigma}{\int D[\sigma]G(\sigma)\Phi(\sigma)}, \tag{5}$$

where $D[\sigma] = \prod_{n,\alpha} d\sigma_\alpha(\tau_n)$ is the volume element and $G(\sigma) = \exp\left[-\frac{1}{2}\Delta\beta\sum_{\alpha,n}|V_\alpha|\sigma_{\alpha n}^2\right]$ is the Gaussian factor.

By using Eq. (5), in the limit of low temperature ($T \to 0$ or $\beta \to \infty$), the properties of the ground state can be extracted. Although this expression is exact, the fluctuations in $\Phi(\sigma) = \frac{\zeta(\sigma)}{|\zeta(\sigma)|}$, which is defined as the sign of the Monte Carlo weight function, determines how precise the Monte Carlo evaluation can be. When the average sign is smaller than its uncertainty, this leads to the sign problem: the sign of the integrand fluctuates among samples and the integral is a result of cancellations that cannot be reproduced with a finite number of samples. When considering finite temperature, canonical ensemble, the sign decreases exponentially with the temperature. Some Hamiltonians are free from sign problems, e.g. pairing plus quadrupole. These Hamiltonians include correctly the dominant collective components of the effective interaction. In addition, the sign problem is related to the time-reversal properties of the ne-body of the Hamiltonian in Eq. (4) [1, 7]. It has been shown that all realistic interactions suffer form the sign problem and the Monte Carlo method in this case fails.

The extrapolation method [3] has been considered as a 'practical' solution to the problem and has made possible the use of realistic interactions in Monte Carlo evaluations.

2. SHIFTED-CONTOUR METHOD FOR SHELL MODEL MONTE CARLO CALCULATIONS

Our goal is to modify the HS transformation so that the formation of large-amplitude oscillations in the integrand are reduced. For this purpose, let us consider the two-body Hamiltonian, Eq. (1), in which the density operator is shifted by an arbitrary density, η:

$$\hat{H} = \sum_{\alpha\beta}(\varepsilon_{\alpha\beta} + W_{\alpha\beta})a^{\dagger}_{\alpha}a_{\beta} + \frac{1}{2}\sum_{\alpha\beta\gamma\delta}V_{\alpha\beta\gamma\delta}(a^{\dagger}_{\alpha}a_{\delta} - n_{\alpha\delta})(a^{\dagger}_{\beta}a_{\gamma} - n_{\beta\gamma}) + const, \qquad (6)$$

where $W_{\alpha\beta} = \sum_{\gamma\delta}V_{\alpha\beta\gamma\delta}n_{\alpha\beta}$ is the change in the one-body part. The Hamiltonian in Eq. (6) is equivalent to the original Hamiltonian, Eq. (1). The effect of the shift in $\hat{H}$ is equivalent to a shift of the auxiliary density σ to a line with the density η. Formally, the contour of integration passes through a stationary point. The HS transformation applied to Eq. (6) becomes:

$$e^{\frac{1}{2}\Lambda\Theta^2} = const \int d\sigma \, e^{-\frac{1}{2}\Lambda\sigma^2 + s(\sigma - s\eta)\Lambda\Theta}, \qquad (7)$$

As in case of electrons, for our calculations we also consider the arbitrary density, $\eta_{\alpha\beta}$, to be chosen as the expectation value of the Hartree-Fock density in the ground state. Due to the change in the integrand, Eq. (7), the oscillating part, $e^{s(\sigma - s\eta)\Lambda\Theta}$, is damped by η and the fluctuations are highly reduced.

This approach was first introduced and successfully applied to electron-structure calculations [5]. Currently, we apply this technique to nuclear systems in sd-shell and consider their properties at finite temperature (canonical ensemble). In Figure 1, we plot the behavior of the sign versus different values of β. Two examples are presented, ^{28}Mg and ^{28}Si. The change of the sign when increasing β for ^{28}Mg based on the shifted-contour method is compared with the SMMC without applying the shift. Our results show a significant delay of the sign due to the mean-field shift. Stabilization is achieved and an overwhelming part of the sign problem is removed. The calculations are performed up to larger β (in these cases, $\beta \approx 2MeV^{-1}$), or lower temperature, as compared to the SMMC calculations.

The shifted-contour method leads to a powerful technique for alleviating the sign problem in the SMMC method. The sign delay opens up the possibility of a more detailed description of different properties of nuclei in spaces where the exact diagonalization is prohibited or the extrapolation method is not valid. While this work exhibits the power of the SC-SMMC method, questions regarding the validity of the method in larger model spaces remain to be answered.

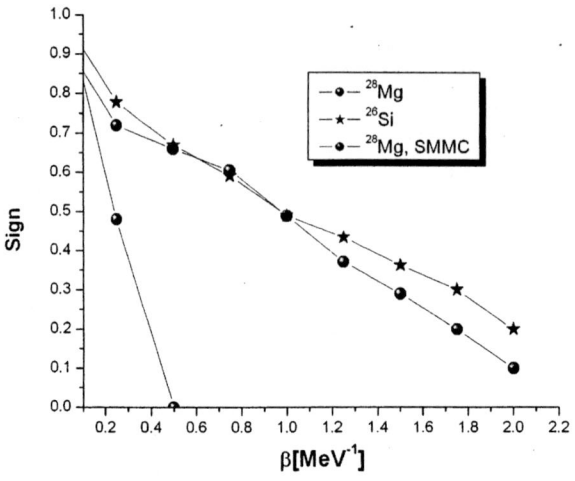

FIGURE 1. The sign as a function of β.

ACKNOWLEDGMENTS

This work has been partly supported by the Department of Energy through the Scientific Discovery through Advanced Computing Program. Oak Ridge National Laboratory is managed by UT-Battelle, LLC for the U.S. Department of Energy under Contract No. DE-AC05-00OR22725.

REFERENCES

1. G. H. Lang, C. W. Johnson, S. E. Koonin, and W. E. Ormand, Phys. Rev. C **48**, 1518 (1993).
2. S. E. Koonin, D. J. Dean, and K. Langanke, Phys. Rep. **278**, 2 (1997).
3. Y. Alhassid, D. J. Dean, S. E. Koonin, G. Lang, and W. E. Ormand, Phys. Rev. Lett. **72**, 613 (1994).
4. N. Rom, D.M. Charutz, and D. Neuhauser, Chem. Phys. Lett. **270**, 382 (1997).
5. S. Jacobi and R. Baer, Journal of Chem. Phys. **120**, 43 (2004).
6. J. Hubbard, Phys. Rev. Lett. **3**, 77 (1959); R. L. Stratonovich, Dokl. Akad. Nauk. S.S.S.R. **115**, 1097 (1957).
7. C.W. Johnson and D.J. Dean, Phys. Rev. C **61**, 044327 (2000)

Nuclear shell model frontiers

T. Papenbrock

Physics Division, Oak Ridge National Laboratory, Oak Ridge, TN 37831, USA
Department of Physics and Astronomy, University of Tennessee, Knoxville, TN 37996, USA

Abstract. I present two recent advances in the nuclear shell model. The first concerns the selection of the most relevant basis states out of the vast number of available configurations. Here, the density matrix renormalization group (DMRG) and the wave function factorization are two approaches that approximate low-lying states. The second part deals with a puzzle of the shell model itself and addresses the preponderance of spin-0 ground states in shell models with random two-body interactions.

The shell model is the dynamical model of the atomic nucleus. In recent years, considerable progress has been made on various aspects and applications of this model. This concerns the determination of the effective interaction itself as well as applications to weakly bound systems. New ideas have also been developed to deal with the enormous size of the configuration space. Here, the density matrix renormalization group (DMRG) and the wave function factorization are two approaches that will be discussed in Sect. 1. Another interesting aspect of the shell model deals with the structure that is imposed by the rotational symmetry alone. The recently found preponderance of spin-0 ground states in a shell model with random two-body interactions has been a theoretical puzzle, and will be partly unraveled in Sect. 2.

1. FACTORIZATION OF SHELL MODEL GROUND STATES

Shell model spaces grow factorial in size as the number of valence nucleons and/or the number of single-particle orbitals is increased. Exact shell model diagonalizations [1, 2, 3, 4] have made impressive progress over the last decade, but the largest shell model problems are presently solved with more favorably scaling stochastic methods like the Shell Model Monte Carlo [5] and the Monte Carlo Shell Model [6]. Recently, several approximation methods have been proposed that truncate the Hilbert space and select the most relevant basis states [7, 8, 9, 10, 11, 12]. Among these methods, the DMRG [10, 11] and the wave function factorization [12] select an optimal basis that is formed by products of correlated states from two different subsets of single-particle orbitals. Possible choices of the subsets are, for instance, particle and hole orbitals, or proton and neutron orbitals. Details can be found in the recent review [13].

CP726, *Nuclear Physics, Large and Small: International Conference on*
Microscopic Studies of Collective Phenomena, edited by R. Bijker, R. F. Casten, and A. Frank
© 2004 American Institute of Physics 0-7354-0207-8/04/$22.00

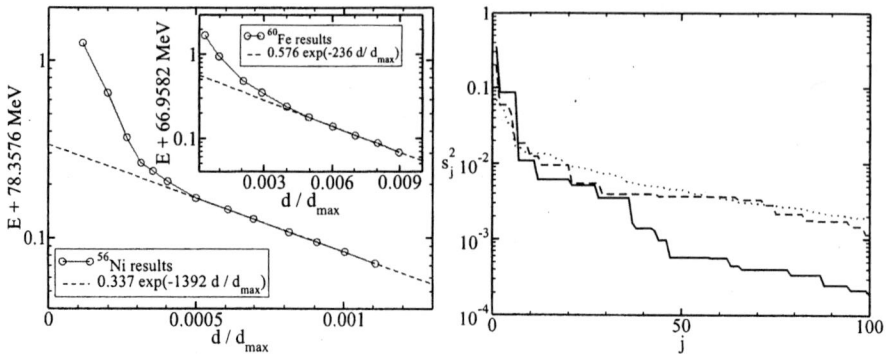

FIGURE 1. Left: Ground state energy as a function of the relative dimension for pf-shell nuclei ^{56}Ni and ^{60}Fe (inset); taken from Ref. [14]. Right: Largest norms s_j^2 for the ground state of the sd-shell nucleus ^{24}Mg for three different factorizations. The two subsets of single-particle orbitals are proton vs. neutron orbitals (solid line), $d_{5/2}$ vs. $d_{3/2}s_{1/2}$ (dashed line), and orbitals with positive angular momentum projection vs. orbitals with negative angular momentum projection (dotted line).

Within the wave function factorization, one expands the shell model ground state $|\psi\rangle$ as

$$|\psi\rangle = \sum_{j=1}^{\Omega} |\alpha_j\rangle |\beta_j\rangle. \tag{1}$$

Here $|\alpha_j\rangle$ and $|\beta_j\rangle$ are many-body states based on single-particle orbitals of subsystem A and B, respectively. The truncation is governed by Ω. The optimal states $|\alpha_j\rangle$ and $|\beta_j\rangle$ are solutions of eigenvalue problems of dimension d. This dimension typically is considerably smaller than the dimension $d_{\max}$ of the untruncated shell model problem. Let subset A and B consist of proton and neutron single-particle orbitals, respectively. One then finds that the ground-state energy and the ansatz (1) converges exponentially as the number Ω is increased. The left part of Fig. 1 shows the ground-state energy as a function of the relative dimension $d/d_{\max}$ for the pf-shell nuclei ^{56}Ni and ^{60}Fe, respectively. (We used the KB3 interaction.) Note that the full m-scheme dimensions are 10^9 and 10^8 for ^{56}Ni and ^{60}Fe, respectively.

It is interesting to compare the proton-neutron factorization with other schemes. Let $s_j^2 = \langle \alpha_j | \alpha_j \rangle \langle \beta_j | \beta_j \rangle$ denote the norms of the optimal basis states obtained in the factorization (1). We consider the sd-shell nucleus ^{24}Mg (USD interaction) and compute the ground-state factorization for different choices of the subsets A and B. As subsets we use (i) proton orbitals vs. neutron orbitals, (ii) $d_{5/2}$ orbital vs. $d_{3/2}s_{1/2}$ orbitals, and (iii) orbitals with positive angular momentum projection vs. orbitals with negative angular momentum projection. Case (i) is motivated by the dominance of proton-proton and neutron-neutron pairing over proton-neutron correlations. Case (ii) separates the energetically favored $d_{5/2}$ orbital and is motivated by the usual shell model configuration truncation. Case (iii) is not expected to be an effective factorization since angular momentum conservation and pairing yield considerable entanglement between states with opposite angular momentum projection. The right part of Fig. 1 shows the largest norms s_j^2 for the three cases. The ground state is least strongly entangled for case (i). Here, $\Omega = 83$ states yield 99% overlap with the ground state. The same accuracy requires 420 and 449 states for case (ii) and case (iii), respectively.

Thus for $0\hbar\omega$ shell model problems considered in this section, a factorization based on proton and neutron states is very efficient. Interestingly, this factorization converges somewhat less rapid for strongly deformed systems due to the strong proton-neutron correlations in these systems [15]. The $N = Z$ nucleus ^{48}Cr is a good example for such a case [14]. The factorization based on energetically ordered subshells seems less efficient for the $0\hbar\omega$ problem considered in this work. However, the recent DMRG calculation suggests that a separation of bound and resonance states vs. scattering states is quite promising for continuum shell-model problems [16].

2. STRUCTURE FROM RANDOM INTERACTIONS

Recently, Johnson, Bertsch and Dean [17] computed the ground states for shell model Hamiltonians with random interactions and found a preponderance of spin-0 ground states in even number systems; i.e., the probability of finding a spin-0 ground state is much higher than the fraction of spin-0 states in the Hilbert space. This surprising result suggests that regularities known in nuclear structure might to some degree be the consequence of symmetries and the two-body character of the interaction, and not solely caused by the specific details of the nucleon-nucleon interaction. Similar results were found in interacting electronic systems [18] and in the interacting boson model [19], thus making the spin-0 preponderance of ground states a robust phenomenon for systems with random two-body interactions.

Several authors have addressed this problem [20, 21, 22, 23, 24, 25, 26, 27, 28, 29] with reviews in Refs. [30, 31]. In systems of spin-1/2 fermions, an explanation can be based on the ensemble averaged width and spectral shape [24, 28]. For the shell model problem, however, the ensemble averaged spin-0 width is not sufficiently large to explain the ground state predominance [20, 25, 29, 30, 31]. However, we will see that the width is indeed the key to the understanding of the spin-0 preponderance of ground states [32].

The left (right) part of Fig. 2 shows data for a system of $n = 7$ ($n = 8$) fermions in a single $j = 19/2$ shell. The data points show the probability that the ground state has spin J. There is a sizable probability for a ground state with minimum spin, while there is only a very small fraction of spin-0 states in the Hilbert space. This is the puzzle observed in Ref. [17]. The solid line in Fig. 2 shows the probability that the largest width has spin J. This probability is sizable only for minimum and maximum spin. Given the large probability of a largest width with minimum spin, the preponderance of ground states with minimum spin is not really surprising. To understand this behavior, we study fluctuations of the width distribution function.

Let us consider a single $j = 19/2$ shell. There are $a = j + 1/2 = 10$ two-body matrix elements (TBME), as two fermions can have spins $0, 2, 4, \ldots, 2j - 1$. The Hamiltonian

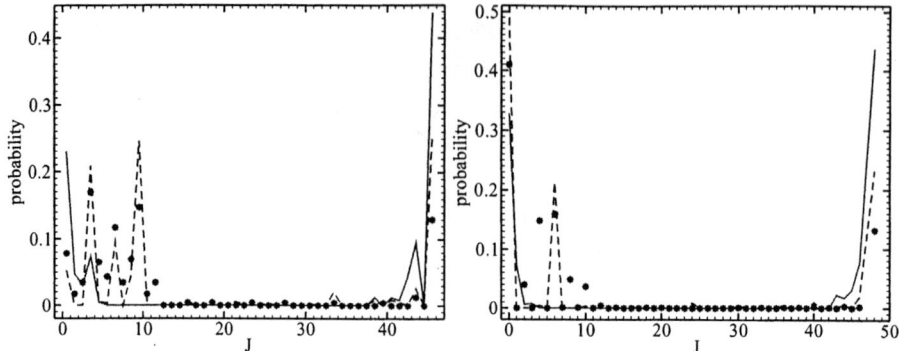

FIGURE 2. Left: 7 fermions in a $j = 19/2$ shell. Data points: probability that ground state has spin J. Solid line: probability that width σ_J is largest. Dashed line: probability that $r_J\sigma_J$ is largest. Right: same as left except for 8 fermions.

of the n-body system with total spin J is given by

$$H(J) = \sum_{\alpha=1}^{a} v_\alpha C_\alpha(J). \tag{2}$$

Here, v_α denotes the TBME, and the operators $C_\alpha(J)$ transport this two-body interaction into the d_J-dimensional Hilbert space of the n-fermion problem. They are complicated objects based on coefficients of fractional parentage and vector addition coefficients. The TBME are independent, random Gaussian variables with zero mean $\langle v_\alpha \rangle = 0$ and unit variance $\langle v_\alpha v_\beta \rangle = \delta_{\alpha\beta}$. Here, $\langle . \rangle$ denotes the ensemble average.

The spin-dependent width σ_J is defined as

$$\sigma_J^2 = d_J^{-1} \mathrm{Tr} H^2(J) = d_J^{-1} \sum_{\alpha,\beta} v_\alpha S_{\alpha\beta} v_\beta. \tag{3}$$

The matrix

$$S_{\alpha\beta} = d_J^{-1} \mathrm{Tr}(C_\alpha(J)C_\beta(J)), \tag{4}$$

contains the geometric information concerning the width. This matrix is symmetric and positive semi-definite. Its diagonalization $S = UsU^T$ yields the eigenvalues $s_1^2 \geq s_2^2 \geq \ldots \geq 0$. The eigenvectors of S can be used to construct two-body operators $B_\alpha(J) = \sum_\beta U_{\beta\alpha} C_\beta(J)$ which are orthogonal under the trace and have width s_α. The random matrix model (2) is equivalent to a Gaussian random matrix model based on the two-body operators $B_\alpha(J)$. This latter model has the advantage that the squared width of the Hamiltonian (3) becomes a diagonal quadratic form of the random variables. The width distribution function is

$$P_J(\sigma) = 2\sigma \langle \delta(\sigma^2 - \sigma_J^2) \rangle. \tag{5}$$

$P_J(\sigma)d\sigma$ gives the probability that spin J has a width between σ and $\sigma + d\sigma$. Writing the δ-function as a Fourier integral, the ensemble average can be performed, and one obtains

$$P_J(\sigma) = \frac{\sigma}{\pi} \int\limits_{-\infty}^{\infty} dt\, e^{it\sigma^2} \prod_{\alpha=1}^{a} \frac{e^{-\frac{i}{2}\arctan 2ts_\alpha^2}}{(1+4s_\alpha^4 t^2)^{1/4}}. \tag{6}$$

Here, s_α^2 denote the eigenvalues of the matrix (4). To understand the width distribution function (6), we consider two limiting cases for orientation. For the first case we assume that there is only one nonzero eigenvalue s_1^2 while all remaining eigenvalues vanish. In this case, the width distribution function is a Gaussian with width s_1. Then $\langle \sigma^2 \rangle = s_1^2$ and the r.m.s. variance·is $\sqrt{2}s_1^2$. For the second extreme, we assume that all eigenvalues are equal, i.e. $s_\alpha^2 = s^2$. In this case, the width distribution function becomes $P(\sigma) \propto (\sigma/s)^{a-1}\exp(-(\sigma/s)^2/2)$. The average squared width is $\langle \sigma^2 \rangle = as^2$ with r.m.s. variance $\sqrt{2}as^2$. Thus, the relative size of fluctuations is suppressed by a factor $a^{-1/2}$ in the second case when compared to the first case.

For our model, one finds that maximum spin has the largest ensemble averaged width. However, there is only one nonzero eigenvalue s_1^2, and width fluctuations are also largest. For minimum spin, all eigenvalues s_α^2 exceed its low-spin neighbors. Thus, minimum spin has a relatively large ensemble averaged width and relatively small width fluctuations. This explains why the width of minimum spin may exceed the width of maximum spin in many realizations of the random Hamiltonian. One also finds that there are strong correlations of the orthogonal transformation matrices U belonging to different spins; i.e., the J-dependence of these matrices is rather weak, and matrices that differ by only a few units of spin are almost identical. This causes correlations of the widths belonging to different spins. For low spins, for instance, the widths are simultaneously large or small for a given realization of the random Hamiltonian. These correlations further enhance the probability that minimum (maximum) spin has the largest width among the low (high) spins. This finding qualitatively explains the full line in Fig. 2.

The J-dependent spectral radius R_J is defined as $R_J = \max_i |E_i(J)|$, where $E_i(J)$ are the energies of the random Hamiltonian with spin J. The spectral radius is either given by the ground state or the highest excited state, and both cases have the same probability. For each realization of the Hamiltonian the spectral radius is related to the width by $R_J \geq \sigma_J$. However, we find that

$$R_J \approx r_J \sigma_J \tag{7}$$

is approximately valid. Here, r_J is a constant and encodes information about the shape of the spectral density. In can be determined by fit to data points (σ_J, R_J) obtained numerically from a large number of realizations of our random Hamiltonian. Once r_J is determined, we compute the probability that the product $r_J\sigma_J$ is maximal. The result, shown as the dashed line in Fig. 2, is a reasonable approximation for the probability that the ground state has spin J.

ACKNOWLEDGMENTS

This research was supported in part by the U.S. Department of Energy under Contract Nos. DE-FG02-96ER40963 (University of Tennessee) and DE-AC05-00OR22725 with UT-Battelle, LLC (Oak Ridge National Laboratory).

REFERENCES

1. E. Caurier, A. P. Zuker, A. Poves, and G. Martínez-Pinedo, Phys. Rev. C **50**, 225 (1994), nucl-th/9307001.
2. E. Caurier, G. Martínez-Pinedo, F. Nowacki, A. Poves, J. Retamosa, and A. P. Zuker, Phys. Rev. C **59**, 2033 (1999), nucl-th/9809068.
3. P. Navrátil, J. P. Vary, and B. R. Barrett, Phys. Rev. Lett. **84**, 5728 (2000), nucl-th/0004058; Phys. Rev. C **62**, 054311 (2000).
4. M. Honma, T. Otsuka, B. A. Brown, and T. Mizusaki, Phys. Rev. C **69**, 034335 (2004).
5. S. E. Koonin, D. J. Dean, and K. Langanke, Phys. Rep. **278**, 1 (1997), nucl-th/9602006.
6. M. Honma, T. Mizusaki, and T. Otsuka, Phys. Rev. Lett. **75**, 1284 (1995).
7. S. R. White, Phys. Rev. Lett. **69**, 2863 (1992); Phys. Rev. B **48**, 10345 (1993).
8. M. Horoi, B. A. Brown, and V. Zelevinsky, Phys. Rev. C **50**, R2274 (1994), nucl-th/9406004.
9. F. Andreozzi and A. Porrino, J. Phys. G **27**, 845 (2001).
10. J. Dukelsky and S. Pittel, Phys. Rev. C **63**, 061303 (2001), nucl-th/0101048.
11. J. Dukelsky, S. Pittel, S. S. Dimitrova, and M. V. Stoitsov, Phys. Rev. C **65**, 054319 (2002), nucl-th/0202048.
12. T. Papenbrock and D. J. Dean, Phys. Rev. C **67**, 051303(R) (2003), nucl-th/0301006.
13. J. Dukelsky and S. Pittel, Rep. Prog. Phys. **67**, 513 (2004), cond-mat/0404212.
14. T. Papenbrock, A. Juodagalvis, and D. J. Dean, Phys. Rev. C **69**, 024312 (2004), nucl-th/0308027.
15. P. Federman and S. Pittel, Phys. Rev. C **20**, 820 (1979).
16. N. Michel, W. Nazarewicz, M. Ploszajczak, and J. Rotureau, nucl-th/0401036.
17. C. W. Johnson, G. F. Bertsch, and D. J. Dean, Phys. Rev. Lett. **80**, 2749 (1998), nucl-th/9802066.
18. P. Jacquod and A. D. Stone, Phys. Rev. Lett. **84**, 3938 (2000), cond-mat/9909067.
19. R. Bijker and A. Frank, Phys. Rev. Lett. **84**, 420 (2000), nucl-th/9911067.
20. R. Bijker, A. Frank, and S. Pittel, Phys. Rev. C **60**, 021302 (1999), nucl-th/9906046.
21. R. Bijker and A. Frank, Phys. Rev. C **64**, 061303 (2001), nucl-th/0111009.
22. D. Kusnezov, Phys. Rev. Lett. **85**, 3773 (2000), nucl-th/0009076.
23. D. Mulhall, A. Volya, and V. Zelevinsky, Phys. Rev. Lett. **85**, 4016 (2000), nucl-th/0005014.
24. L. Kaplan, T. Papenbrock, and C. W. Johnson, Phys. Rev. C **63**, 014307 (2001), nucl-th/0007013.
25. Y. M. Zhao and A. Arima, Phys. Rev. C **64**, 04L301(R) (2001), nucl-th/0108052.
26. P. Chau Huu-Tai, A. Frank, N. A. Smirnova, and P. Van Isacker, Phys. Rev. C **66**, 061302(R) (2002), nucl-th/0301061.
27. Y. M. Zhao, A. Arima, and N. Yoshinagá, Phys. Rev. C **66**, 034302 (2002), nucl-th/0112075.
28. V. K. B. Kota and K. Kar, Phys. Rev. E **65**, 026130 (2002).
29. N. Yoshinaga, A. Arima, and Y. M. Zhao, J. Phys. A **35**, 8575 (2002).
30. V. Zelevinsky and A. Volya, Phys. Rep. **391**, 311 (2004), nucl-th/030907.
31. Y. M. Zhao, A. Arima, and N. Yoshinaga, nucl-th/0311050.
32. T. Papenbrock and H. A. Weidenmüller, nucl-th/0404022.

Many-body systems in the presence of the random interactions and the J pairing interaction

Akito Arima

The House of Councilors, 2-1-1 Nagatacho, Chiyodaku, Tokyo 100-8962, Japan

Abstract. In this talk I shall discuss some regularities of many-body systems in the presence of random interactions and regularities of a single-j shell for the J pairing interaction which works only when two particles are coupled to spin J. I shall first explain an empirical rule to predict the spin I ground state probability. Then I shall present some interesting results of a single-j shell under the J pairing interaction. Last I shall discuss some preliminary results of binding energies in the presence of random two-body interactions.

It is my great pleasure for me to talk to you here. I would like to thank the organizers of this conference and say "congratulations" to Professor Stuart Pittel. I am also very glad to see many of my friends.

My talk includes four subjects. (1) spin 0 ground state dominance; (2) Some regularities under the J pairing interaction; (3) ground state parities and binding energies.

SPIN 0 GROUND STATE DOMINANCE

The spin 0^+ ground state (0 g.s.) dominance was discovered and first studied by Johnson, Bertsch, Dean, and Talmi [1, 2]. This phenomenon has attracted much attention [3, 4].

In Ref. [5] we proposed a simple approach to predict the probability (denoted as $P(I)$) for a certain spin I to be the ground state spin without using random interactions: We first set one of the two-body matrix elements of the problem to -1 and all the rest to zero, and then see which angular momentum I gives the lowest eigenvalue among all eigenvalues of this many-body system. If the number of independent two-body interaction matrix elements is N, the above procedure is repeated N times. After all N calculations have been done, we count how many times (denoted as $\mathcal{N}_I$) the angular momentum I gives the lowest eigenvalue. We then predict $P(I)$ to be $\mathcal{N}_I/N$.

This approach was applied to predict the $P(I)$ for various systems, such as fermions in a single-j shell or many-j shells, sd bosons or sdg bosons. Here we take a single-j shell with $j = 9/2$ as an example. The Hamiltonian is defined as follows.

$$H = \sum_J G_J A^{J\dagger} \cdot A^J \equiv \sum_J \sqrt{2J+1} \left(A^{J\dagger} \times A^J \right)^{(0)} ,$$

CP726, *Nuclear Physics, Large and Small: International Conference on Microscopic Studies of Collective Phenomena*, edited by R. Bijker, R. F. Casten, and A. Frank
© 2004 American Institute of Physics 0-7354-0207-8/04/$22.00

$$A^{J\dagger} = \frac{1}{\sqrt{2}} \left[a_j^\dagger \times a_j^\dagger \right]^{(J)}, \quad A^J = -(-1)^M \frac{1}{\sqrt{2}} \left[\tilde{a}_j \times \tilde{a}_j \right]^{(J)}. \tag{1}$$

where G_J is given by $\langle j^2 J | V | j^2 J \rangle$, and V is a two-body interaction. Here the J pairing interaction is defined as the interaction which has $G_{J'} = -\delta_{JJ'}$.

One can obtain that $\mathcal{N}_0 = 3$ and $\mathcal{N}_4 = \mathcal{N}_{I_{max}} = 1$ for four fermions ($n = 4$, n labels the particle number) and $\mathcal{N}_j = 2$ and $\mathcal{N}_{3/2} = \mathcal{N}_{5/2} = \mathcal{N}_{I_{max}} = 1$ for $n = 5$. Here $N = j + 1/2 = 5$. Thus one predicts that $P(0) = 60\%$ and $P(4) = P(I_{max}) = 20\%$ for $n = 4$, and $P(j) = 40\%$ and $P(3/2) = P(5/2) = P(I_{max}) = 20\%$ for $n = 5$. The I g.s. probabilities obtained by using the random interactions are as follows: $P(0) = 66.4\%$, $P(4) = 11.8\%$, $P(I_{max}) = 17.9\%$ and other $P(I)$'s are close to zero for $n = 4$; $P(j) = 33.9\%$ and $P(3/2) = 20.5\%$, $P(5/2) = 15.5\%$, $P(I_{max}) = 18.4\%$ and other $P(I)$'s are close to zero for $n = 5$. The agreement between the predicted values and those obtained by the random interactions is good.

There are two other conclusions based on our empirical rule. First, $P(I_{max})$ is considerably large for a single-j shell, with the predicted $P(I_{max}) = 1/N = 1/(j + \frac{1}{2})$ which is independent of n. This prediction has been confirmed quantitatively. Second, through the process of finding the spin I of the ground state of the system with $G_{J'} = -\delta_{JJ'}$, one is able to predict which interactions are essential for a spin I g.s. probability.

REGULARITIES OF MANY-BODY SYSTEMS UNDER THE J PAIRING INTERACTION

The above empirical rule stimulated our studies of a single-j shell under the J pairing interaction. To avoid confusion, here we use superscript (n) to specify the particle number n for spin $I^{(n)}$ and the eigenvalue $E_{I^{(n)}, J(j)}^{(n)}$ for a single-j shell with the J pairing interaction.

We recently showed in Ref. [6] that a system of three fermions in a single-j shell is exactly solvable for any J pairing interaction. There is at most one non-zero eigenvalue for a fixed J and for any $I^{(3)}$. The non-zero eigenvalue is given by

$$E_{I^{(3)}, J(j)}^{(3)} = -1 - 2(2j+1) \left\{ \begin{array}{ccc} J & j & I^{(3)} \\ J & j & j \end{array} \right\}. \tag{2}$$

for $G_{J'} = -\delta_{JJ'}$.

The summation of all eigenvalues over J for a fixed $I^{(3)}$ is equal to the number of spin $I^{(3)}$ states multiplied by a factor $\frac{n(n-1)}{2} = 3$. Combining this with the result of Eq. (2), one can obtain a number of new sum rules for six-j symbols. These sum rules were found and discussed in Ref. [6].

The case of four fermions for the J pairing interaction is not soluble except for $J = 0$. Here we concentrate on the J_{max} pairing. In Ref. [7] we found that many of eigenvalues $E_{I^{(4)}, J_{max}(j)}^{(4)}$ are asymptotically integers (0, -1 and -2), and related them with the number of pairs with spin J_{max} in their wavefunctions. Besides these asymptotic integer eigenvalues, there are many $E^{(4)}_{I^{(4)},J_{max}(j)}$ which are not close to integers. We called them "non-integer" eigenvalues, and noticed that they are very close to the non-integer eigenvalues of $n = 3$. Therefore, it might be anticipated the wavefunction corresponding to the non-integer $E^{(4)}_{I^{(4)},J_{max}(j)}$ could be approximated by that of the corresponding $E^{(3)}_{I^{(3)},J_{max}(j)}$, to which $E^{(4)}_{I^{(4)},J_{max}(j)}$ is asympotically equal, coupled to a single-j particle. This anticipation was confirmed by calculating their overlaps which are 1.0 within a high precision. This last particle is called a "spectator" because of its weak coupling with the other strongly bound three particles which can be regarded as a cluster.

It is noted that the similar pattern holds for $n = 5$ and 6, and also for bosons with spin l. The non-integer eigenvalues and integer eigenvalues can be unified by the following picture of clusters which include both J_{max} pairs and spectators: The J_{max} pairing interaction favors clusters with n_1 particles ($n_1 \le n$), with $I^{(n_1)} \sim I^{(n_1)}_{max}$.

As is well known, the existence of degeneracy indicates that the Hamiltonian has a certain symmetry. The degeneracy for J_{max} pairing interaction, however, is not exact. It would be interesting to explore the broken symmetry hidden in the J_{max} pairing interaction discussed in this paper. It would be also interesting to discuss the modification of the J_{max} pairing interaction in order to recover the exact degeneracy.

GROUND STATE PARITIES AND BINDING ENERGIES

There are many interesting patterns of many-body systems in the presence of random interactions. Some characteristic features such as the spin 0 g.s. dominance and the odd-even staggering of binding energies have been numerically well known under the random interactions. Below I discuss two other regular patterns.

The first is related to the parity distribution. In Ref. [8] we studied many systems filling different shells in the presence of two-body random ensemble, and found that positive parity is dominant in the ground states of even-even nuclei, and the possibility for the ground states of odd-mass and odd-odd nuclei to have either positive or negative parity is comparable in general. This feature is very similar to that for the 0 g.s. dominance. Because parity is a much simpler quantity, the positive parity dominance might be well understood in the near future. This may help us to obtain a deeper understanding of the 0 g.s. dominance.

The second is related to binding energies obtained by the random interactions. The binding energy is a very important and fundamental quantity not only for various aspects of nuclear physics but also for other branches of physics such as astrophysics. For a recent review of nuclear masses, see Ref. [9]. Recently Bohigas and Leboeuf claimed in Ref. [10] that there is a lower limit on the accuracy of theoretical predictions of binding energies due to chaotic components in the nuclear mass. Here I present some preliminary results for binding energies of a few neighboring nuclides in the sd shell which are obtained by using the random interactions. We are interested in the Garvey-Kelson relations which connect the masses of neighboring nuclides. Our calculated results seem to satisfy the algebraic Garvey-Kelson relations within a reasonable precision in the presence of random interactions. For instance, we calculated the deviations from the

Garvey-Kelson relation for example around ^{22}Na, which involves of binding energies of Na^{21,22}, Mg^{21,23} and Al^{22,23}, and found that these deviations are small. More work is necessary to clarify the indications of Ref. [10].

SUMMARY

In this talk I have discussed a few interesting aspects concerning the regularities of many-body systems under the random interactions and a single-j shell under the J pairing interaction. First, I explained our empirical rule which is useful in predicting the I g.s. probabilities and understanding the spin 0 g.s. dominance. Second, I showed that three fermions in a single-j shell for any J pairing interaction is solvable. I also discussed more fermions among which the J_{max} pairing interaction works, and suggested that the J_{max} pairing interaction favors clusters, where $I^{(n_1)}$ ($n_1 \leq n$) for each cluster is very close to or equal to the maximum spin for n_1 particles. Last, I showed two other interesting features for the ground state parities and binding energies by using the random interactions: (1) the positive parity dominates in the ground states of even-even nuclei, and (2) the calculated binding energies satisfy the Garvey-Kelson relations reasonably well.

ACKNOWLEDGMENTS

I would like to thank Dr. Y. M. Zhao for his collaboration in this work. I also thank Dr. I. Talmi for his valuable comments on some results of Sec. 2.

REFERENCES

1. C. W. Johnson, G. F. Bertsch, and Dean, D. J., *Phys. Rev. Lett.*, **80**, 2479 (1998).
2. C. W. Johnson, G. F. Bertsch, D. J. Dean., and I. Talmi, *Phys. Rev. C*, **61**, 014311 (2000).
3. Y.M. Zhao, A. Arima, and N. Yoshinaga, *nucl-th/0311050, to be published*.
4. V. Zelevinsky and A. Volya, *Phys. Rep.*, **391**, 311 (2004).
5. Y. M. Zhao, A. Arima, and N. Yoshinaga, *Phys. Rev. C*, **66**, 034302 (2002).
6. Y. M. Zhao and A. Arima, *nucl-th/0404041, to be published*.
7. Y. M. Zhao, A. Arima, J. N. Ginocchio, and N. Yoshinaga, *Phys. Rev. C*, **68**, 044320 (2003).
8. Y.M. Zhao, A. Arima, N. Shimizu, K. Ogawa, N. Yoshinaga, and O. Scholten, *nucl-th/0404042, to be published*.
9. D. Lunney, J. M. Pearson., and C. Thibault, *Rev. Mod. Phys.*, **75**, 1021 (2003).
10. O. Bohigas and P. Leboeuf, *Phys. Rev. Lett.*, **88**, 092502 (2002).

A short review on recent advances in cluster physics

P. O. Hess

Instituto de Ciencias Nucleares, UNAM, A.P. 70-543, 04510 México D.F., Mexico

Abstract. A short review on recent advances in cluster physics is given.

Cluster Physics is a rather broad subject which includes the Cluster Radioactivity (CR), nuclear molecules and fission (for an excellent review, concerning all topics mentioned in this contribution, see [1, 2]).

In the first part I will concentrate on models which observe the Pauli exclusion principle. In the last part, shortly some phenomenological developments are mentioned.

In a microscopic treatment, a particular rôle in this subject plays the spectroscopic factor. In the last 30 years the $SU(3)$ shell model was used as a tool to calculate the spectroscopic factors (see as examples the Refs. [3, 4, 5, 6]).

Here, I will focus on some particular new developments. Concerning the CR, in [7] a scaling law for the spectroscopic factor in the emission of a light cluster was given, which helped to understand the probability of the emission of certain clusters. Extending the scalar law to heavier fragments, in [8] it was shown that CR and fission can be treated on the same footing.

About ten year ago, the Semimicroscopic Cluster Model (SACM) was developed [9] in which the individual clusters are treated in the $SU(3)$ model and the irreducible representations (irreps) of the cluster $SU(3)$'s are coupled with the relative motion to total $SU(3)$ irreps. The list of all resulting $SU(3)$ irreps is matched to those allowed by the shell model, i.e. the SACM observes the Pauli exclusion principle. The SACM is able to describe states with positive and negative parity at the same time. It has been applied to various nuclei with success (see, e.g., [9] and for the most recent application [10]). Efforts have been made to understand within the SACM the allowed clusterizations and the dependence of the emission of clusters as a function on the mass number and deformation [11]. Efforts are currently done to parameterize the spectroscopic factor within the SACM. In [8] an attempt was made to use the algebraic parameterization of the spectroscopic factor in order to describe CR and fission on the same footing.

Another interesting procedure predict the preferred clusterizations, as it was proposed by the Oxford group [12]. Maximizing the binding energy of the two clusters determines a function whose maxima are at the position of preferred clusterizations, in accordance to experiment. It is a very simple and effective procedure for the prediction of preferred clusterizations.

Next, I would like to mention the *Antisymmetrized Molecular Dynamics* model [13]. It assumes gaussian wave packets for the individual nuclei and constructs from them

CP726, Nuclear Physics, Large and Small: International Conference on Microscopic Studies of Collective Phenomena, edited by R. Bijker, R. F. Casten, and A. Frank
© 2004 American Institute of Physics 0-7354-0207-8/04/$22.00

anti-symmetric many-particle wave functions. Using a microscopic Hamiltonian and a *Molecular Dynamics* formulation, physical clusterizations for light systems can be described.

In [14] a method is described how to construct Pauli allowed states in a multi cluster system. It was applied to multi-α clusters and to core+$n\alpha$ clusters. In [15] an interesting description of cluster models via the Two-Center-Shell model is given. With this nice model, e.g., the ^{10}Be nucleus can be pictured as two α-particles which are bound by two valence neutrons.

Finally, I would like to mention some developments in phenomenological models. The most famous one is the nuclear vibron model [16], where the relative motion of two clusters are described within a U(4) group and the basic degrees of freedom are vector bosons and a scalar boson. This model was extended to various clusters, with one deformed [17] and extended two and three deformed clusters in [18].

Another type of model is based on a geometrical description of cluster systems, like nuclear molecules. There exist applications to heavy two-cluster systems [19] and three-cluster systems [20]. An extension of [19] to light nuclei is given in [21] and applied to the particular case of ^{24}Mg $\to$ ^{12}C+^{12}C. The same system was also considered in [22].

Unfortunately, I can not report on further interesting developments, due to lack of space. However, I hope, that the references given here, though not complete, will serve as a continuation for a further investigation.

REFERENCES

1. W. Greiner, J. Y. Park and W. Scheid, *Nuclear Molecules* (Singapore: World Scientific, 1995).
2. R. Gupta and W. Greiner, Int. J. Mod. Phys. (Suppl.) (1994), 335.
3. K. T. Hecht and D. Braunschweig, Nucl. Phys. A **244** (1975), 365.
4. K. T.Hecht, E. J. Reske, T. H. Seligman and W. Zahn, Nucl. Phys. A **356** (1981), 146.
5. J. P. Draayer, Nucl. Phys. A**237** (1975), 157.
6. K. Kato and H. Bando, Prog. Theo. Phys. **59** (1978), 774.
7. R. Blendowske and H. Walliser, Phys. Rev. Lett. **61** (1988), 1930
8. S. Mişicu and P. O. Hess, submitted for publication
9. J. Cseh, Phys. Lett. B **281** (1992), 173;
    J. Cseh and G. Lévai, Ann. Phys. (N.Y.) **230** (1994), 165.
10. L. Hernández de la Peña, P. O. Hess, G. Lévai and A. Algora, J. Phys. G **27** (2001), 2019.
11. A.Algora, J.Cseh and P.O.Hess, J. of Phys. G **24** (1998), 2111;
    A. Algora, J. Cseh and P.O. Hess, J. Phys. G **25** (1999), 775;
    A. Algora, J. Cseh, P.O. Hess and M. Hunyadi, Heavy Ion Phys. **13** (2001), 145.
12. B. Buck, A. C. Merchant, M. J. Horner and S. M. Perez, Phys. Rev. C **61** (2000), 024314;
    B. Buck, A. C. Merchant and S. M. Perez, Few Body Syst. **29** (2000), 53.
13. Y. Kanada-En'yo, H. Horiuchi and A. Doté, Phys. Rev. C **60** (1999), 064304 (and references therein).
14. K. Katō, K. Fukatsu and H. Tanaka, Progr. Theor. Phys. **80** (1988), 663 (and references therein).
15. H. G. Bohlen et al., Nuovo Cimento **111** (1998), 6 (and references therein).
16. F. Iachello, Phys. Rev. C **23** (1981), 2778.
17. H. J. Daley and F. Iachello, Ann. Phys. (N.Y.) **167** (1986), 73.
18. H. Yépez-Martínez, P. O. Hess and Ş. Mişicu, Phys. Rev. C **68** (2003), 014314.
19. P. O. Hess and W. Greiner, Il Nuovo Cimento, **83** (1984), 76.
20. P. O. Hess, Ş. Mişicu, W. Greiner and W. Scheid, J. Phys. G **26** (2000), 957.
21. R. Maass and W. Scheid, J. Phys. G **18** (1992), 707.
22. P. O. Hess and P. Pereyra, Phys. Rev. C **42** (1990), 1632.

STUART PITTEL AND THE $f_{7/2}$ SHELL REVISITED: MAGNETIC MOMENTS IN THE Ca ISOTOPES.

N. Benczer-Koller*, G. Kumbartzki*, T. J. Mertzimekis†, Y.Y. Sharon*, K.-H Speidel**, M. J. Taylor‡ and L. Zamick*

*Department of Physics and Astronomy, Rutgers University, New Brunswick, NJ 08903, USA
†NSCL, Michigan State University, East Lansing, MI 48824, USA
**Universität, Bonn, D-53115 Bonn, Germany
‡School of Engineering, University of Brighton, Brighton, BN2 4GJ, UK

Abstract.
The study of the structure of the Ca isotopes occupied Stuart Pittel for many years at the beginning of his career. In spite of many experiments and theoretical investigations over the last thirty years, the Ca isotopes still provide considerable surprises. Recently, magnetic moments of 2_1^+ states in ^{42,44,46}Ca have been measured and will be discussed in terms of a model encompassing a two-component wave function representing particle-hole excitations of the ^{40}Ca core and single particle configurations.

INTRODUCTION

I have known Stuart Pittel for about 3.5 decades during which our interests have often intersected. His early papers were concerned with understanding the structure of light nuclei in the Ca region. As ^{40}Ca is doubly magic it was expected that shell model calculations would reproduce the experimental evidence available at the time, energy levels, B(E2)'s, and spectroscopy factors in stripping reactions.

It became obvious early on that one had to introduce configurations with a very high number of particle-hole configurations. The analysis of ^{40}Ca included 0p-0h, 2p-2h and 4p-4h with the particle and holes restricted to the $f_{7/2}$ and $d_{3/2}$ orbitals respectively. That model reproduced all the known 0^+ levels in ^{40}Ca with remarkable accuracy. Similarly the calculation in ^{42}Ca included only the 2p-0h and 4p-2h configurations, and the same matrix elements used in the analysis of Ca^{40} [1, 2, 3, 4]. In fact, the 0_2^+ state in Ca^{40} supported a regular rotational band.

Since then a wealth of experiment has yielded precision data on lifetimes and magnetic moments. Magnetic moments are particularly sensitive to the single particle components of the wave function and therefore provide a critical challenge to the model, in particular aspects of the interplay between single particle configurations and collective excitations in the form of particle-hole contributions to the wave function or deformation.

CP726, Nuclear Physics, Large and Small: International Conference on
Microscopic Studies of Collective Phenomena, edited by R. Bijker, R. F. Casten, and A. Frank
© 2004 American Institute of Physics 0-7354-0207-8/04/$22.00

Precision measurements of the g factors of 2^+_1 states in ^{46,48}Ti and ^{50,52}Cr [5] have been evaluated in terms of full fp shell model calculations. Deviations of the g factors and BE(2) values from the predictions were attributed to excitations from the ^{40}Ca core. Furthermore, MonteCarlo calculations [6] support the assertion that ^{48}Ca is a better inert core than ^{40}Ca.

In order to examine the nature of particle-hole excitations in the fp shell, the simpler system of the Ca isotopes, with the protons nominally closing the sd shell and the neutrons nominally occupying the fp, shell was investigated. The magnetic moments of the ground states of the odd Ca isotopes have been known for a long time. In the present paper, measurements of the g factors of the 2^+_1 states of the three even Ca isotopes by the Rutgers and Bonn groups are presented. The experimental results [7, 8, 9, 10] will be discussed in terms of various shell model calculations.

EXPERIMENTAL PROCEDURES

These studies required the application of newly developed experimental techniques. These involve Coulomb excitation of a heavy ion beam by a light nucleus, in inverse kinematics. The excited heavy nuclei traverses a ferromagnetic material where it undergoes a transient hyperfine interaction that causes a precession of the state magnetic moment which was originally aligned by the reaction [11, 12, 13]. Finally the excited ions stop in a copper foil where they experience no further hyperfine interactions. Often the *same* target can used for several isotopes, thus reducing systematic uncertainties.

The light target ions are scattered forward and are recorded in a detector located at 0° with respect to the beam. The γ-rays are detected in four detectors placed at angles where the angular correlation has optimum slope. The details of the experiments discussed in this paper are described in several papers on Ca isotopes [7, 8, 9, 10].

RESULTS AND DISCUSSION

Fig. 1 displays the measured g factors in the Ca isotopes and Table 1 summarizes the data.

Several different approaches can be used to describe these results. Calculations can be carried out by considering pure $(f_{7/2})_\nu$ configurations. These can be expanded to consider the whole fp shell [7]. These calculations have used different nucleon-nucleon interactions such as KB3 and FPD6. In addition, a full large scale shell model (LSSM) [9] calculation was carried out, by including excitations of protons and neutrons from the $2s_{1/2}2d_{3/2}$ shell into the fp shell. Finally, one may rely on a model that doesn't require specific knowledge of the wave function, but considers a two-components wave function with amplitudes C and D where C represents the amplitude of the single particle configurations while D represents the amplitude of the deformed component of

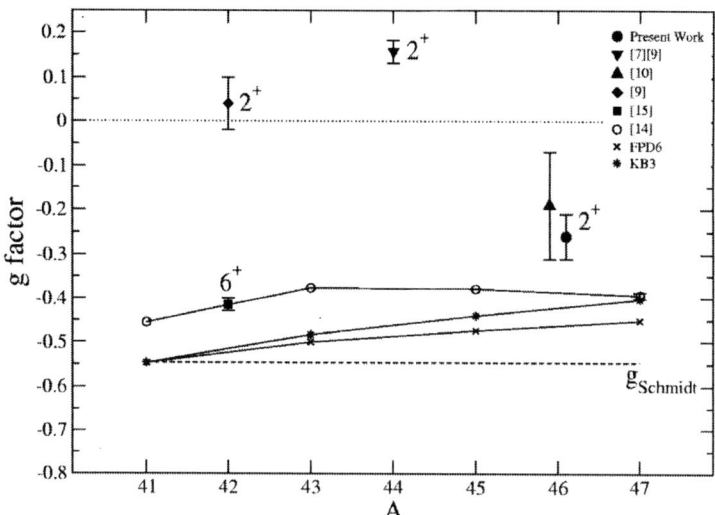

FIGURE 1. Measured g factors of the ground states of the odd Ca isotopes [14], of 2_1^+ states of even Ca isotopes as well as of the 6_1^+ state in ^{42}Ca [15]. $(fp)_\nu$ shell model calculations for the ground state g factors of the odd A Ca isotopes using the KB3 and FPD6 interactions are shown for comparison with experiment. $g_{Schmidt}(f_{7/2})_\nu = -0.5465$ is indicated by the dashed line. The solid lines are drawn to guide the eye.

TABLE 1. Measurements of g factors of 2_1^+ states in the ^{42,44,46}Ca isotopes. For ^{44}Ca and ^{46}Ca the average g factors from Refs. [7, 9, 10] and this work, are also shown.

Laboratory	$g(2_1^+)$	^{42}Ca	^{44}Ca	^{46}Ca
Bonn		+0.04(6)	+0.17(3)	−0.19(12)
Rutgers			+0.12(5)	−0.26(5)
	$<g>$	+0.04(6)	+0.16(3)	−0.25(5)

the wave function, a term that includes core excitations [7, 16].

The $(fp)_\nu$ shell model calculations for the ground state g factors of the odd Ca nuclei using the KB3 and FPD6 interactions are shown for comparison with experiment in Fig. 1. These calculations were performed using effective proton and neutron charge values of $e_\pi = 1.5$ and $e_\nu = 0.5$. The bare spin and angular momentum g factor values for protons and neutrons were taken as $g_s(\pi) = 5.586$, $g_s(\pi) = -3.826$, $g_l(\pi) = 1.0$ and $g_l(\nu) = 0$.

It is clear from the figure that the g factors of the $(7/2)^-$ ground states of the odd A Ca isotopes are all smaller than the Schmidt value for a pure $f_{7/2}$ neutron configuration. Whereas in the $f_{7/2}$ model the g factors of all the Ca isotopes would be the same, in the $(fp)^n$ model there is a steady decrease in magnitude of the g factors of the $(7/2)^-$

ground states of the the odd Ca isotopes in the progression from ^{41}Ca to ^{47}Ca.

The positive g factors for the ^{42,44}Ca isotopes contrast with the negative g factors expected if the wave function of the 2^+_1 state consisted of pure $f_{7/2}$ or $(fp)_v$ neutron configurations with an inert ^{40}Ca core.

Pure $(f_{7/2})_v$ shell model calculations for the even ^{42,44,46}Ca isotopes yield simple results since the 2^+_1 states of definite seniority are unique. The calculations utilized KB3 and mass independent FPD6 interactions, the free neutron value for $(g_s)_v = -3.826$ and an effective neutron charge, $e_v = 0.65$, as used previously for even mass Ca isotopes [7, 16]. The calculated excitation energy, $E(2^+_1)$, of the first excited 2^+_1 state in all three nuclei is 0.839 MeV with KB3 and 1.380 for FPD6. With either interaction, the $B(E2;0^+_1 \rightarrow 2^+_1)$ in (eb)^2 are 0.00471, 0.00643, and 0.00495 for ^{42}Ca, ^{44}Ca, and ^{46}Ca respectively. The corresponding quadrupole moments of the 2^+_1 states, in eb, with either interaction are +0.04855, 0, and –0.04976 for ^{42}Ca, ^{44}Ca (at midshell), and ^{46}Ca, respectively.

The calculations performed with the full $(fp)_v$ shell using again KB3 and FPD6 interactions are compared to experimental data in Fig. 2.

There are large discrepancies between these model calculations and the experimental values for the $g(2^+_1)$'s for ^{42}Ca and ^{44}Ca, but the measured $g(2^+_1)$ for ^{46}Ca agrees with the calculations based on the FPD6 interaction. This agreement confirms that there is very little or no contribution to the excited state wave function from excitations from the ^{48}Ca core. In addition, the B(E2) is much smaller for ^{46}Ca than for ^{44}Ca and the energy of the 2^+_1 state is higher. These two properties also reflect lower collectivity in ^{46}Ca than in ^{44}Ca.

LSSM calculations of level energies, B(E2)'s and g factors for even Ca isotopes have also been carried out and are described in detail in Ref. [10]. Whereas the results of these calculations for the g factors and B(E2)'s are reasonable for ^{42}Ca and ^{44}Ca, they lie very far from the data for ^{46}Ca. The calculated B(E2) is much larger than experiment and the g factor is large and positive, $g(^{46}$Ca$,2^+_1) = +0.89$ [10]. The problem seems to be that with the interaction that was chosen in that calculation both the ^{40}Ca and ^{48}Ca shells are badly broken which contradicts the evidence from the present experiment that requires that only the ^{40}Ca shell be broken but that the ^{48}Ca core be a reasonably good closed shell.

Finally, following earlier work by Gerace and Green [18] and by Towsley, Cline and Horoshko [19] using the "co-existence" approach, Sharon and Zamick used the newly measured experimental $g(2^+_1)$ values to estimate the intensity of the core-excited component in the 2^+_1 state wave functions of ^{42,44}Ca [7, 16]. This approach, which does not require a detailed knowledge of the particle-hole excitations that make up the core-excited components in the wave function, is summarized by Eq.1. The 2^+_1 state wave function for the even Ca isotopes with n neutrons beyond ^{40}Ca consists of two components, one of amplitude C corresponding to a spherical shell model wave function and the other of amplitude D representing many particle-hole excitations ($2p - 2h$, $4p - 4h$, etc...),

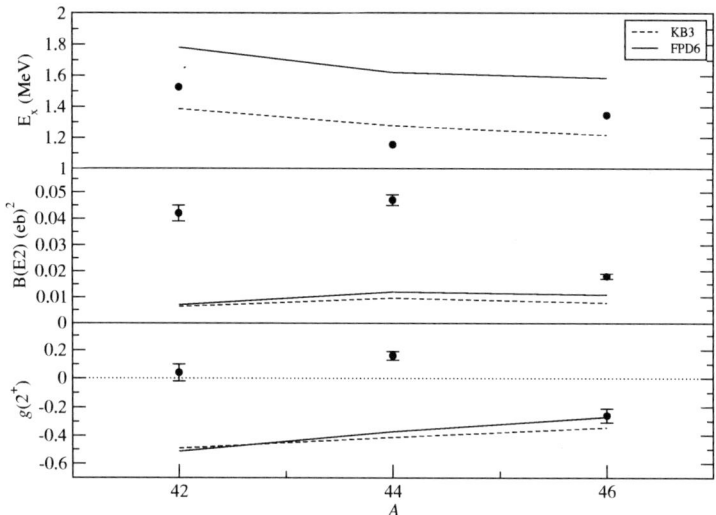

FIGURE 2. Experimental values for the excitation energies $E(2_1^+)$, $B(E2;0_1^+ \rightarrow 2_1^+)$ values [17] and $g(2_1^+)$ factors in ^{42,44,46}Ca. Also shown are the results from full fp shell model calculations using the KB3 and FPD6 interactions with connecting lines drawn to guide the eye.

$$\langle \Psi(2_1^+) \rangle = C[(fp)_\nu^n]_{J=2} + D[\Psi_{def}]_{J=2} \qquad (1)$$

The g factor becomes:

$$g(2_1^+)_{meas} = C^2[g(fp)_\nu^n] + D^2[g(\Psi_{def})] \qquad (2)$$

In Eq. 2, $g(\Psi_{def}) = Z/A$ while $g[(fp)_\nu^n]$ is the average of $g(2_1^+)$ obtained in the full $(fp)_\nu$ shell model calculation with the KB3 and FPD6 interactions.

The resulting intensities of the two components of the wave function are displayed in Table 2 for the three even Ca isotopes. For ^{42,44}Ca the intensity of the deformed component in the wave function is large. Deformation is nearly negligible in ^{46}Ca. These data are in agreement with the known deformation effects in ^{40}Ca where a well deformed band is supported by the second 0^+ state whereas no such effects are known in ^{48}Ca.

It should be noted that in Refs. [18, 19] smaller core-excited components ranging from 17% to 27% were deduced for the 0_1^+ ground states of ^{42,44}Ca. Similarly, a smaller core-excited component of 15% was obtained for the $(7/2)^-$ ground state of ^{41}Ca [18].

TABLE 2. Relative intensities of the single particle C^2 and deformed D^2 components of the 2_1^+ wave functions in the ^{42,44,46}CaCa isotopes. For ^{44}Ca and ^{46}Ca the average g factors from Refs. [7, 9, 10] and this work are shown.

Nucleus	Method	$<g(2_1^+)>$	C^2	D^2
^{42}Ca	Gerace and Green; Ref.[18]		~ 0.50	~ 0.50
	$Q_{meas}=-0.19(8)$eb; Ref. [19]		~ 0.42	~ 0.58
		+0.04(6)	0.45(6)	0.55(6)
^{44}Ca	Stripping reactions; Refs. [20, 19]		0.47	0.53
	$Q_{meas}=-0.14(7)$eb; Ref. [19]		~ 0.43	~ 0.57
		+0.16(3)	0.35(4)	0.65(4)
^{46}Ca		−0.26(5)	0.93(7)	0.07(7)

ACKNOWLEDGMENTS

Support for this work was provided in part by the U.S. National Science Foundation and the U.S. Department of Energy.

REFERENCES

1. P. Federman and S. Pittel, *Phys. Rev.*, **186**, 1106 (1969).
2. S. Pittel and B. F. Bayman, *Phys. Rev.*, **187**, 1398 (1969).
3. P. Federman and S. Pittel, *Nuc. Phys.*, **A139**, 108 (1969).
4. P. Federman and S. Pittel, *Nuc. Phys.*, **A155**, 161 (19670).
5. R. Ernst, K.-H. Speidel, O. Kenn, U. Nahum, J. Gerber, P. Maier-Komor, N. Benczer-Koller, G. Kumbartzki, L. Zamick, *et al.*, *Phys. Rev. Lett.*, **84**, 416 (2000).
6. T. Otsuka *et al.*, *Phys. Rev. Lett.*, **81**, 1588 (1998).
7. M. J. Taylor, N. Benczer-Koller, G. Kumbartzki, T. J. Mertzimekis, S. Robinson, Y.Y. Sharon, L. Zamick, A. Stuchbery, C. Hutter, C. Beausang, *Phys. Lett.*, **B559**, 187 (2003).
8. M. J. Taylor, N. Benczer-Koller, L. Bernstein, J. Cooper, K. Hiles, D. S. Judson, G. Kumbartzki, P. Maier-Komor, M. McMahan, T. J. Mertzimekis, L. Phair, Y.Y. Sharon, K.-H. Speidel, A. E. Stuchbery, and L. Zamick, *to be submitted to Phys. Rev. C* (2004).
9. S. Schielke, *et al.*, *Phys. Lett.*, **B571**, 29 (2003).
10. K.-H. Speidel, S. Schielke, O. Kenn, J. Leske, D. Hohn, H. Hodde, J. Gerber, P. Maier-Komor, O. Zell, Y.Y. Sharon, *et al.*, *Phys. Rev.*, **C68**, 061302(R) (2003).
11. N. Benczer-Koller, M. Hass and J. Sak, *Annual Rev. Nucl. Part. Sci.*, **30**, 53 (1980).
12. K.-H. Speidel, N. Benczer-Koller, G. Kumbartzki, C. Barton, A. Gelberg, J. Holden, G. Jakob, N. Matt, R. H. Mayer, M. Satteson, *et al.*, *Phys. Rev.*, **C57**, 2181 (1998).
13. N. K. B. Shu *et al.*, *Phys. Rev.* (1980).
14. P. Raghavan, *Atomic Data and Nuclear Data Tables*, **42**, 189 (1989).
15. L. E. Young, R. Brown, S. K. Bhattacherjee, D. B. Fossan and G. D. Sprouse, *Phys. Rev. Lett.*, **35**, 497 (1975).
16. N. Benczer-Koller, *et al.*, "Magnetic Moments from the Mediterranean to Mt. Fuji," in *Symmetries in Nuclear Structure*, World Scientific Publishing Co., Singapore, 2003.
17. S. Raman, C. W. Nestor Jr., P. Tikkanen, *Atomic Data and Nuclear Data Tables*, **78**, 1 (2001).
18. W. J. Gerace and A. M. Green, *Nuc. Phys.*, **A93**, 110 (1967).
19. C.W. Towsley, D. Cline and R. N. Horoshko, *Nuc. Phys.*, **A204**, 574 (1973).
20. J. H. Bjerregard and O. Hansen, *Phys. Rev.*, **155**, 1229 (1967).

Regularities vs. chaos in nuclear masses

Jorge G. Hirsch*, Alejandro Frank*, José Barea *, Piet Van Isacker † and
Víctor Velázquez **

*Instituto de Ciencias Nucleares, Universidad Nacional Autónoma de México, AP 70-543, 04510
México DF, Mexico
†GANIL, BP 55027, F-14076 Caen Cedex 5, France
**Departamento de Física, Facultad de Ciencias, Universidad Nacional Autónoma de México, AP
70-348, 04511 México DF, México

Abstract. It has been suggested that there might be an inherent limit to the accuracy with which
nuclear masses can be calculated, chaotic motion inside the atomic nucleus being responsible for
this lack of predictability. However, a thorough application of a set of parameter-free relationships
among neighboring nuclei, known as the Garvey-Kelson relations, to an up-to-date compilation of
more than 2500 nuclear masses allows to set an upper bound for the proposed chaotical component,
which turns out to be almost an order of magnitude smaller than previously suggested. This is good
news for astrophysicists attempting to understand the processes that fuel the stars.

INTRODUCTION: CLASSICAL AND QUANTUM CHAOS

Does chaos exist in the quantum world? The issue of quantum chaos is not settled and
diverse points of view coexist in the literature [1, 2]. It would appear that quantum
uncertainty can actually blur the basic requirement of sensitive dependence on initial
conditions, smoothing the expected characteristic signals of chaos. However, although
the actual mathematical structure of quantum mechanics would seem to forbid the
presence of chaos, several recent experiments, prominently those of Rydberg atoms
subject to strong magnetic fields [3] and others involving quantum dots [4], clearly
display the footprints of chaos. These signals emerge not in any particular state but rather
in the distribution of the energy levels and other spectral characteristics of a quantum
system. It is in this context that the recent proposal that there might be an inherent limit
to the accuracy with which nuclear masses can be calculated, due to the presence of
chaotic motion inside the atomic nucleus [5] should be analyzed. This suggestion could
have important consequences to the fields of nuclear physics and astrophysics [6].

NUCLEAR MASSES

Atomic nuclei consist of a collection of A nucleons (protons and neutrons) bound
together by the nuclear interaction. Nuclei range from a single proton to more than two
hundred nucleons and thus they constitute particularly complex systems, far too large for
a detailed microscopic treatment but too small for statistical methods. They are not easily
subject to theoretical scrutiny. Each nucleus displays a rich variety of spectral behavior,

CP726, Nuclear Physics, Large and Small: International Conference on
Microscopic Studies of Collective Phenomena, edited by R. Bijker, R. F. Casten, and A. Frank
© 2004 American Institute of Physics 0-7354-0207-8/04/$22.00

but its most basic property is its mass. The importance of an accurate knowledge of nuclear masses to understand the processes occurring in astrophysical phenomena has been abundantly stressed [6]. Though great progress has been made in the challenging task of measuring the mass of exotic nuclei, theoretical models are necessary to *predict* their mass in regions far from stability [7]. The efforts to calculate these masses have been hampered by the absence of an exact theory of the nuclear interaction and by the difficulties inherent to quantum many-body calculations, so diverse models which attempt to bring forth the fundamental physics of the atomic nucleus have been devised. The simplest approach is that of the liquid drop model. It incorporates the essential macroscopic terms, which means that the nucleus is at this stage pictured as a very dense, charged liquid drop (LDM). Including the discrete character of the nucleons and their basic interactions requires more sophisticated treatments. The finite range droplet model (FRDM) [8], which combines the macroscopic effects with microscopic shell and pairing corrections, has become the *de facto* standard for mass formulas. A microscopically inspired model has been introduced by Duflo and Zuker (DZ) [9, 10] with good results. Among the mean field methods it is also worth mentioning the Skyrme-Hartree-Fock approach [11]. These mass formulas can calculate and predict the masses (and often other properties) of as many as 8979 nuclides [7], but it is in general difficult to match theory to experiment (for all known nuclei) to an average precision better than about .7 million electron volts [7]. This minute quantity, corresponding to a mass slightly larger than that of a single electron and less than a part in 10^5 of the mass of a typical nucleus, still represents a significant fraction of the energy released in nuclear decays, strongly affecting the extrapolations required in astrophysical processes and nuclear beta decays near the lines of stability [6]. There is thus a permanent search for better theoretical models that reduce the difference with the experimental masses and produce reliable predictions for unstable nuclei. The distribution of the mass deviations predicted by each of these procedures is quite characteristic and different from the others. These distributions may conceal some useful and remarkable information, as we proceed to discuss.

CHAOS IN NUCLEAR MASSES

It has been postulated that the presence of chaotic motion in highly excited nuclear systems can be determined through the statistics of the high-lying energy levels or resonances [12, 13]. The analysis of level distributions and their comparison with random matrix results has become a standard gauge for quantum chaos [14]. We can then say that the basic corollary of Random Matrix Theory [13] is that near or around the neutron absorption energies predictability is hopeless and only a statistical analysis makes sense. One may then ask whether at lower energies any remnant of chaos subsists, perhaps even at the ground state of nuclei.

The question of the remaining mass deviations observed in the nuclear mass formulas was addressed in [5] from a novel perspective. The errors among experimental and calculated masses in [8] were interpreted in terms of two types of contributions. The first one is associated with a regular part, related to the underlying collective dynamics (LD model), plus the shell energy correction, while the other was assumed to arise from some inherent dynamics, possibly higher order interactions among nucleons [5], that lead to chaotic behavior. According to [15] the latter could be interpreted as remaining signals of the chaotic dynamics occurring at higher energies. Their magnitude suggests that we have essentially already achieved the maximum accessible precision in the calculation of the masses in mean-field theories [5]. If this is the case, chaotic motion could be interpreted as comprising an ultimately unpredictable component of nuclear mass, thus setting a lower limit on how accurately nuclear masses can be theoretically predicted [15].

GARVEY-KELSON RELATIONS

Besides the "global" formulas of which the FDRM method has become the standard, there are a number of "local" mass formulas. These local methods are usually effective when we require the calculation of the mass of a nucleus, or a set of nuclei, which are fairly close to a number of other nuclei of known mass, exploiting the relative smoothness of the masses as a function of proton (Z) and neutron (N) numbers, conforming the M(Z,N) mass surface, to deduce systematic trends. Among these methods there are a set of algebraic relations for neighboring nuclei, known as the Garvey-Kelson (GK) relations [16, 17].

There are a number of different GK relations [18], but the two most common, each involving the masses of six neighboring nuclei, are written below [17, 18]:

$$-M(N+1,Z-2)+M(N+1,Z)-M(N+2,Z-1) \tag{1}$$
$$+M(N+2,Z-2)-M(N,Z)+M(N,Z-1) \quad =0,$$
$$M(N+2,Z)-M(N,Z-2)+M(N+1,Z-2) \tag{2}$$
$$-M(N+2,Z-1)+M(N,Z-1)-M(N+1,Z) \quad =0.$$

These relations do not have any free parameters and can be derived from an independent particle picture. Equation (2) is satisfied for essentially all values of N and Z, while equation (1) is valid only for $A \geq 16$ and $N \geq Z$ and , if N=Z, N may not be odd [17]. A separate equation that interchanges N and Z can be used instead of (1) for $N \leq Z$, but there are not many nuclei in this category. These innocent-looking equations are based on a very clever idea. The combinations are such that the number of neutron-neutron, neutron-proton and proton-proton interactions cancel. In addition to having the correct number of interactions, the single-particle energies and the residual interactions within each level, to a first approximation, cancel too [17].

TABLE 1. $\sigma_{r.m.s.}$ mass differences, in keV for the LDM, FRDM, DZ and GK calculations, and for different GK calculations.

model	LDM	FRDM	DZ	GK	
$\sigma_{r.m.s.}$	3447	669	346	189	
GK relations	1-12	4-12	7-12	10-12	12
A $\geq$ 16	189	162	117	95	86
A $\geq$ 60	123	102	87	81	80

ANALYSIS OF THE DIFFERENCES BETWEEN MEASURED AND CALCULATED MASSES

We apply the GK procedure to all nuclei in the 2003 compilation [19] where at least one of the relations (1) or (2) is applicable. Using the GK procedure we obtain a very specific prediction, determined by that of its neighbors, which guarantees that any mass calculated in this form does not involve a chaotic component. In this procedure there are no free parameters and there is no fit to the data, just a prediction of nuclear masses arising from those of its neighbors. The masses calculated in this way must thus represent a regular, non-chaotic component of the true mass in a very fundamental sense. This is to be contrasted with the "global" formulas that we discussed before, where a fitting procedure involving many parameters, or the determination of the effective interaction among nucleons must be achieved. In these cases it is difficult to separate *a priori* a regular and a chaotic part in the calculation, since this division depends on the particular methodology applied. In what follows we compare the mass deviations found in three of the global methods (LDM, FRDM, DZ) and our GK studies. The corresponding $\sigma_{r.m.s.}$ deviations, defined as

$$\sigma_{r.m.s.} = \left[\frac{1}{N} \sum_{i=1}^{N} (M_{exp}^i - M_{th}^i)^2 \right]^{1/2} \tag{3}$$

are displayed in Table 1, where we also include the smaller samples GK-n which involve the application of n or more GK relations, for which the average deviation is also quoted. Note the systematic decrease in the errors as a consequence of a better determination of the masses, proportional to the number of GK relations applied. In Fig. 1 we show a color-coded depiction of the distribution of mass deviations for each of the four methods employed. In our best scenario, that of GK-12, we find an r.m.s. deviation of 80 keV, almost an order of magnitude smaller than the FDRM one.

The LDM, FRDM and DZ plots display systematic errors which can be analyzed in linear arranges of data [20, 21]. To do this we first need to map the two-dimensional information present in figures (3) into one-dimensional arrays. The direct 2D study is difficult because of the irregular form of the data displayed in the nuclear charts, and cuts along fixed N, Z or A lines have a small number of nuclei, making it difficult to extract definite conclusions [20].

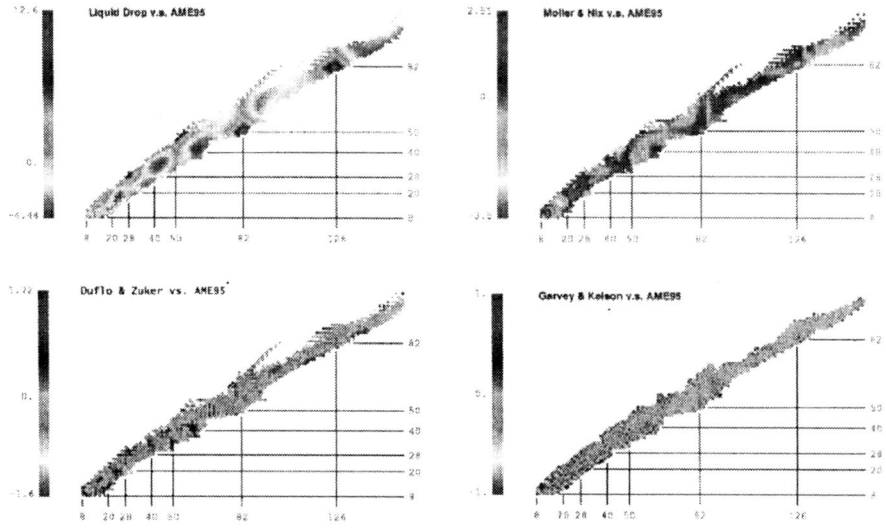

FIGURE 1. Mass differences from the LDM, FRDM, DZ and our GK studies , in MeV, as functions of N and Z.

CONCLUSIONS

We arrive at the conclusion that chaotic components in the nuclear masses, as suggested in [5] and analyzed in [15], if they exist at all, must contribute a rather small amount, with an upper bound of less than 100 keV (or one part in 10^6 for an A=100 nucleus), according to the criteria put forward in this paper. The order of magnitude of these components is consistent with the recent theoretical estimates of Ref. [22], although not with the A-dependence predicted in that study. Note however, that our approach can be considered model independent. The remaining correlations found by means of a power spectrum analysis seem to be consistent with zero, i.e. behave like white noise. Deviations between experimental masses and LDM, FDRM and DZ, respectively, show diminishing but significant remaining correlations. The results presented in this paper thus imply that nuclear masses can in principle be predicted with an average accuracy of better than 100 keV. Although the presence of chaos at this small scale cannot be discarded by this study, this is good news for the continued efforts to predict unknown masses with increasing accuracy.

ACKNOWLEDGMENTS

Our conversations with R. Bijker, O. Bohigas, J. Dukelsky, J.Flores, J.M. Gomez, P. Leboeuf, F. Leyvraz, R. Molina, S. Pittel, A. Raga and A. Zuker are gratefully acknowledged. This work was supported in part by Conacyt, México.

REFERENCES

1. Gutzwiller, M., *Chaos in Classical and Quantum Mechanics*, Springer-Verlag, Berlin, 1990.
2. Stöchmann, H.-J., *Quantum Chaos: An Introduction*, Cambridge University Press, Cambridge, 1999.
3. Blümel, R., and Reinhardt, W., *Chaos in Atomic Physics*, Cambridge University Press, Cambridge, 1997.
4. Nakamura, K., and Harayama, T., *Quantum Chaos and Quantum dots*, Oxford University Press, Oxford, 2004.
5. Bohigas, O., and Leboeuf, P., *Phys. Rev. Lett.*, **88**, 92502 (2002).
6. Rolfs, C., and Rodney, W., *Cauldrons in the Cosmos*, University of Chicago Press, Chicago, 1988.
7. D.Lunney, Pearson, J., , and Thibault, C., *Rev. Mod. Phys.*, **75**, 1021 (2003).
8. Möller, P., Nix, J., Myers, W., and Swiatecki, W., *At. Data Nucl. Data Tables*, **59**, 185 (1995).
9. Duflo, J., *Nucl. Phys. A*, **576**, 29 (1994).
10. Zuker, A. P., *Nucl. Phys. A*, **576**, 65 (1994).
11. Goriely, S., Tondeur, F., and Pearson, J., *Atom. Data Nucl. Data Tables*, **77**, 311 (2001).
12. Wigner, E., *Statistical Theories of Spectra: Fluctuations*, Academic, New York, 1965.
13. Mehta, M., *Random Matrices*, Academic Press, London, 1990.
14. Bohigas, O., Haq, R., and Pandey, A., *Phys. Rev. Lett.*, **54**, 1645 (1985).
15. Aberg, S., *Nature*, **417**, 499 (2002).
16. Garvey, G., and Kelson, I., *Phys. Rev. Lett.*, **16**, 197 (1966).
17. Garvey, G., Gerace, W., Jaffe, R., Talmi, I., and Kelson, I., *Rev. Mod. Phys.*, **41**, S.1 (1969).
18. Comay, E., I.Kelson, and Zidon, A., *At. Data Nucl. Data Tables*, **39**, 235 (1988).
19. Audi, G., Wapstra, A., and Thibault, C., *Nucl. Phys. A*, **729**, 337 (2003).
20. Hirsch, J. G., Frank, A., and Velázquez, V., *Phys. Rev. C*, **69**, 37304 (2004).
21. Velázquez, V., Frank, A., and Hirsch, J. G., 'Systematic correlations and chaos in mass formulae," in *Computational and Group Theoretical Methods in Nuclear Physics*, edited by J. Escher, O. C. nos, J. Hirsch, G. Stoitcheva, and S. Pittel, World Scientifi c, Singapore, 2004.
22. Molinari, A., and Weidenmüller, H., *arXiv:nucl-th/*, p. 0403028 (2004).

Probing additional dimensions in the universe with neutron experiments

Alejandro Frank*, Piet Van Isacker† and Joaquín Gómez-Camacho**

*Instituto de Ciencias Nucleares, Universidad Nacional Autónoma de México, Apartado Postal
70-543, 04510 México, D.F., México
†GANIL, B.P. 55027, F-14076 Caen Cedex 5, France
**Departamento de Física Atómica, Molecular y Nuclear, Facultad de Física, Universidad de
Sevilla, Apartado 1065, 41080 Sevilla, Spain

Abstract. We investigate the modifications of the gravitational force in the context of recent work on large supplementary dimensions and deduce a formula for the expected compactification radius for the n additional dimensions in the universe, as a function of the Planck and the electroweak scales. We argue that the correspondingly modified gravitational force gives rise to effects that, although very small, might be within the detection range of dedicated neutron scattering experiments. We discuss the characteristics of an experimental setup to observe these effects.

INTRODUCTION

Why gravity is so weak? This question may sound strange to the man in the street, which experiences gravity as a very strong force. However, when the gravitational force is compared with the electromagnetic force for two elementary particles, like two electrons, the difference is more than 40 orders of magnitude [1]. The deep reason behind this difference arises from the difference in size of two fundamental constants. On one side, there is Newton's gravitational constant G, which determines the strength of gravity. On the other hand, there is Fermi's constant G_F, which is mainly associated with the weak interaction. Both constants have dimensions, in natural units, of the square of a length. The corresponding lengths are Planck length R_p and the electroweak length R_e, given by

$$R_p = \sqrt{\hbar G/c^3} = 1.62 \cdot 10^{-35} m, \quad R_e = \sqrt{G_F/\hbar c} = 6.73 \cdot 10^{-19} m$$

How many dimensions are there in the universe? The obvious answer to this is that there are 3 space dimensions, plus time. However, string theorists have been arguing that there should be as many as 10 or 11 dimensions, in order to have a consistent understanding of gravity and the standard model of particle physics. The additional dimensions are curved (compactified), and have a characteristic extension given by an unknown compactification radius R_c which is generally assumed to be of the order of the Planck length R_p.

These two questions have been related in recent works by Arkani-Hamed, Dimoupoulos, and Dvali (ADD) [2] In their papers, ADD conjecture the existence of two or more additional dimensions in which gravity, but not the strong or the electro-weak forces,

CP726, Nuclear Physics, Large and Small: International Conference on
Microscopic Studies of Collective Phenomena, edited by R. Bijker, R. F. Casten, and A. Frank
© 2004 American Institute of Physics 0-7354-0207-8/04/$22.00

might be acting, diluting itself by spreading its lines of force into these extra dimensions. Essentially, this would explain its apparent weakness [2]. The proposal arises from a bold modification of pre-existing many-dimensional string theories and the more recent M-theories (M for membrane) which encompass the former [3, 4]. In these theories only gravitons are able to traverse the extra dimensions, whereas other particles are fixed to our observable 3-D world, since the former are described as closed strings free to wander while the latter are open strings with their ends fixed to our 'brane'. Additional dimensions are 'compactified', i.e., they are closed on themselves with a characteristic radius of compactification R_c (which for simplicity is assumed to be the same for all additional dimensions). For ranges smaller than R_c, we thus expect a modified gravitational force. In this scenario, instead of catching up with the other forces at Planck's length scale, the N-D gravitational force actually joins the other interactions at a distance about 10^{16} times larger, namely at the electro-weak unification scale of R_e. As will be shown below, this conjecture fixes the strength of the N-D gravitational force and the value of R_c. The most important consequence of the LED hypothesis is the possible transition from the entirely Platonic, inaccessible situation posed by the Planck-scale compactification, to one where it is conceivable that experimental measurements may actually test these ideas. The experiments envisioned to date are predominantly of two kinds. The ones involving submillimeter torsion-balance experiments [5], and high-energy collider experiments in the TeV energy region, where diverse theoretical predictions exist for the indirect observation of additional dimensions, such as the occurrence of missing energy carried away by undetected gravitons [6]. The Casimir effect has also been proposed as a test of new forces at micrometer distances [7]. Here we explore the possible effects of N-D gravity in experiments with neutrons.

GRAVITATIONAL FORCE IN N DIMENSIONS

We start in the spirit of ADD [2] by carrying out a classical analysis of gravity in $N = n + 3$ dimensions. If space would have $n + 3$ (extended) dimensions, Gauss' law implies that the force of gravity would be of the form

$$F_n = -\frac{m_1 m_2 G_n}{r^{n+2}}, \tag{1}$$

where G_n is a constant which reduces to Newton's gravitational constant G for $n = 0$. This expression should be valid for distances below R_c. In general, one should assume a soft transition from N-D gravity to 3-D gravity at distances around R_c. We shall instead impose the equality of forces at the compactification length R_c, $F_n(R_c) = F_0(R_c)$, which implies that $G_n = G R_c^n$.

We then implement the ADD conjecture that the N-D force at the electro-weak length R_e is as strong as the 3-D force at the Planck length R_p, $F_n(R_e) = F_0(R_p)$. This leads to the desired equation

$$\left(\frac{R_c}{R_e}\right)^n = \left(\frac{R_e}{R_p}\right)^2 = \frac{c^2 G_F}{\hbar^2 G} = 1.73 \cdot 10^{33} \tag{2}$$

which relates the number of additional dimensions n, the radius of compactification R_c, and the electro-weak scale R_e, to the ratio of two fundamental constants in nature: Fermi's and Newton's constants. A similar result can be obtained [8] requiring that the N-D force be comparable to the electromagnetic interaction. at the electroweak length scale. This is an indication of the robustness of formula (2). In Table 1 we display the values of R_c as a function of n, which turn out to be close to the ones evaluated by more sophisticated arguments [9]. Note that $n = 1$ can be readily discarded since it leads to a value of R_c larger than the size of the solar system and hence to unstable planetary orbits. For $n = 2$ we find R_c of a few cm. Deviations from Newton's law at this scale seem to be discarded by experiment. Nevertheless, it should be noted that our analysis can only be expected to give rough estimates with considerable uncertainties in the prediction of R_c. It is the range below 1 mm, which is very difficult to explore with macroscopic gravity experiments, that we would like to investigate by means of neutrons.

From the expression of the force, one can derive the expression of the gravitational potential, valid at distances smaller than R_c

$$V_n(r) = \frac{m_1 m_2 G_F c / \hbar^3}{n+1} \left(\frac{R_e}{r} \right)^n \frac{\hbar c}{r}. \tag{3}$$

Note that the strength of this potential depends on the Fermi constant G_F, which describes not only the electroweak interaction, but also the enhanced gravitational force.

ENHANCED GRAVITATIONAL EFFECTS IN NEUTRON SCATTERING

To investigate the possible gravitational effects in the scattering of particles, the strong and long range Coulomb force should be avoided. Thus, one should consider neutral heavy particles. This leaves neutrons as the obvious choice for projectiles, colliding with heavy nuclei as targets.

In order to attempt neutron N-D gravity experiments at short ranges we face problems from the outset. The more obvious one is the strong nuclear force, present at range scales of the order of 10^{-15} m. Other problems are related to the magnetic interaction with the spin of the neutron [10] and the dipole polarizability of the neutron [11], which could generate weak long range interactions. We will focus here in reducing the effect of the strong and short range nuclear force.

Let us consider the interaction of a neutron beam with a heavy nucleus such as ^{208}Pb. Although the constant $m_1 m_2 G_F c / \hbar^3 = 0.0021$ is not too small, the typical separations r are much larger than R_e giving rise to very small values of the potential energy. We will discuss what are the optimal experimental conditions which could allow observation of this tiny effect in neutron-scattering experiments and will consider what is the adequate energy and angular momentum so that the phase shift due to the gravitational force is as large as possible.

The nuclear potential can be parametrized with a Woods-Saxon shape, so that

$$V_{\text{nucl}}(r) = \frac{V_0}{1 + \exp(r - R)/a}, \tag{4}$$

95

TABLE 1. Estimates of R_c, R_m (fm), gravitational potential (MeV), optimal energy (MeV) and maximum gravitational phaseshift as a function of the number of large extra dimensions n.

n	1	2	3	4	5	6	7
R_c	$1.2\,10^{30}$	$2.8\,10^{13}$	$8.1\,10^{7}$	$1.4\,10^{5}$	$3.0\,10^{3}$	$2.3\,10^{2}$	38
R_m	18	25	32	39	47	54	62
$V_n(R_m)$	$4.4\,10^{-7}$	$4.1\,10^{-12}$	$3.1\,10^{-17}$	$1.9\,10^{-22}$	$9.0\,10^{-28}$	$4.2\,10^{-33}$	$1.5\,10^{-38}$
E_{opt}	4.6	2.3	1.5	0.98	0.68	0.51	0.39
ϕ_{opt}	$4.1\,10^{-7}$	$7.6\,10^{-12}$	$8.8\,10^{-17}$	$8.2\,10^{-22}$	$5.6\,10^{-27}$	$3.5\,10^{-32}$	$1.6\,10^{-37}$

Reasonable parameters are $V_0 = 50$ MeV, $R = 1.2\,A^{1/3}$ fm, and $a = 0.6$ fm. This gives, for distances below $r = 10$ fm, values of the potential in the MeV range and any gravitational effect at that distance would be drowned by the uncertainties in the nuclear potential. Instead, one must probe distances at which the nuclear and gravitational potential are of the same order. In Table 1 we indicate the distances R_m at which the nuclear and gravitational potential are equal, as a function of n. These values are, for $n \leq 6$, larger than R_c, and this indicates that there is some range of distances $R_m < R < R_c$ in which gravitational effects, although very small, are larger than the nuclear force, and thus could, in principle, be investigated.

The scattering observables are also affected by the fact that the neutron has bound states in the nuclear potential generated by ^{208}Pb. From the shell structure of this nucleus one knows that the single-particle potential supports bound states up to angular momentum $L = 7$ (the $1j_{15/2}$ orbital). Thus, the scattering of neutrons with $L \leq 7$ is affected by the nuclear potential, even if the neutron energy is very small, because the scattering wave functions have to be orthogonal to the bound states. Consequently, to obtain scattering observables free of nuclear contamination, we need to consider $L > 7$.

Estimates of the phase shifts due to the gravitational interaction can be found in the eikonal approximation [8]. In this way, we find

$$\phi(L,E) \simeq \frac{m_1 m_2 G_F c/\hbar^3}{n+1} \left(\frac{R_e}{b} \right)^n \frac{\hbar c}{\hbar v}. \tag{5}$$

From this expression we see that, in order to enhance the scattering effects of the gravitational force, one would need to have, in principle, small impact parameter b and small velocity v which are related through $bvm_n = (L+1/2)\hbar$. To maximize the phase shift $\phi(L,E)$, the best choice is to take the minimum angular momentum $L_{min} = 8$ and the minimum impact parameter which corresponds to the distance R_m at which the nuclear force may start to play a role. This gives the following optimal energy for scattering:

$$E_{opt} = \frac{(L_{min}+1/2)^2 \hbar^2}{2 m_n R_m^2}. \tag{6}$$

These energies are shown in Table 1 and are of the order of 1 MeV. At these energies the optimal phase shifts (for $L = L_{min}$), also shown in Table 1, can be evaluated to give

$$\phi_{opt} \simeq \frac{V_n(R_m)}{2 E_{opt}} (L_{min} + 1/2). \tag{7}$$

96

It should be noticed that the factor $(L_{min} + 1/2)$ was missing in eq. (14) of ref. [8]. For $n > 2$ the phase shift is extremely small and decreases as n grows. However, we believe that the case of $n = 2$ may be within reach of a dedicated scattering experiment. For this case, the elastic scattering amplitude turns out to be energy independent and can be written in the closed form

$$A_g(E, \theta) = iA_g f_g(\theta) \tag{8}$$

$$A_g = \frac{2\hbar c (m_1 m_2 G_F c / \hbar^3)^2}{3(m_1 + m_2)c^2} \tag{9}$$

$$f_g(\theta) = \sum_L (L + 1/2)^{-1} P_L(\cos\theta) \tag{10}$$

where we have made use of Eq. 2. The modified gravitational amplitude can now be evaluated for the scattering of neutrons on ^{208}Pb, and we find $A_g = 0.298 \cdot 10^{-8} fm$, to be compared with the typical scattering amplitudes for the nuclear force, which for a range of energies of a few MeV, are of the order of $A_n = 7 fm$, although the amplitude can strongly fluctuate with energy as resonances are crossed. It would seem that it is impossible to observe such a tiny gravitational effect, being so small compared to the nuclear amplitude. However, the angular dependence of these amplitudes is quite different. In contrast to the nuclear part, which is essentially independent of the scattering angle for $\theta \ll 1/L_{min}$, the gravitational amplitude involves the contribution of a significant number of angular momenta. In addition, we find that the gravitational amplitude diverges for small scattering angles. It is the combination of these characteristics which may open a window to observe an interference effect. More specifically, we have derived that $f_g(\theta) = K_0(\theta/2) + \delta_g(\theta)$, where K_0 is the Bessel function, which diverges logarithmically as $\theta \to 0$, and $\delta_g(\theta)$ is a smooth function of the angle.

Our analysis then implies that at very small scattering angles, the neutron-nucleus differential cross section is given by

$$\frac{d\sigma}{d\Omega} =. |A_n(E, \theta) + A_g(E, \theta)|^2 \tag{11}$$

$$\simeq |A_n(E, 0)|^2 + 2|A_n(E, 0)|A_g \sin(\arg(A_n(E, 0)))K_0(\theta/2) \tag{12}$$

Note that while $|A_n(E, 0)|^2$ is about 5 b/sr, $2|A_n(E, 0)|A_g$ is about 0.4 nb/sr. Discerning such a faint whisper in the midst of the nuclear background roar can be a formidable task. But this feat may be accomplished by carefully monitoring both the angular and energy dependences of the cross section. As the phase of the nuclear amplitude changes as resonances are crossed, the diminutive interference between nuclear and gravitational amplitudes changes from constructive to destructive interference. In this sense, the application of interferometric techniques, as used in the scattering of slow neutrons [12], that would allow to reduce the angle-independent nuclear amplitude would be extremely useful.

An experimental verification of these ideas would involve measurements with a neutron detector array measuring small angles (below 10 degrees), as well as a neutron source which provides neutron beams in the MeV range, with energy resolution sufficient to see the resonances in n+^{208}Pb scattering. For purely nuclear scattering the ratio

97

of the cross sections (the ratio of detected neutrons) measured at different small angles should remain constant as energy is changed. The presence of any long range interaction between the neutron and the target would be signalled by small fluctuations associated to the interference between the two amplitudes. Moreover, these fluctuations would not be random, but should correlate with the scattering energy as resonances are crossed.

A careful analysis of these fluctuations may isolate a gravitational signature from higher order electromagnetic effects such as magnetic interactions or neutron dipole polarizability.

ACKNOWLEDGMENTS

We wish to thank H.G. Börner, R.F. Casten, A. Villari and V. Nesvizhevsky for encouraging discussions. AF is supported by CONACyT, Mexico and JGC by the spanish DGICyT project FPA2002-04181-C04-04.

REFERENCES

1. M. Kaku, *Quantum Field Theory. A Modern Introduction* (Oxford University Press, New York, 1993).
2. N. Arkani-Hamed, S. Dimopoulos and G. Dvali, Phys. Lett. B **429** (1998) 263; Phys. Rev. D **59**, 086004 (1999).
3. P. Horava and E. Witten, Nucl. Phys. B **460** (1996) 506; Nucl. Phys. B **475** (1996) 94.
4. L. Randall and R. Sundrum, Phys. Rev. Lett. **83** (1999) 3370;
    L. Randall, Science **296** (2002) 1422.
5. C.D. Hoyle *et al.*, Phys. Rev. Lett. **86** (2001) 1418.
6. S. Cullen and M. Perelstein, Phys. Rev. Lett. **83** (1999) 268.
7. D.E. Krause and E. Fischbach, Phys. Rev. Lett. **89** (2002) 190406.
8. A. Frank, P. Van Isacker, J. Gómez-Camacho Physics Letters B582 (2004) 15-20, nucl-th/0305029.
9. J. Blum *et al.*, Phys. Rev. Lett. **85** (2000) 2426;
    Z. Chacko and E. Perazzi, hep-ph/0210254.
10. Yu.A. Alexandrov, Phys. Part. Nucl. **32** (2001) 708.
11. J. Schmiedmayer, P. Riehs J. A. Harvey and N. W. Hill, Phys. Rev. Lett. 66 (1991) 1015.
12. J. Byrne, *Neutrons, Nuclei and Matter. An Exploration of the Physics of Slow Neutrons* (Institute of Physics, Bristol, 1994).

Nuclear abundances in the high-energy cosmic radiation

Thomas K. Gaisser

Bartol Research Institute, University of Delaware, Newark, DE 19716

Abstract. This paper is based on a talk presented April 21st at the International Conference on Nuclear Physics, Large and Small, Cocoyoc, Mexico. The conference, in honor of my colleague at Bartol, Stuart Pittel, celebrated his work in nuclear theory, which reflects a long tradition of nuclear physics at our institution. Here I briefly review some current aspects of another traditional field at Bartol, cosmic ray physics. I describe three examples in which the energy-dependence of the relative abundances of different nuclei in the cosmic radiation is used to understand their origin. The emphasis is on the high-energy end of the cosmic-ray spectrum, which is accessible only to large, ground-based air shower experiments.

INTRODUCTION

The cosmic-ray spectrum extends over some 11 decades of energy from below 1 GeV to above 10^{20} eV. The intensity decreases by a factor of 50 or more per factor of ten increase in energy. As a consequence, direct measurements of the primary cosmic rays with detectors on high-altitude balloons or spacecraft before they interact in the atmosphere are not practical for energies above 10^{15} eV because the flux is too low. Above this energy it is necessary to resort to large arrays of detectors on the ground which sample the atmospheric cascade initiated by the primaries high in the atmosphere.

With direct measurements above the atmosphere it is possible to identify the charge and energy of each nucleus (and, in the GeV range, the mass as well). Such detailed information forms the basis of our understanding of the likely sources and acceleration mechanisms that produce the cosmic rays. In particular, from the the energy content of galactic cosmic radiation, together with the characteristic dwell time of particles inside the Galaxy before they escape, one can estimate the power required of the sources to maintain the observed intensity of cosmic-rays.

With air shower experiments, in contrast, the best one can hope for is to assign a rough probability for the primary mass and to determine the relative contribution of major groups of nuclei to the total event sample in broad bins of energy (e.g. $\Delta E/E \sim .25$). The goal, however, remains the same–to make inferences about the origin of the highest energy cosmic particles. In this talk I review two aspects of the high-energy spectrum as inferred from air shower experiments, making use of analogies with results of direct measurements at lower energy. I begin with a brief reminder of the main conclusions from direct observations of primary cosmic rays.

CP726, Nuclear Physics, Large and Small: International Conference on
Microscopic Studies of Collective Phenomena, edited by R. Bijker, R. F. Casten, and A. Frank
© 2004 American Institute of Physics 0-7354-0207-8/04/$22.00

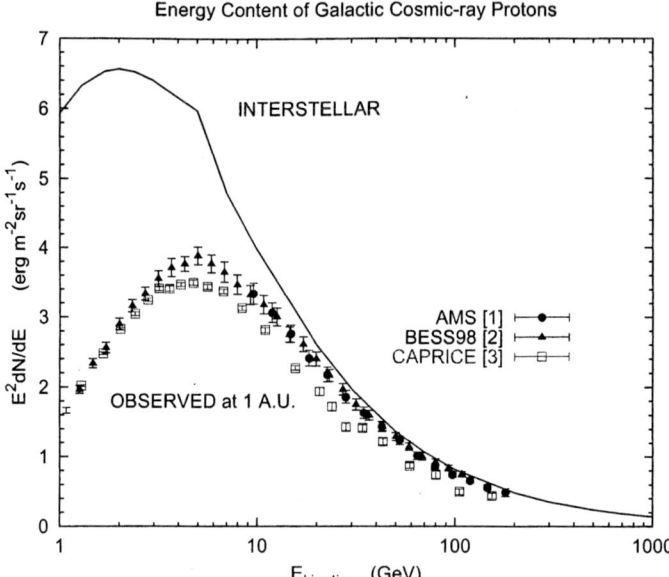

FIGURE 1. Three measurements of the protons spectrum at Earth[1, 2, 3]. The line is an estimate of the interstellar flux corrected for the effect of solar modulation.

DIRECT MEASUREMENTS

Most of the energy in the cosmic radiation is carried by particles with energy less than 100 GeV. This is illustrated in Fig. 1. The figure shows energy content per logarithmic interval of energy on a linear scale so that energy content is proportional to the geometrical area under the curve. (This is like the νF_ν plot familiar from multiwavelength astronomy.) The line corrects the low energy part of the spectrum for the effects of the solar wind, which decelerates incoming particles and excludes the lowest energy particles from the inner heliosphere. The energy content is obtained by integrating the curve in Fig 1 over energy and converting from flux to energy density:

$$\rho_E = \frac{4\pi}{c} \times \int \frac{E\, dN}{dE}\, dE \approx 10^{-12}\, \mathrm{erg\, cm^{-3}}. \tag{1}$$

The source power required to maintain the observed spectrum is

$$P_{CR} \sim \rho_E / \tau_{esc} \approx 10^{-26}\, \mathrm{erg\, cm^{-3} s^{-1}}, \tag{2}$$

where $\tau_{esc} \sim 3 \times 10^6$ yrs is an estimate of the typical time cosmic rays diffuse in the Galaxy before escaping. Eq. 2 is a simplified form of a detailed model of cosmic-ray propagation. There are several more or less elaborate models with parameters based on observations of spectra of individual elements and of diffuse gamma radiation. The general picture is that cosmic-ray sources are distributed in the disk of the Galaxy. The sources emit particles which diffuse in the disk for a characteristic time τ_{esc}. An estimate of the characteristic time is based on the relatively large amounts of the elements Li, Be and B in the cosmic radiation as compared to their very low abundance in general. They are produced as spallation products of abundant C and O nuclei. The known spallation cross sections, when convolved with the density of gas in the interstellar medium (ISM), lead to a numerical value of the order used in Eq. 2.

Assuming that the energy density in cosmic radiation near the Solar system is typical of the Galactic disk in general, the total power requirement is

$$P_{CR,tot} = P_{CR} \times V_{Galaxy} \approx 3 \times 10^{40} \text{ erg/s.}$$

In comparison, the power available in kinetic energy of supernova ejecta, assuming one supernova per century and 10^{51} ergs per supernova, is 3×10^{41} erg/s. Thus a $\sim 10\%$ efficiency is adequate for supernovae to power the observed cosmic radiation. This fact, together with a theory of acceleration by supernova-driven shocks, accounts for the prominence of the supernova model of cosmic-ray origin.

A characteristic feature of acceleration by supernova shocks is a maximum accessible energy per particle. In general, shock acceleration takes time because the process involves cycling particles back and forth across the shock many times. In the case of acceleration at a shock driven by rapid expansion of a supernova remnant, the shock weakens when the remnant slows down significantly. This occurs after it has swept up a mass of interstellar gas comparable to the original ejected mass. Estimates of the maximum energy depend on models used for the explosion and for the surrounding medium, but they are generally in the range of 10^{14} to 10^{15} eV for protons [4].

THE HIGH-ENERGY SPECTRUM

It is natural to ask how much of the entire cosmic-ray spectrum could be explained by supernovae or other sources inside our Galaxy. Fig. 2 shows the high-energy portion of the spectrum plotted as $E^{2.7} \times dN/dE$. Plotting the spectrum in this way amplifies deviations from the low-energy behavior, which is described by a power law with differential spectral index -2.7. The steepening of the spectrum between 10^{15} and 10^{16} eV is known as the *knee* of the spectrum. The range of values from different experiments indicates underlying systematic uncertainties characteristic of the indirect nature of air shower experiments. One possibility is that the knee could reflect the end of the supernova acceleration mechanism.

At higher energy, somewhere above 10^{18} eV, the spectrum again flattens, a feature known as the *ankle*. The flux above 10^{18} eV is about 30 per square kilometer per year. To explore this energy region therefore requires giant air shower experiments with effective areas of thousands of square kilometers. Major current activity in this field by the AGASA [6], HiRes [7] and Auger [8] experiments aims at resolving the question of whether the highest energy particles are suppressed by energy losses as they propagate through the cosmological black body radiation. An ankle feature in the spectrum is naturally interpreted as a transition to a different population of particles with a harder spectrum. The spectrum presumably continues to low energy underneath

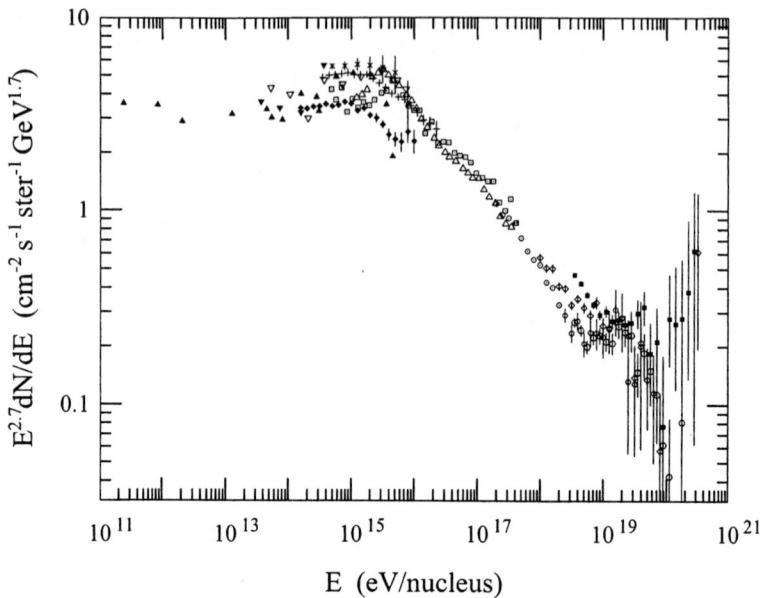

FIGURE 2. Overview of the high-energy cosmic-ray spectrum. [5]

the more abundant population with the steeper spectrum. The gyro-radius of a 10^{18} eV proton in the typical few μ-Gauss field of the ISM is comparable to the thickness of the galactic disk. It is therefore natural to interpret the ankle as the transition from galactic to extra-galactic cosmic rays. Systematic uncertainties blur the location of the transition energy, which is currently a point of considerable interest.

Knee of the spectrum

Acceleration and propagation of particles in an astrophysical setting both depend on magnetic fields and hence on gyro-radius, $r_L = R/B$. Here $R = pc/Ze$ is magnetic rigidity and $pc \approx E$ is the total momentum ($\approx$ energy) per nucleus of charge Ze. (Nuclei are assumed to be fully ionized.) Thus if the spectrum (either as a consequence of acceleration or of propagation) has a feature at a characteristic rigidity R_c, one expects the feature to show up at systematically different energies per particle as a function of nuclear charge:

$$E_c(Z) \sim Z R_c. \tag{3}$$

Total energy per particle is the relevant quantity for air shower experiments which essentially use the atmosphere as a calorimeter.

Since the major components of the cosmic radiation are the elements extending from protons ($Z = 1$) to iron ($Z = 26$), one expects any rigidity-dependent feature to extend over at most a factor of 30 in total energy per particle. Moreover, the change should occur with a systematic pattern, first for protons, then successively for helium, carbon/oxygen, Mg/Si and finally for the iron group [9]. (If another source does not intervene, it is conceivable that much rarer trans-iron elements could become important at higher energy [10].) This pattern of successively heavier elements dominating the spectrum with increasing energy has been observed in at least one air shower experiment in the energy region up to nearly 10^{17} eV [11]. The results are based on measurements of ratios of different components of the showers at ground level (e.g. ratio of low energy muons to the dominant e^+/e^- component).

Although the structure of the knee itself may be confined within a factor of 30, the spectrum continues smoothly for another order of magnitude or more before the ankle. Explanation of this fact requires either a conspiracy or an accident. One example of a conspiracy would be a change in the properties of particle diffusion in the interstellar medium such that τ_{esc} decreases more rapidly with energy above the knee. Since the observed equilibrium spectrum in the Galaxy is a product of the source spectrum and τ_{esc}, such a transition would be consistent with the smooth continuation of the spectrum to the ankle provided the change in diffusion produced a change in power law consistent with with $\delta\gamma \approx 0.3$ as observed. In addition, such an explanation assumes the existence of galactic accelerators that produce particles with energy up to 10^{18} eV. The compositional signature of this explanation would be one sequence of change of slope, starting with protons and ending with the heaviest abundant element, iron.

Another possibility is that the knee is signaling the upper limit of the acceleration mechanism. In the case of shock acceleration, any particular source is expected to have an exponential cutoff at its maximum energy. In general, one would expect the observed spectrum to consist of the contribution of many individual sources. An explanation of the knee as an accident could be that a large number of sources contribute at low energy, and one or two sources with larger E_{max} produce the spectrum continuing up to $\sim 10^{18}$ eV. Such accidental explanations have been advocated by Erlykin & Wolfendale [12]. The compositional signature of such an explanation would be repeated sequences of composition cycles as each successively higher energy source reached its maximum energy. The power required of the high-energy sources depends on the assumed spectrum of the "accidental" component, which would by assumption be harder at low energy than the supernova component. It also depends on the energy dependence of τ_{esc}. Reasonable assumptions lead to an estimate of $\sim 2 \times 10^{39}$ erg/s, less than 10% of the total power requirement for all galactic cosmic-rays. For comparison, the micro-quasar SS433 at 3 kpc distance has a jet power estimated as 10^{39} erg/s [13].

A more orthodox explanation requires a conspiracy of acceleration mechanisms so that the population of particles injected by the dominant low energy sources (usually taken to be supernova shocks) is re-accelerated to higher energy by a different acceleration mechanism capable of reaching a higher energy. Axford [14] suggests extended acceleration by multiple supernova remnants. Jokipii& Morfill [15] and Völk & Zirakashvili [16] suggest re-acceleration to higher energy by shocks in a galactic wind (analogous to shocks of various types in the solar wind). This explanation would have a different signature in composition, more like that of the change of diffusion mentioned above.

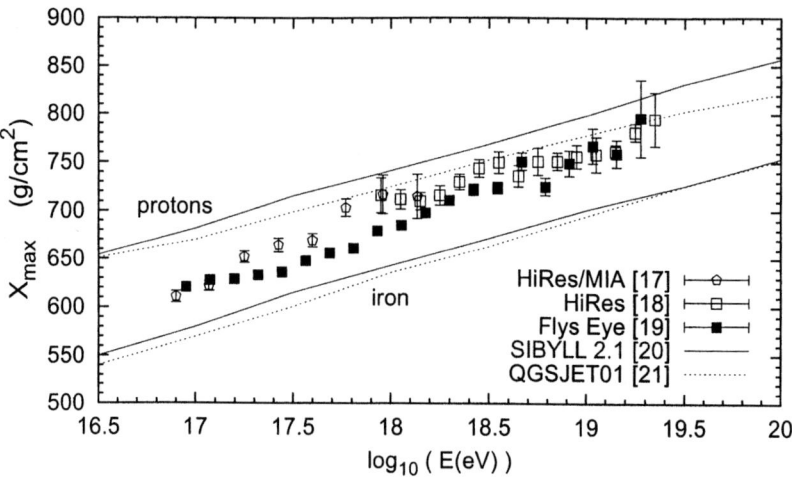

FIGURE 3. Depth of shower maximum from HiRes [17, 18] and Fly's Eye [19] compared with two theoretical calculations for primary protons and two for primary iron nuclei.

Ankle of the spectrum

Whatever the explanation of the knee, one expects the highest energy cosmic rays of galactic origin to be enriched in heavy nuclei for the reasons explained above. If so, the compositional signature of the transition from Galactic to extragalactic particles would naturally be a transition back to predominantly protons.

Fig. 3 shows data from three experiments [17, 18, 19] compared to calculations [20, 21] of average depth of shower maximum. The data come from air shower detectors that use the fluorescence technique to track shower development through the atmosphere. For a given total energy, cascades initiated by heavy nuclei develop higher in the atmosphere (smaller X_{max}) than those initiated by protons. The depth (in g/cm^2) along the shower axis until the shower reaches maximum number of particles is related to the mass (A) of the primary nucleus by

$$X_{max} = K_1 \ln(E_0/A) + K_2, \qquad (4)$$

where E_0 is the total energy of the incident nucleus. The parameters in Eq. 4 depend on the nature of high-energy nuclear interactions in the atmosphere.

The data in Fig. 3 show the expected transition from heavy to light. Two measurements made with the HiRes detector [17, 18] indicate that the transition occurs between 10^{17} and 10^{18} eV, while the original Fly's Eye experiment shows the transition at a higher energy [19]. Uncertainty in the location of the transition translates into a corresponding uncertainty in the power required for the extragalactic cosmic rays, as explained in the next paragraph.

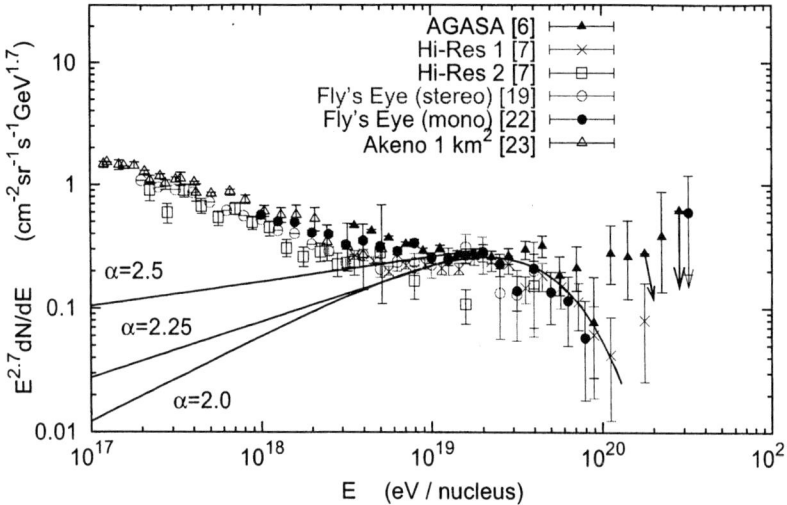

FIGURE 4. The cosmic-ray spectrum above 10^{17} eV. The curves show several possibilities for an extragalactic component.

Following the pattern of analysis for galactic cosmic-ray sources the next step is to estimate the power required for the extragalactic component. The answer depends on the energy at which one chooses to normalize this component to the data and on the spectral index assumed to extrapolate it to lower energy. Fig. 4 illustrates the problem. The curves show several possibilities for the spectrum of the extragalactic component superimposed on an expanded plot of measurements [6, 7, 19, 22, 23] of the spectrum above 10^{17} eV. [1] Assuming a differential spectral index of -2 and normalizing at 10^{19} eV, the estimated energy density is 2×10^{-19} erg/cm^3. For a cosmological distribution of sources the relevant time is the Hubble time, 10^{10} years, which leads to an estimated power requirement of $\sim 10^{37}$ erg/Mpc^3/s. Shifting the normalization point down in energy would increase the power requirement, as would assuming a differential spectral index < -2.

It is instructive to compare this order of magnitude estimate of total power required to what it implies for the power per source for various possibilities. The numbers in Table 1 follow from the known densities (or rate) of the objects (events). In all cases, the power that would be required in cosmic rays is comparable to the power observed in other wavelengths. Active galaxies [25, 26] and gamma-ray bursts (GRB) [27, 28] are

---

[1] Note that the curves in Fig. 4 all show the suppression of the flux above 5×10^{19} eV expected as a consequence of energy losses as ultra-high energy cosmic rays propagate through the cosmological black body radiation. The extremely interesting question of whether this suppression is actually consistent with the data is not addressed here. See [24] for a recent review.

among the sites often considered as possible sources of the highest energy cosmic rays.

TABLE 1.   Densities and power requirements for various possible cosmologically distributed sources of high-energy cosmic rays.

Object (event)	Density (rate)	Power required
Galaxies	$3 \times 10^{-3}/\text{Mpc}^3$	5×10^{39} erg/s/galaxy
Clusters of galaxies	$3 \times 10^{-6}/\text{Mpc}^3$	4×10^{42} erg/s/cluster
AGN	$10^{-7}/\text{Mpc}^3$	10^{44} erg/s/AGN
(GRB)	(1000/yr)	3×10^{52} erg/GRB

CONCLUSION

The common theme of the examples discussed here is the use of primary composition to understand the origin of high-energy cosmic radiation. In the case of the knee region around 3×10^{15} eV, there is a considerable variation among experimental results [10, 29]. The Kascade experiment shows the beginning of the progression of cutoffs expected for a rigidity-dependent effect, although the fits show a particularly low relative abundance of protons throughout the knee region. As with other small air shower detectors, the energy region explored with good statistics is limited to just above 10^{16} eV.

At higher energy, comparison of measurements of $X_{\max}$ with calculations shows a transition from heavy back toward lighter primary composition. This is an expected signature of a transition from cosmic rays of galactic origin to extragalactic origin. Energies around 10^{18} eV are, however, in the threshold region for air fluorescence detectors. The exact location of the change toward lighter primaries is uncertain, with the hybrid HiRes-MIA (fluorescence + muon array) detector indicating that the transition begins already at 10^{17} eV.

A novel approach to primary composition is to measure the ratio of shower size at the surface to the number of penetrating muons observed deep underground. This has been realized with the EASTOP over the deep underground MACRO [30] and with the South Pole Air Shower Experiment (SPASE) above the Antarctic Muon and Neutrino Detector Array (AMANDA) [31]. Both experiments find an increase in the average primary mass in the knee region.

The SPASE-AMANDA configuration is a forerunner of the IceCube experiment, now under construction at the South Pole. IceCube [32] is a kilometer-scale neutrino telescope between 1.45 and 2.45 kilometer depth in the Antarctic ice sheet together with a kilometer-square air shower array on the surface. The primary function of the surface array (IceTop) is for tagging downward background events in the deep detector for study, calibration and partial veto. As a consequence of its large size and high altitude, this three-dimensional air shower array is also expected to have excellent sensitivity to primary mass with good energy resolution up to 10^{18} eV [33]. With a trigger threshold of 3×10^{14} eV, the combined detector will give a measure of the primary composition from below the knee to the beginning of the transition to extragalactic cosmic rays with a single technique.

ACKNOWLEDGMENTS

The South Pole Air Shower Experiment is supported by a grant from the Office of Polar Programs of the National Science Foundation (OPP-9980801). The IceCube construction project is supported by the Office of Polar Programs of the National Science Foundation.

REFERENCES

1. J. Alcarez *et al.*, Phys. Lett. B490 (2000) 27-35.
2. T. Sanuki *et al.*, Ap.J. 545 (2000) 1135-1145.
3. M. Boezio *et al.*, Ap.J. 518 (2999) 457-472.
4. E.G. Berezhko in *Invited, Rapporteur and Highlight papers of ICRC2001* (Copernicus Gesellschaft, 2002) 226-233.
5. T.K. Gaisser & Todor Stanev in *Reviews of Particle Properties*, K. Hagiwara *et al.*, Phys. Rev. D (2002) 010001-182-188.
6. M. Takeda *et al.*, Astropart Phys. 19 (2003) 447-462.
7. T. Abu-Zayyad *et al.*, astro-ph/0208243. Also, D. Bergman, Proc. 28th Int. Cosmic Ray Conf. (Tsukuba) 1 (2003) 397-400.
8. http://www.auger.org/
9. B. Peters, Nuovo Cimento 22 (1961) 800.
10. J.R. Hörandel, Astropart. Phys. 21 (2004) 241-265.
11. M. Roth *et al.*, Proc. 28th Int. Cosmic Ray Conf. (Tsukuba) 1.(2003) 139-142.
12. A.D. Erlykin & A.W. Wolfendale, J.Phys. G27 (2001) 1005.
13. C. Distefano, D. Guetta, E. Waxman & A. Levinson, Ap.J. 575 (2002) 378.
14. W.I. Axford, Ap.J. Suppl. 90 (1994) 937.
15. J.R. Jokipii & G.E. Morfill, Ap.J. 312 (1987) 170.
16. H.J. Völk & V.N. Zirakashvili, Proc. 28th Int. Cosmic Ray Conf. (Tsukuba) 4 (2003) 2031-3024.
17. T. Abu-Zayyad et al. Ap.J. 557 (2001) 686-699.
18. G. Archbold & P. Sokolsky, Proc. 28th Int. Cosmic Ray Conf. (Tsukuba) 1 (2003) 405-408.
19. D.J. Bird et al., Phys. Rev. Letters 71 (1993) 3401.
20. R. Engel, T.K. Gaisser, P. Lipari & T. Stanev, Proc. 26th Int. Cosmic Ray Conf. (Salt Lake City) 1 (1999) 415-418; R. Engel, T.K. Gaisser & T. Stanev, Proc. 27th Int. Cosmic Ray Conf. (Hamburg) 2 (2001) 431-434.
21. N.N. Kalmykov, S.S. Ostapchenko & A.I. Pavlov, Nucl. Phys. B (Proc. Suppl.) 52B (1997) 17-28.
22. D.J. Bird *et al.*, Ap.J. 424 (1994) 491-502.
23. M. Nagano *et al.*, J. Phys. G10 (1984) 1295.
24. A.V. Olinto, astro-ph/0404114.
25. K. Mannheim, Astron. Astrophys. 269 (1993) 67.
26. V. Berezinsky, A. Gazizov & S. Grigorieva, astro-ph/0210095.
27. E. Waxman, Phys. Rev. Letters 75 (1995) 386.
28. M. Vietri, Ap.J. 453 (1995) 883. Also, M. Vietri, D. De Marco & D. Guetta, astro-ph/0302144.
29. S. Swordy *et al.*, Astropart. Phys. 18 (2002) 129-150.
30. M. Aglietta *et al.*, Astropart. Phys. 20 (2004) 641-652.
31. J. Ahrens *et al.*, Astropart. Phys. (to be published).
32. J. Ahrens *et al.*, Astropart. Phys. 20 (2004) 507-532.
33. T.K. Gaisser *et al.*, Proc. 28th Int. Cosmic Ray Conf. (Tsukuba) 2 (2003) 1117-1120.

Open Questions in Stellar Nuclear Physics: I [1]

Moshe Gai

Laboratory for Nuclear Science at Avery Point,
University of Connecticut, 1084 Shennecossett Road, Groton, CT 06340.
moshe.gai@yale.edu, http://www.phys.uconn.edu

Abstract. No doubt, among the most exciting discoveries of the third millennium thus far are oscillations of massive neutrinos and dark energy that leads to an accelerated expansion of the Universe. Accordingly, Nuclear Physics is presented with two extraordinary challenges: the need for precise (5% or better) prediction of solar neutrino fluxes within the Standard Solar Model, and the need for an accurate (5% or better) understanding of stellar evolution and in particular of Type Ia super nova that are used as cosmological standard candle. In contrast, much confusion is found in the field with contradicting data and strong statements of accuracy that can not be supported by current data. We discuss an experimental program to address these challenges and disagreements.

THE STANDARD SOLAR MODEL

The Standard Solar Model [1] is dependent on nuclear inputs and the most critical ones are cross sections of nuclear reactions [2] at solar conditions of central temperature of 15.7 MK and central density of approximately 150 g/cm^3. The two most important reaction cross sections that must be measured with an accuracy of 5% or better are the $^7Be(p,\gamma)^8B$ reaction and the $^4He(^3He,\gamma)^7Be$ reaction and the corresponding $S_{17}(0)$ and $S_{34}(0)$ defined in Ref. [2].

A major effort on measuring $S_{17}(0)$ with high accuracy was carried out in several labs and agreement among high precision data collected at GSI [3], Weizmann [4] and Seattle [5] was found. Most amazing is the excellent agreement between the Weizmann data that were measured with a ^7Be target and the GSI data that employed the Coulomb dissociation method. However as shown in Fig. 1 the slopes of these three results are sufficiently different. The d-wave correction to $S_{17}(0)$ on the other hand is directly related to this slope, and thus it is ill determined. Since the d-wave correction reduces $S_{17}(0)$ by as much as 15%, it precludes an accurate extrapolation of $S_{17}(0)$. This conclusion contradicts the strong statement of the Seattle group [5] that $S_{17}(0)$ has been determined with a theoretical uncertainty of 2.5%. This issue must be resolved by future high precision measurements of the slope, most likely with ^7Be beams [6], so as to allow accurate (5% or better) extrapolation of $S_{17}(0)$.

In contrast to the intensive work on $S_{17}(0)$, no progress what-so-ever was achieved on measuring $S_{34}(0)$ with high precision, and it is still poorly known with an error of 9% [2]. This inadequate situation must be improved in the near future as we expect the direct

---

[1] Work Supported by USDOE Grant No. DE-FG02-94ER40870.

CP726, Nuclear Physics, Large and Small: International Conference on
Microscopic Studies of Collective Phenomena, edited by R. Bijker, R. F. Casten, and A. Frank
© 2004 American Institute of Physics 0-7354-0207-8/04/$22.00

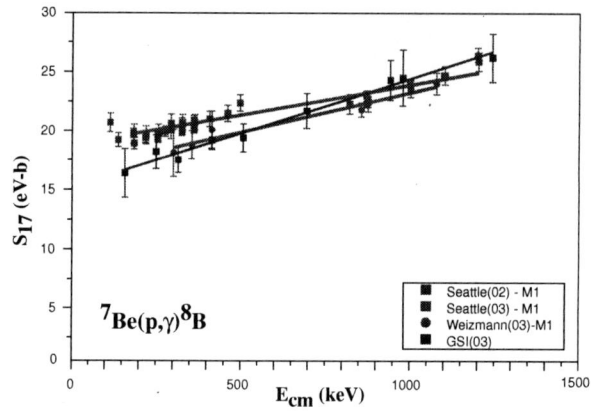

FIGURE 1. A comparison of the GSI [3], Weizmann [4] and Seattle [5] measurements of the astrophysical cross section factor S_{17} of the $^7Be(p, \gamma)^8B$ reaction, as defined in [2] and discussed in the text.

detection of ^7Be solar neutrinos. These measurements will conclude a four decade long quest by Nuclear Physicists for the nuclear inputs to the SSM. When the controversy on the composition of sun (Z/X) will also be resolved [7], it will allow high precision prediction of all solar neutrino fluxes including the ^8B neutrino flux. The high precision on one hand may provide a strong evidence for the SSM, but may also allow for a study of fundamental neutrino processes including oscillation to sterile neutrinos.

REFERENCES

1. J.N. Bahcall, Neutrino Astrophysics, Cambridge University Press, New York 1989.
2. E.G. Adelberger et al.; Rev. of Modern Phys. **70**(1998)1265.
3. F. Schumann et al.; Phys. Rev. Lett. **90**(2003)232501.
4. L.T. Baby et al., Phys. Rev. Lett. **90**(2003)022501, ibid Phys. Rev. **C67**(2003)065805.
5. A.R. Junghans et al.; Phys. Rev. Lett **88**(2002)041101, ibid Phys. Rev. **C68**(2003)065803.
6. M. Gai, M. Hass and Th. Delbar, for the UConn-Weizmann-LLN-ISOLDE collaboration; Letter of Intent to INTC, I37 INTC 2001-007.
7. J.N. Bahcall, and A.M. Serenelli, astro-ph/0403604. S. Basu and H. M. Antia; astro-ph/0403485.

Shape Phase Transitions in Nuclei

F. Iachello

Center for Theoretical Physics, Sloane Physics Laboratory, Yale University, New Haven, Connecticut 06520-8120, USA

Abstract. The theory of shape phase transitions in nuclei is briefly reviewed with particular emphasis to new results.

INTRODUCTION

The study of shape phase transitions in nuclei initiated in the early 80's [1, 2, 3] following some previous work of Gilmore [4]. Experimental manifestation of these phase transitions was found [5]. The field laïd dormant until the late 1990's, when Casten et al. [6] noted that accurate remeasurements of some electromagnetic matrix elements in ^{152}Sm showed major departures from accepted values. The new values were consistent with ^{152}Sm being very close to the critical point of the spherical to axially deformed $[U(5) - SU(3)]$ shape phase transition. The last three years have seen a renaissance in the study of shape phase transitions in nuclei with the introduction of novel concepts, such as that of critical point symmetry (spectroscopic signatures of shape phase transitions). In this note, dedicated to Stuart Pittel on the occasion of his 60th birthday, a brief review of some of the novel developments will be given. A review of the subject up to 2002 is given in [7].

ONE-FLUID SYSTEMS

Consider first the case of a one-fluid (or one-component) system described by the Interacting Boson Model-1 [5]. Gilmore's algorithm to study shape phase transitions within algebraic models is:

(i) Choose a Hamiltonian constructed of Casimir operators of each dynamical symmetry of the model. For the Interacting Boson Model-1, this gives [1, 2]

$$H = \sum_i C_i = \alpha_1 C_1(U5) + \alpha_2 C_2(SU3) + \alpha_3 C_2(O6). \tag{1}$$

Instead of (1), one can use an equivalent form [8]

$$H = \varepsilon \hat{n}_d + \kappa \hat{Q}^\chi \cdot \hat{Q}^\chi \tag{2}$$

that emphasizes the physical meaning of the operators.

Define control parameters $\left(\frac{\alpha_2}{\alpha_1}, \frac{\alpha_3}{\alpha_1} \right)$ or $\left(\frac{\kappa}{\varepsilon}, \chi \right)$.

CP726, *Nuclear Physics, Large and Small: International Conference on Microscopic Studies of Collective Phenomena*, edited by R. Bijker, R. F. Casten, and A. Frank
© 2004 American Institute of Physics 0-7354-0207-8/04/$22.00

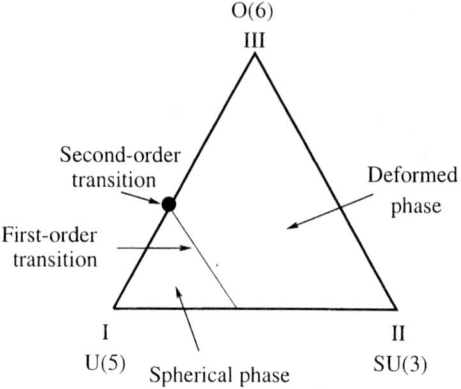

Second-order
transition

Deformed
phase

First-order
transition

I

II

U(5)   Spherical phase   SU(3)

FIGURE 1. Phase diagram of IBM-1.

(ii) Introduce coherent states and define classical variables, β, γ [1, 9, 10]. Construct the energy functional

$$E(N; \beta, \gamma) = \frac{\langle N; \beta, \gamma \mid H \mid N; \beta, \gamma \rangle}{\langle N; \beta, \gamma \mid N; \beta, \gamma \rangle} \qquad (3)$$

and the scale-free energy per particle $V(N; \beta, \gamma) = E/\varepsilon N$, or $E/\alpha_1 N$.

(iii) Minimize V as a function of β, γ and determine V_{min} and the order parameters β_e and γ_e. For the Hamiltonian (1) or (2), $\gamma_e = 0°, 60°$ only, and $V = V(N; \beta)$.

(iv) Study the behavior of V_{min} as a function of the control parameters, ξ (Erhenfest classification). A transition is called zeroth order if V_{min} is discontinuous, first order if $\frac{\partial V_{min}}{\partial \xi}$ is discontinous, etc.

Use of this algorithm produces the phase-diagram of nuclei in IBM-1 shown in Fig.1. Experimental evidence for phase transitions in nuclei, especially the $U(5) - SU(3)$ transition, was put forward in the 1980's [5]. This evidence was based on several observables. Some key quantities are:

(i) $B(E2; 2_1^+ \to 0_1^+)$.

(ii) $B(E0; 0_2^+ \to 0_1^+)$ and isomer shift $\delta \langle r^2 \rangle = \langle r^2 \rangle_{2_1^+} - \langle r^2 \rangle_{0_1^+}$.

(iii) Two-neutron separation energies $S_{2n}(N) = E_B(N+1) - E_B(N)$.

(iv) Isotope shift $\Delta \langle r^2 \rangle = \langle r^2 \rangle^{(N+1)} - \langle r^2 \rangle^{(N)}$.

The $B(E2; 2_1^+ \to 0_1^+)$ and two-neutron separation energies, $S_{2n}(N)$, in the Nd-Sm-Gd region are shown in Figs. 2 and 3.

In is interesting to note at this stage, that because of the inner automorphism $SU(3) \to \overline{SU(3)}$ [5] one can construct an extended phase diagram with Z_2 symmetry $\chi \to -\chi$ (prolate-oblate), as shown in Fig.4. All quantities are either symmetric or antisymmetric under Z_2. This symmetry has been studied recently by Jolie et al. [11].

Nuclei are finite quantal systems. Shape phase transitions are examples of 'phase transitions' in mesoscopic systems. In recent years, the study of phase transitions in mesoscopic systems has received new impetus. In addition to the classical (mean-field) method given above, shape phase transitions have been studied by direct computation of

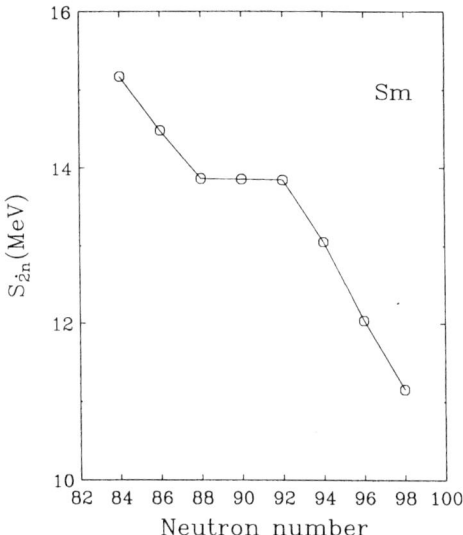

FIGURE 2. Two neutron separation energies in $_{62}$Sm as a function of neutron number, showing that the $U(5) - SU(3)$ transition is first order.

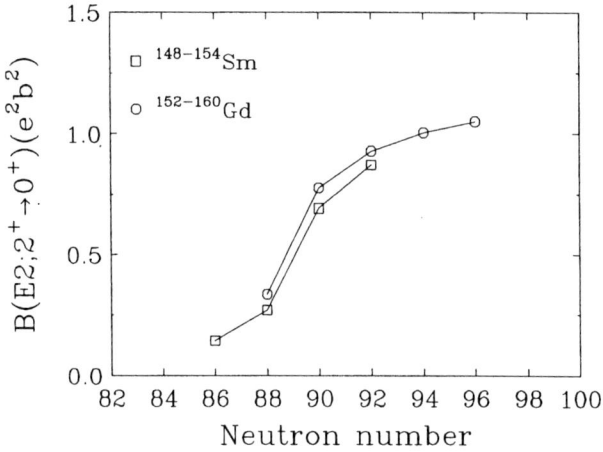

FIGURE 3. $B(E2; 2_1^+ \rightarrow 0_1^+)$ values in $_{62}$Sm and $_{64}$Gd.

the order parameter. The algorithm used to study 'phase transitions' in finite systems is:

(i) Choose a Hamiltonian.

(ii) Define quantal order parameters and evaluate them for finite N.

(iii) Extrapolate to N$\rightarrow \infty$.

The description of shape phase transitions in nuclei by order parameters is currently being developed [12]. For second order $[U(5) - SO(6)]$ transitions this description is

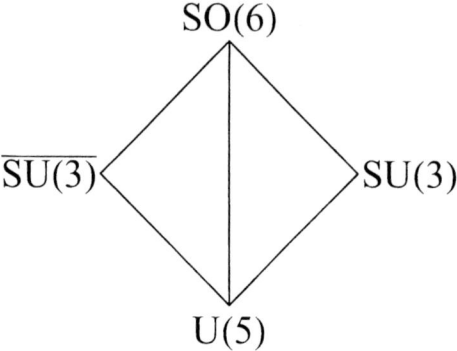

FIGURE 4. Extended phase diagram of IBM-1 including the automorphism $SU(3) \rightarrow \overline{SU(3)}$, $\chi \rightarrow -\chi$.

well known [3]. One needs one order parameter. A convenient quantal order parameter is

$$v_1 = \langle N;0_1^+ \mid \hat{n}_d \mid N;0_1^+ \rangle / N. \tag{4}$$

The quantal order parameter, v_1, is related to the classical order parameter, β_e, by $\beta_e = \sqrt{v_1/(1-v_1)}$. In this case there is only one critical value of the control parameter, ξ_c. The critical exponent for the classical order parameter β_e can be evaluated to be

$$\beta_e \propto (\xi - \xi_c)^\mu, \quad \xi \geq \xi_c \ ,$$
$$\mu = \frac{1}{2} \text{ mean field,}$$
$$\mu = 0.53 \pm 0.05 \text{ numerical simulation.} \tag{5}$$

The behavior of the order parameter v_1 as a function of the control parameter ξ is shown in Fig. 5.

For first order $[U(5) - SU(3)]$ transitions, it has been recently suggested to use two order parameters [12]

$$v_1 = \langle N;0_1^+ \mid \hat{n}_d \mid N;0_1^+ \rangle / N,$$
$$v_2 = [\langle N;0_2^+ \mid \hat{n}_d \mid N;0_2^+ \rangle - \langle N;0_1^+ \mid \hat{n}_d \mid N;0_1^+ \rangle] / N. \tag{6}$$

The behavior of the two order parameters v_1 and v_2 as a function of the control parameter ξ is shown in Fig. 6. There are now three important values of the control parameter ξ^*(spinodal), ξ_c(critical), ξ^{**}(antispinodal). The spinodal exponent has been evaluated for the classical order parameter β_e to be

$$(\beta_e - \beta_e^*) \propto (\xi - \xi^*)^\mu, \quad \xi \geq \xi^*$$
$$\mu = \frac{1}{2} \text{ mean field}$$
$$\mu = 0.52 \pm 0.07 \text{ numerical simulation.} \tag{7}$$

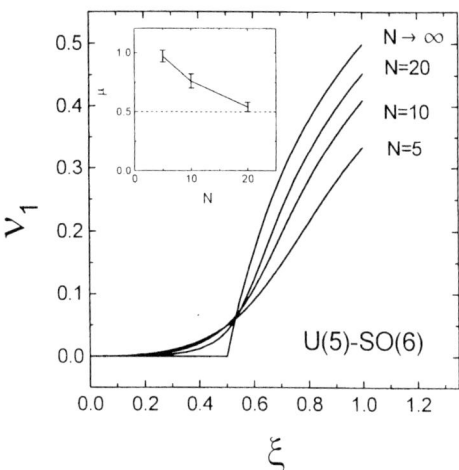

FIGURE 5. The order parameter v_1 as a function of the control parameter ξ for the $U(5) - SO(6)$ second order transition. In the insert the critical exponent. From [12].

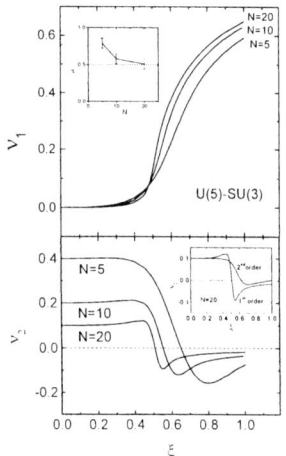

FIGURE 6. The behavior of the two order parameters v_1 and v_2 as a function of the control parameter ξ for the first order transition $U(5) - SU(3)$. From [12].

The most important new development in the study of phase transitions in nuclei has been the discovery of spectral signatures of critical points. These signatures, called critical point symmetries, have been used to analyze data in the transition regions. To find these symmetries, one needs to complete the algorithm described above and

(v) Study the behavior of the potential V as a function of the classical variables β, γ at the critical point (Landau approach). Solve the corresponding differential equation

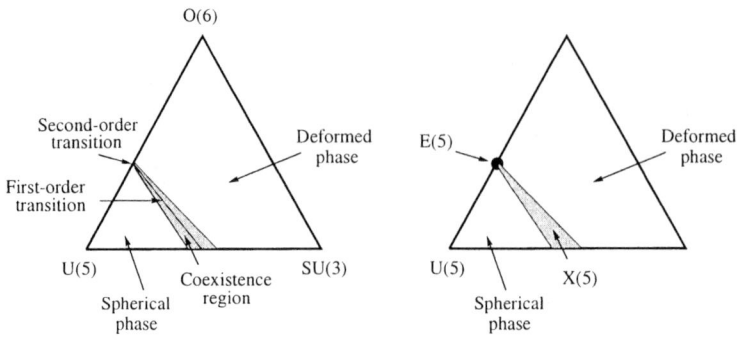

FIGURE 7. Phase diagram of IBM-1 showing the location of the solutions $E(5)$ and $X(5)$.

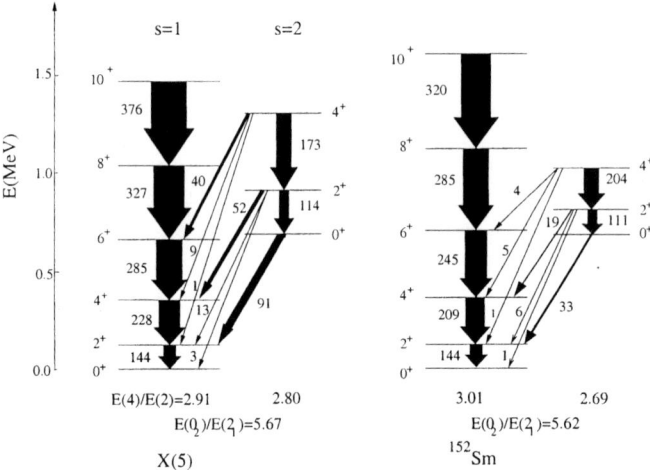

FIGURE 8. Comparison between the experimental spectrum of ^{152}Sm and the spectrum of $X(5)$.

$$D\psi = E\psi.$$

In the case in question, the differential operator D is the Bohr Hamiltonian. The solutions [13] and [14] have been called:

E(5) $[U(5) - SO(6)$ transition] and X(5) $[U(5) - SU(3)$ transition], Fig.7.

Experimental evidence for critical points has been found [15, 16]. One example is shown in Fig.8.

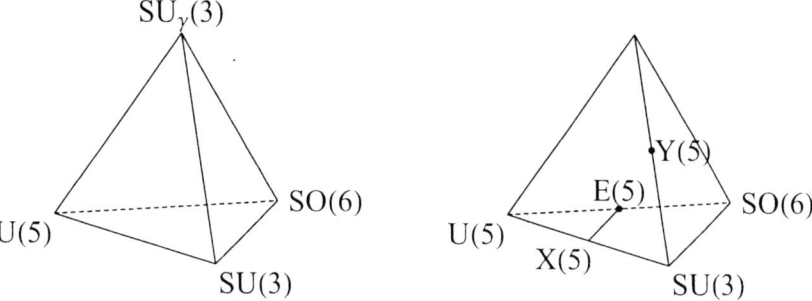

FIGURE 9. Left panel: Phase diagram of extended IBM-1 [17]. The fourth vertex is denoted $SU_\gamma(3)$. It corresponds to a situation in which the ground state is an $SU(3)$ representation (λ, μ) with $\mu \neq 0$. Right panel: Phase diagram of extended IBM-1 [19]. The fourth vertex is not defined in terms of symmetries. The location of the $Y(5)$ solution is also schematically shown.

ONE-FLUID SYSTEMS WITH INTERACTIONS HIGHER THAN ORDER TWO

These systems can be studied with extensions of the Interacting Boson Model. Consider first the extension with higher order Casimir operators [17]. The Hamiltonian is now

$$
\begin{aligned}
H = {} & \alpha_1 C_1(U5) + \alpha_2 C_2(SU3) + \alpha_2'[C_2(SU3)]^2 \\
& + \alpha_3 C_2(O6)
\end{aligned}
\tag{8}
$$

and the control parameters are $(\frac{\alpha_2}{\alpha_1}, \frac{\alpha_2'}{\alpha_1}, \frac{\alpha_3}{\alpha_1})$ with phase diagram, given in Fig. 9 (left). The phase transitions in this model have been recently investigated by Caprio [18]. An interesting property of these phase transitions is that in the classical limit the ground state of the triaxial phase has an infinite degeneracy. This infinite degeneracy must broken if one wants to have a realistic description of triaxial nuclei. Work in this direction is in progress.

An alternative extension was suggested long ago by Chen and van Isacker [19]

$$
H = \varepsilon \hat{n}_d + \kappa \hat{Q}^\chi \cdot \hat{Q}^\chi + \eta \hat{H}_3,
\tag{9}
$$

where $\hat{H}_3$ denotes a combination of cubic terms. The control parameters are $(\frac{\kappa}{\varepsilon}, \chi, \frac{\eta}{\varepsilon})$. One difficulty here is that the fourth vertex is not defined in terms of symmetries, since $\hat{H}_3$ cannot be simply written in terms of Casimir operators. Nonetheless, spectral signatures of critical points for this case have been found [20]. The corresponding solution has been called Y(5), Fig.9 (right). The spectrum of the $Y(5)$ solution is shown in Fig. 10.

TWO-FLUID SYSTEMS

The shape phase transitions of two-fluid (two-component systems) were, to some extent, investigated in the 1980's by Dieperink and Bijker [21]. In view of the availability of

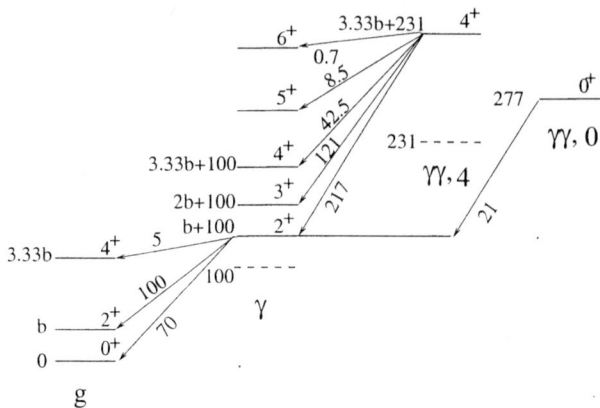

FIGURE 10. Schematic representation of the spectrum of the $Y(5)$ solution (critical point of the axially deformed to triaxially deformed shape phase transition in the IBM-1 with cubic interactions). From [20].

new nuclear species at Radioactive Beam Facilities, this problem has returned to be of interest. The framework here is the Interacting Boson Model-2. The situation is much more complex than in the case of IBM-1. The algorithm is:

(i) Choose a Hamiltonian either in terms of Casimir operators

$$
\begin{aligned}
H = {} & \alpha_1 C_1(U_{\pi+\nu}5) + \alpha_2 C_2(SU_{\pi+\nu}3) \\
& + \alpha_2^* C_2(SU_{\pi+\nu}3^*) + \alpha_3 C_2(SO_{\pi+\nu}6)
\end{aligned}
\tag{10}
$$

or in terms of quadrupole operators

$$
H = \varepsilon\left(\hat{n}_{d_\pi} + \hat{n}_{d_\nu}\right) + \kappa\left(\hat{Q}_\pi^{\chi_\pi} + \hat{Q}_\nu^{\chi_\nu}\right) \cdot \left(\hat{Q}_\pi^{\chi_\pi} + \hat{Q}_\nu^{\chi_\nu}\right)
\tag{11}
$$

and define control parameters $\left(\frac{\alpha_2}{\alpha_1}, \frac{\alpha_2^*}{\alpha_1}, \frac{\alpha_3}{\alpha_1}\right)$ or $\left(\frac{\kappa}{\varepsilon}, \frac{\chi_\pi + \chi_\nu}{2}, \frac{\chi_\pi - \chi_\nu}{2}\right)$.

(ii) Introduce coherent states and define classical variables $\beta_\pi, \gamma_\pi, \beta_\nu, \gamma_\nu, \varphi_1, \varphi_2, \varphi_3$, where $\varphi_1, \varphi_2, \varphi_3$ are the three Euler angles that determine the orientation in space of the neutron intrinsic system relative to the proton intrinsic system. Construct the energy functional

$$
E(N_\pi, N_\nu; \beta_\pi, \gamma_\pi, \beta_\nu, \gamma_\nu, \varphi_1, \varphi_2, \varphi_3)
\tag{12}
$$

and the scale-free energy per particle, V. A simplification occurs here since one can take $\varphi_1 = \varphi_3 = 0$ and $\varphi_2 = \varphi$.

(iii) Minimize V as a function of the classical variables and determine $V_{\min}$ and the order parameters. For the Hamiltonian (11), $\gamma_{\pi e} = 0°, 60°$ and $\gamma_{\nu e} = 0°, 60°$ only, and $V = V(N_\pi, N_\nu; \beta_\pi, \beta_\nu, \varphi)$.

(iv) Study the behavior of $V_{\min}$ (Erhenfest classification).

The phase diagram of nuclei in IBM-2 is a tetrahedron [22] with a symmetry at each vertex, $U_{\pi+\nu}(5), SU_{\pi+\nu}(3), SO_{\pi+\nu}(6)$, and $SU_{\pi+\nu}(3)^*$, Fig.11. A detailed study of phase transitions in IBM-2 has been initiated by Caprio [23] and by Arias, Dukelsky,

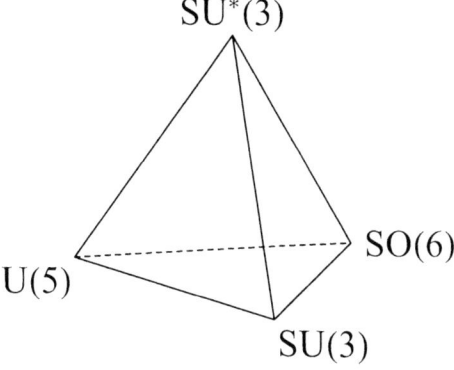

FIGURE 11. Phase diagram of IBM-2 [22].

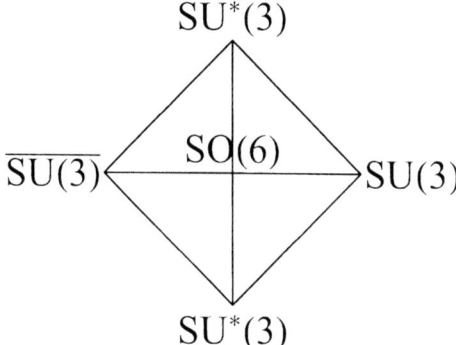

FIGURE 12. A portion of the extended phase diagram of IBM-2 including the automorphisms $SU(3) \rightarrow \overline{SU(3)}$ and $SU(3)^* \rightarrow \overline{SU(3)^*}$.

and García-Ramos [24]. The corresponding phase diagrams, obtained under different assumptions, are shown in [23] and [24].

It should be noted that because of the inner automorphisms $SU(3) \rightarrow \overline{SU(3)}$ and $SU(3)^* \rightarrow \overline{SU(3)^*}$, the extended phase diagram has additional symmetries. A portion of this extended diagram is shown in Fig. 12.

CONCLUSIONS

(a) The study of shape phase transitions in nuclei, initiated in 1980, is undergoing a revival. Several new results have been obtained. Of particular interest are: (i) Spectral signatures of critical points in finite quantal systems and (ii) Phase transitions in two-fluid systems.

(b) Shape phase transitions in nuclei are an example for all mesoscopic systems.

ACKNOWLEDGMENTS

This talk is dedicated to Stuart Pittel on the occasion of his 60th birthday. The Federman-Pittel mechanism is the foundation of the microscopic origin of shape phase transitions in nuclei.

This work was supported in part under USDOE Contract DE-FG-02-91ER-40608.

REFERENCES

1. A.E.L. Dieperink, O.Scholten, and F. Iachello, Phys. Rev. Lett. **44**, 1747 (1980).
2. D.H. Feng, R. Gilmore, and S.R. Deans, Phys. Rev. C **23**, 1254 (1981).
3. O.S. Van Roosmalen, *Algebraic Description of Nuclear and Molecular Rotation-Vibration Spectra*, Ph.D. Thesis, University of Gröningen, The Netherlands(1982).
4. R. Gilmore and D.H. Feng, Nucl. Phys. A **301**, 189 (1978); R. Gilmore, J. Math. Phys. **20**, 89 (1979).
5. For a review see, F. Iachello and A. Arima, *The Interacting Boson Model*, Cambridge University Press, Cambridge (1987).
6. R.F. Casten, M. Wilhelm, E. Radermacher, N.V. Zamfir, and P. von Brentano, Phys. Rev. C **57**, R1553 (1998).
7. F. Iachello, in Proc. Int. School of Physics "Enrico Fermi", Course CLIII, A. Molinari, L. Riccati, W.M. Alberico and M. Morando, eds., IOS Press, Amsterdam (2003), p.1.
8. D.D. Warner and R.F. Casten, Phys. Rev. C **28**, 1798 (1983).
9. J. Ginocchio and M. W. Kirson, Phys. Rev. Lett. **44**, 1744 (1980).
10. A. Bohr and B.R. Mottelson, Physica Scripta **22**, 468 (1980).
11. J. Jolie, R.F. Casten, P. von Brentano, and V. Werner, Phys. Rev. Lett. **87**, 162501 (2001); J. Jolie, P. Cejnar, R.F. Casten, S. Heinze, A. Linneman, and V. Werner, Phys. Rev. Lett. **89**, 182502 (2002)
12. F. Iachello and N.V. Zamfir, Phys. Rev. Lett., in press (2004).
13. F. Iachello, Phys. Rev. Lett. **85**, 3580 (2000).
14. F. Iachello, Phys. Rev. Lett. **87**, 052502 (2001).
15. R.F. Casten and N.V. Zamfir, Phys. Rev. Lett. **85**, 3584 (2000).
16. R.F. Casten and N.V. Zamfir, Phys. Rev. Lett. **87**, 052503 (2001).
17. Y.F. Smirnov, N.A. Smirnova and P. van Isacker, Phys. Rev. C **61**, 041302 (2000).
18. M. A. Caprio, private communication (2004).
19. P. Van Isacker and J.Q.Chen, Phys. Rev. C **24**, 684 (1981).
20. F. Iachello, Phys. Rev. Lett. **91**, 132502 (2003).
21. A.E.L. Dieperink and R. Bijker, Phys. Lett. B **116**, 77 (1982).
22. A. E. L. Dieperink, in *Progress in Particle and Nuclear Physics,* Vol.9, D. Wilkinson ed., Pergamon (1983).
23. M. A. Caprio, These Proceedings (2004).
24. J. M. Arias, J. Dukelsky, and J. E. García-Ramos, These Proceedings (2004).

Geometry of coexistence
in the interacting boson model

P. Van Isacker*, A. Frank†** and C.E. Vargas‡

*Grand Accélérateur National d'Ions Lourds, BP 55027, F-14076 Caen Cedex 5, France
†Instituto de Ciencias Nucleares, UNAM, Apdo. Postal 70-543, 04510 México, D.F. Mexico
**Centro de Ciencias Físicas, UNAM, Apdo. Postal 139-B, 62251, Cuernavaca, Mor., Mexico
‡Facultad de Física e Inteligencia Artificial, UV, Sebastián Camacho; Centro, Xalapa, Ver., 91000, Mexico

Abstract. The Interacting Boson Model (IBM) with configuration mixing is applied to describe the phenomenon of coexistence in nuclei. The analysis suggests that the IBM with configuration mixing, used in conjunction with a (matrix) coherent-state method, may be a reliable tool for the study of geometric aspects of shape coexistence in nuclei.

1. NUCLEAR SHAPE COEXISTENCE

Nuclear coexistence and shape phase transitions have been a matter of major theoretical and experimental interest for many years [1]. In the context of the nuclear shell model the origin of coexistence can be traced back to many-particle many-hole excitations across shell gaps, which become energetically favourable as a result of the interplay between shell effects and the neutron-proton interaction [2].

In the neutron-deficient lead isotopes shape coexistence was theoretically anticipated by May et al. [3] in a Nilsson framework including shell corrections. Subsequent deformed mean-field calculations have predicted the occurrence in the energy surface of close-lying oblate and prolate minima near the ground-state spherical configuration, specifically in lead nuclei with neutron number close to $N = 104$ (see, e.g., Ref. [4]). These predictions have been confirmed by a series of experiments on the neutron-deficient lead isotopes, culminating in the recent observation of three close low-lying 0^+ levels in ^{186}Pb [5], interpreted as having spherical, oblate, and prolate shapes. Present-day sophisticated mean-field calculations [6] can account for these three minima in a region of energy that is consistent with experimental findings. Although these theoretical studies are very impressive, they have focussed mainly on the properties of the potential energy surface of ^{186}Pb. A description beyond mean field that includes spectroscopic properties requires simplifying assumptions such as axial symmetry and even then it represents a major computational effort [6]. Shell-model calculations, on the other hand, are very difficult to carry out in this region due to the rapidly increasing size of the model space, particularly for the case of interest, which involves the opening up of shells through particle-hole (p-h) configurations.

The Interacting Boson Model (IBM) [7] provides a simple alternative to describe the phenomenon of nuclear coexistence. The model assumes that low-lying collective excitations of the nucleus can be described in terms of N s and d bosons. The bosons correspond to pairs of nucleons in the valence shell, coupled to angular momentum 0 or 2; N is constant for a given nucleus and equal to half its number of valence nucleons. Core (i.e., non-valence p-h) excitations can be included in a natural way since they correspond to configurations with higher numbers of active nucleons and thus with higher boson numbers. Specifically, 0p-0h, 2p-2h, 4p-4h,... shell-model configurations correspond to systems of $N, N+2, N+4,...$ interacting bosons which are simultaneously treated and possibly mixed in this configuration-mixed version of the IBM [8]. This model was applied recently to describe the evolving properties of the lead isotopes [9].

While both mean-field theory and IBM give a satisfactory description of nuclear coexistence, an obvious connection between them is lacking. In this contribution we argue that the algebraic description of nuclear coexistence can be given a geometric interpretation and hence be connected with mean-field theory [10]. This approach opens the possibility for a study of the relation between coexistence and criticality.

2. THE IBM WITH CONFIGURATION MIXING

We first explain the essential ingredients of the model with specific reference to the lead isotopes. The model space is built from $N, N+2$, and $N+4$ bosons and constitutes a boson representation of the shell-model configurations that are dominant in the low-energy region of the lead isotopes. The N-boson states correspond to excitations of neutrons only, for which the proton shell $Z = 82$ remains closed; they can be characterized as the 0p-0h configuration. The states with $N+2$ and $N+4$ bosons correspond to 2p-2h and 4p-4h excitations of the protons across the $Z = 82$ shell gap coupled the valence neutrons in the $N = 82$-126 shell. The Hamiltonian is

$$H = H_{0p-0h} + H_{2p-2h} + H_{4p-4h} + H_{mix}^{02} + H_{mix}^{24}, \tag{1}$$

where H_{ip-ih} is the Hamiltonian for the ith of the three configurations, taken in the simplified form [11]

$$H_{ip-ih} = \varepsilon_i n_d + \kappa_i Q_i \cdot Q_i. \tag{2}$$

This Hamiltonian provides a simple parametrization of the essential features of nuclear structural evolution in terms of a vibrational term n_d (the number of d bosons) and a quadrupole interaction $Q_i \cdot Q_i$ with $Q_i = (s^\dagger \tilde{d} + d^\dagger s)^{(2)} + \chi_i (d^\dagger \tilde{d})^{(2)}$. The Hamiltonian (2) conserves the number of bosons and does not provide any mixing between the $N, N+2$, and $N+4$ configurations. Configuration mixing arises through the boson-number non-conserving parts $H_{mix}^{ii'}$ which are assumed to have the simple form [8]

$$H_{mix}^{ii'} = \omega_0^{ii'}(s^\dagger s^\dagger + ss) + \omega_2^{ii'}(d^\dagger \cdot d^\dagger + \tilde{d} \cdot \tilde{d}), \tag{3}$$

with $ii' = 02$ or 24. The operator H_{mix}^{02} mixes the states of the 0p-0h and 2p-2h configurations while H_{mix}^{24} does the same for the 2p-2h and 4p-4h configurations. This mixing directly follows from the two-body nature of the shell-model interaction.

3. GEOMETRY OF THE IBM WITH CONFIGURATION MIXING

The algebraic formalism of the previous section does not directly provide a geometry for the IBM Hamiltonian. A way to establish a connection [12] with the geometric Bohr-Mottelson model [13], is obtained by defining a coherent state with fixed boson number [14]

$$|N, \beta\gamma\rangle \equiv \left(s^\dagger + \beta \left[\cos\gamma d_0^\dagger + \sqrt{\tfrac{1}{2}} \sin\gamma (d_{+2}^\dagger + d_{-2}^\dagger) \right] \right)^N |0\rangle, \tag{4}$$

which endows the algebraic model with an intrinsic geometric structure in terms of the quadrupole-shape variables β and γ. For each of the separate Hamiltonians H_{ip-ih} in (2) one obtains from its expectation value in the coherent state an energy surface

$$
\begin{aligned}
E_i(\beta, \gamma) = {} & \frac{N_i \varepsilon_i \beta^2}{1 + \beta^2} + \kappa_i \left[\frac{N_i (5 + (1 + \chi_i^2)\beta^2)}{1 + \beta^2} + \right. \\
& + \left. \frac{N_i(N_i - 1)}{(1 + \beta^2)^2} \left(\tfrac{2}{7} \chi_i^2 \beta^4 + 4\sqrt{\tfrac{2}{7}} \chi_i \beta^3 \cos 3\gamma + 4\beta^2 \right) \right],
\end{aligned}
\tag{5}
$$

where $N_i = N + i$ is the boson number associated with that particular configuration.

For the lead isotopes in particular, a set of parameters describes a potential energy surface for N ranging from 0 to 11 bosons (for neutron numbers from the closed shell $N = 126$ to mid-shell $N = 104$). The 0p-0h configuration corresponds to N bosons whereas the 2p-2h and 4p-4h excitations require 2 and 4 additional bosons, respectively. This leads to a 3×3 *potential energy matrix* [10]

$$
\mathbf{E}_N(\beta, \gamma) = \begin{pmatrix}
E_0(\beta, \gamma) & \Omega_{02}(\beta) & 0 \\
\Omega_{02}(\beta) & E_2(\beta, \gamma) + \Delta_2 & \Omega_{24}(\beta) \\
0 & \Omega_{24}(\beta) & E_4(\beta, \gamma) + \Delta_4
\end{pmatrix}, \tag{6}
$$

where Δ_2 (Δ_4) corresponds to the single-particle energy expended in raising 2 (4) protons from the lower (50-82) to the upper (82-126) shell, corrected for the gain in energy due to pairing, and where $\Omega_{ii'}(\beta) \equiv \langle N + i, \beta\gamma | \hat{H}_{\mathrm{mix}}^{ii'} | N + i', \beta\gamma \rangle$ are the non-diagonal matrix elements

$$
\begin{aligned}
\Omega_{02}(\beta) &= \frac{\sqrt{(N+1)(N+2)}}{1 + \beta^2} \left(\omega_0^{02} + \omega_2^{02} \beta^2 \right), \\
\Omega_{24}(\beta) &= \frac{\sqrt{(N+3)(N+4)}}{1 + \beta^2} \left(\omega_0^{24} + \omega_2^{24} \beta^2 \right).
\end{aligned}
\tag{7}
$$

The eigenpotentials of the matrix (6) are obtained by diagonalization. The eigensolutions depend on the parameters ε_i, κ_i, χ_i, and $\omega_k^{ii'}$ which are taken *without any modification* from Ref. [9] where they are extracted from a comprehensive fit to the energy spectra of several nuclei in the lead mass region. The 0p-0h configuration is found to be spherical ($\kappa_0 = 0$) while the 2p-2h and 4p-4h configurations turn out to be oblate ($\kappa_2 \neq 0$, $\chi_2 > 0$) and prolate ($\kappa_4 \neq 0$, $\chi_4 < 0$), respectively.

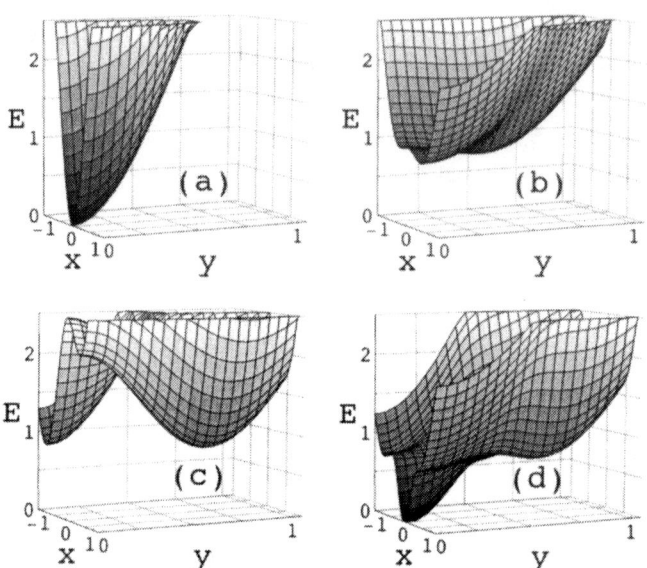

FIGURE 1. Potential energy surfaces in β and γ for ^{186}Pb. On the horizontal axes, $x \equiv \beta \sin(\gamma + 30^o)$ runs from -1 to 1, while $y \equiv \beta \cos(\gamma + 30^o)$ runs from 0 to 1. The vertical axis gives the energy in MeV. The four plots show (a) the 0p-0h configuration before mixing, (b) the 2p-2h configuration before mixing, (c) the 4p-4h configuration before mixing, (d) the lowest eigenpotential of $\mathbf{E}_N(\beta, \gamma)$ in Eq. (6) after mixing.

In Fig. 1 the potential energy surfaces in β and γ resulting from $\mathbf{E}_N(\beta, \gamma)$ are shown for the nucleus ^{186}Pb. Figures 1a-c give the separate 0p-0h, 2p-2h, and 4p-4h configurations before mixing. The first of these exhibits a minimum with spherical shape at $x \equiv \beta \sin(\gamma + 30^o) = 0$ and $y \equiv \beta \cos(\gamma + 30^o) = 0$. The 2p-2h configuration has an oblate minimum at $x = -0.21$, $y = 0.37$ ($\beta = 0.43$, $\gamma = \pm 60^o$) with an energy of 820 keV. A prolate minimum occurs for the 4p-4h configuration at $x = 0.36$, $y = 0.62$ ($\beta = 0.72$, $\gamma = 0^o$) with an energy of 848 keV. These energies are reasonably close to those of the 0^+ excited states observed in ^{186}Pb. Finally, the bottom plot in Fig. 1 shows the potential energy surface of the lowest energy eigenvalue of $\mathbf{E}_N(\beta, \gamma)$ in Eq. (6). The potential energy surface of Fig. 1d exhibits a remarkable similarity with that obtained within a mean-field approach [5] (apart from an overall factor in β). Nevertheless, with the mixing as determined in Ref. [9] only two (spherical and prolate) of the three minima remain. There is, however, still considerable uncertainty in the exact value of the mixing. A reduction of the mixing leads to *three* minima in the lowest eigenpotential of (6) which otherwise is very similar in appearance to the surface shown in Fig. 1d.

4. COEXISTENCE AND CRITICALITY

The geometric interpretation of the IBM with configuration mixing may also improve our understanding of the relation between coexistence and criticality. The different phases of the IBM-1 have been rigorously determined on the basis of catastrophe theory [15] and its phase diagram is well established [16]. In particular, it is known [17] that there is little scope for coexistence and only coexistence of spherical and deformed shapes [U(5)-SU(3)]. The present matrix coherent-state formalism broadens the scope of this investigation: A catastrophe analysis, similar to that in Ref. [15], can be undertaken of the IBM with configuration mixing. A first analysis indicates a wealth of new results such as phase diagrams with spherical-deformed and prolate-oblate coexistence, triaxial instabilities as well as a connection with the critical symmetries proposed by Iachello [18]. The results of this catastrophe analysis will be the subject of a separate paper [19].

5. CONCLUSION

The main aim of this contribution has been to give a geometric interpretation of the IBM with configuration mixing via a matrix coherent-state method. The application to the lead isotopes and the similarities found between the mean-field approach and the geometric interpretation of the IBM, suggest that this geometric interpretation is a reliable tool in the description of shape coexistence phenomena in nuclei. The lowest potential energy surface predicted with matrix coherent states is in agreement with that found using mean-field theory while it is at the same time derived from an IBM Hamiltonian that describes the known spectroscopic properties of the lead isotopes.

The proposed matrix coherent-state method opens up the possibility for a systematic investigation of the relation between coexistence and criticality. In contrast to IBM-1, where coexistence plays a marginal role only and where it is confined to a small U(5)-SU(3) region, coexistence acquires a prominent status in the IBM with configuration mixing and can be studied in full detail by applying a catastrophe theory of critical behaviour to the eigenpotentials obtained from the matrix coherent-state method.

ACKNOWLEDGMENTS

It is a pleasure to dedicate this contribution to Stuart Pittel, our friend and collaborator over many years, who has been a constant inspiration to us, as evident from the preceding discussion. The authors are grateful to O. Castaños, R. Fossion, F. Iachello, and E. Padilla for enlightening conversations. CV is grateful to E. Rojas, N. Bagatella, and R. Hdz.-Montoya from Universidad Veracruzana (UV) in Mexico for stimulating discussions and also to G. Mandujano for computational support at UV. This work was supported in part by CONACyT (Mexico).

REFERENCES

1. K. Heyde, P. Van Isacker, M. Waroquier, J.L. Wood, and R.A. Meyer, Phys. Rep. **102**, 293 (1983); J.L. Wood, K. Heyde, W. Nazarewicz, M. Huyse, and P. Van Duppen, Phys. Rep. **215**, 101 (1992).
2. P. Federman and S. Pittel, Phys. Lett. B **69**, 385 (1977).
3. F.R. May, V.V. Pashkevich, and S. Frauendorf, Phys. Lett. B **68**, 113 (1977).
4. W. Nazarewicz, Phys. Lett. B **305**, 195 (1993).
5. A.N. Andreyev *et al.*, Nature **405**, 430 (2000).
6. T. Duguet, M. Bender, P. Bonche, and P.-H. Heenen, Phys. Lett. B **559**, 201 (2003).
7. F. Iachello and A. Arima, *The Interacting Boson Model*, (Cambridge University Press, Cambridge, 1987).
8. P.D. Duval and B.R. Barrett, Nucl. Phys. A **376**, 213 (1982).
9. R. Fossion, K. Heyde, G. Thiamova, and P. Van Isacker, Phys. Rev. C **67**, 024306 (2003).
10. A. Frank, P. Van Isacker, and C.E. Vargas, Phys. Rev. C, to be published.
11. D.D. Warner and R.F. Casten, Phys. Rev. C **28**, 1798 (1983); P.O. Lipas, P. Toivonen, and D.D. Warner, Phys. Lett. B **155**, 295 (1985).
12. J.N. Ginocchio and M.W. Kirson, Phys. Rev. Lett. **44**, 1744 (1980); A.E.L. Dieperink, O. Scholten, and F. Iachello, Phys. Rev. Lett. **44**, 1747 (1980); A. Bohr and B.R. Mottelson, Phys. Scripta **22**, 468 (1980).
13. A. Bohr and B.R. Mottelson, *Nuclear Structure. II Nuclear Deformations* (Benjamin, New York, 1975).
14. R. Gilmore, J. Math. Phys. **20**, 891 (1979).
15. E. López-Moreno and O. Castaños, Phys. Rev. C **54**, 2374 (1996).
16. J. Jolie, R.F. Casten, P. von Brentano, and V. Werner, Phys. Rev. Lett. **87**, 162501 (2001).
17. F. Iachello, N.V. Zamfir, and R.F. Casten, Phys. Rev. Lett. **81**, 1191 (1998).
18. F. Iachello, Phys. Rev. Lett. **85**, 3580 (2000); *ibid.* **87**, 052502 (2001); *ibid.* **91**, 132502 (2003).
19. P. Van Isacker and A. Frank, to be published.

Critical points in the Interacting Boson Model

J.M. Arias*, J. Dukelsky† and J.E. García-Ramos**

*Departamento de Física Atómica, Molecular y Nuclear, Facultad de Física, Universidad de Sevilla, Apartado 1065, 41080 Sevilla, Spain
†Instituto de Estructura de la Materia, CSIC, Serrano 123, 28006 Madrid, Spain
**Departamento de Física Aplicada, Universidad de Huelva, 21071 Huelva, Spain

Abstract. We study the quantum shape phase transitions in nuclei within the Interacting Boson Model (IBM). The phase diagram for the standard IBM related to axial shapes is revised. The phase diagram of the IBM in relation to the recently proposed Y(5) critical point, connected to the change from axial to triaxial shapes, is investigated.

INTRODUCTION

The study of quantum phase transitions is one of the most difficult but exciting topics in Physics. In Nuclear Physics the introduction of the interacting boson model [1] allowed to study the shape phase transitions in nuclei. This model has four dynamical symmetries (SU(5), O(6), SU(3), and $\overline{SU(3)}$) corresponding to well defined nuclear shapes (spherical, deformed γ-unstable, prolate axial deformed and oblate axial deformed, respectively). The structure of the Hamiltonian allows to study systematically the transition from one shape to another. Although there were some pionering works some time ago [2, 3, 4], it has been the recent introduction of the concept of critical point symmetry [5, 6] what has recall the attention of the community to the topic of phase transitions in nuclei.

The phase diagram of the interacting boson model has been studied from several points of view [2, 3, 7, 8, 9]. In this paper we revise briefly the phase diagram of the IBM related to axial forms. In addition, the new $Y(5)$ symmetry [6] has suggested the importance of considering triaxial shapes in the phase diagram of the IBM. However, the usual IBM with up to two body interactions does not provide triaxial shapes. To get these one has either to include three-body interactions or to go to the version of IBM in which the neutron-proton degree of freedom is explicitly taken into account (IBM-2). Here we present a study of the IBM-2 critical point between axial deformed and triaxial shapes at the mean field level.

MEAN FIELD FOR THE IBM-1

We start by introducing the Hamiltonian. We use here a simplified form of the IBM Hamiltonian that keeps all the main ingredients of the full Hamiltonian, it is called the Consistent-Q Hamiltonian [10]

CP726, Nuclear Physics, Large and Small: International Conference on
Microscopic Studies of Collective Phenomena, edited by R. Bijker, R. F. Casten, and A. Frank
© 2004 American Institute of Physics 0-7354-0207-8/04/$22.00

$$H = x\hat{n}_d - \frac{1-x}{N} Q \cdot Q \qquad (1)$$

where n_d is the d-boson number operator, x is the control parameter and Q is the usual IBM quadrupole operator

$$Q = (s^\dagger \tilde{d} + d^\dagger \tilde{s}) + \chi \, (d^\dagger \tilde{d})^{(2)} . \qquad (2)$$

With this simple Hamiltonian the IBM dynamical symmetries are obtained as follows: U(5) corresponds to x=1 (independent of χ); SU(3) is obtained for x=0 and $\chi = -\frac{\sqrt{7}}{2}$; the selection of the control parameters x=0 and $\chi = 0$ provides the O(6) limit, and $\overline{SU(3)}$ is realized for x=0 and $\chi = \frac{\sqrt{7}}{2}$. For introducing geometry into the IBM the following boson condensate trial function is usually used [11, 12],

$$|g; N, \beta, \gamma\rangle = \frac{1}{\sqrt{N!}} \left(\frac{1}{\sqrt{1+\beta^2}} \left[s^\dagger + \beta \cos \gamma d_0^\dagger + \frac{1}{\sqrt{2}} \beta \sin \gamma (d_2^\dagger + d_{-2}^\dagger) \right] \right)^N |0\rangle . \qquad (3)$$

The energy surface for IBM is easily obtained for the proposed Hamiltonian

$$\begin{aligned} E_N(\beta, \gamma) &= \langle g; N, \beta, \gamma | H | g; N, \beta, \gamma \rangle = -\frac{(5 + \beta^2 + \beta^2 \chi^2)(1-x)}{(1+\beta^2)} \\ &+ \frac{N \beta^2 x}{1+\beta^2} - \frac{(N-1)(1-x)\beta^2}{(1+\beta^2)^2} \left(4 - 4\frac{\sqrt{2}}{\sqrt{7}} \beta \chi \cos[3\gamma] - \frac{2}{7}\beta^2 \chi^2 \right) . \end{aligned} \qquad (4)$$

The IBM-1 phase diagram

i) The critical point in the transition U(5)–O(6).
  This transition can be studied with

$$H = x\hat{n}_d - \frac{1-x}{N} Q^{\chi=0} \cdot Q^{\chi=0} . \qquad (5)$$

The corresponding critical point can be readily obtained with the condition $\left(d^2 E(N,\beta)/d\beta^2 \right)_{\beta=0} = 0$, which gives $x_c = \frac{4N-8}{5N-8}$ and a β^4 dependence in first order. In Ref. [13] the corresponding Bohr equation was solved numerically and it was shown that the corresponding results are different to those obtained in the E(5) critical point symmetry. This phase transition has been shown to be second order [2, 3, 7, 8, 9].

ii) The critical point in the transition U(5)–SU(3).
  This transition is obtained by choosing in our general Hamiltonian, Eq. (1,2), $\chi = -\sqrt{7}/2$ and varying the control parameter x from 0 (SU(3)) to 1 (U(5)). In Fig. 1 it is shown the behavior of the order parameter (β) as a function of the control parameter (x). The characteristic hysteresis cycle of the order parameter when a first order phase transition takes place is observed. In this situation two minima in the energy surface

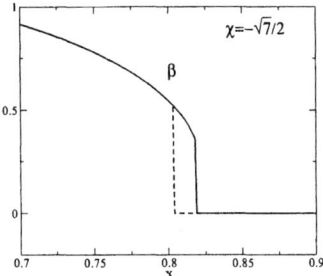

FIGURE 1. The behavior of the order parameter β in the phase transition from U(5) to SU(3). The full line corresponds to a situation in which initially the system is deformed and evolves to a spherical phase. The dashed line illustrates the situation in which initially the system is spherical and evolves to the deformed phase.

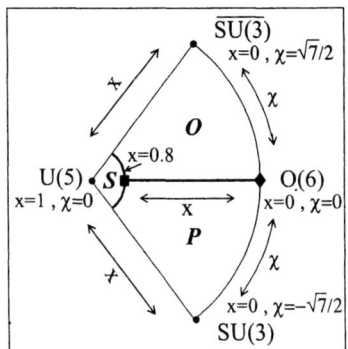

FIGURE 2. Schematic representation of the IBM-1 shape phase diagram.

coexist for a range of values of the control parameter. Within this range the state of the system depends on its previous history. If it comes from the spherical phase (dashed line) it remains spherical ($\beta = 0$) up to x approximately 0.8 and then the system jumps to the deformed minimum. If the system comes from a deformed regime (full line) it stays deformed up to around x=0.82 and then jumps to the spherical minimum. In the range 0.8 –0.82 the two minima coexist, one is stable and the other meta-stable.

With the preceding results and knowing that there is no phase transition from O(6) to SU(3) the phase diagram for IBM-1 is presented in Fig. 2. The transitions from spherical to axial deformed nuclei and from prolate to oblate shapes are first order with a singular point in the transition from U(5) to O(6) which is second order.

PHASE TRANSITIONS IN IBM-2

In this section we are presenting schematically the formalism to study the transition from axial to triaxial shapes in IBM-2. The Hamiltonian proposed is the natural extension of the previously discussed in IBM-1

$$H = x \left(n_{d_\pi} + n_{d_v} \right) - \frac{1-x}{N} Q^{(\chi_\pi, \chi_v)} \cdot Q^{(\chi_\pi, \chi_v)}, \tag{6}$$

where $n_d = \Sigma_\mu d^\dagger_\mu d_\mu$, $Q^{(\chi_\pi, \chi_v)}_\mu = \left(Q^{\chi_\pi}_\pi + Q^{\chi_v}_v \right)_\mu$ with $Q^\chi_\mu = \left[d^\dagger \tilde{s} + s^\dagger \tilde{d} \right]^2_\mu + \chi \left[d^\dagger \tilde{d} \right]^2_\mu$ and
N is the total number of bosons, which is equal to the number of valence nucleon pairs. The IBM phase diagram studied up to now corresponds to the selection $\chi_\pi = \chi_v$ which produces either spherical or axial shapes. In this section we explore the phase transition from axial shapes to triaxial shapes within a mean field formalism. One of the forms this can be done is by introducing a boson condensate [14]

$$|g; N_\pi, N_v, \beta_\pi, \gamma_\pi, \beta_v, \gamma_v\rangle = \frac{1}{\sqrt{N_\pi! N_v!}} (\Gamma^\dagger_g)^{N_\pi}_\pi (\Gamma^\dagger_g)^{N_v}_v |0\rangle \tag{7}$$

with

$$(\Gamma^\dagger_g)_\rho = \frac{1}{\sqrt{1+\beta^2_\rho}} \left[s^\dagger_\rho + \beta_\rho \cos\gamma_\rho d^\dagger_{\rho 0} + \frac{1}{\sqrt{2}} \beta_\rho \sin\gamma_\rho (d^\dagger_{\rho 2} + d^\dagger_{\rho -2}) \right] \tag{8}$$

with $\rho = \pi, v$. The equilibrium values of the order parameters $(\beta_\pi, \gamma_\pi, \beta_v, \gamma_v)$ and the energy of the system for given values of the control parameters in the Hamiltonian (x, χ_π, χ_v) can be obtained by minimizing the expected value of the Hamiltonian (6) in the intrinsic state (7)

$$\delta \langle g; N_\pi, N_v, \beta_\pi, \gamma_\pi, \beta_v, \gamma_v | H | g; N_\pi, N_v, \beta_\pi, \gamma_\pi, \beta_v, \gamma_v \rangle = 0. \tag{9}$$

In the limit of large N_π, N_v the energy to be minimized is easily obtained to be

$$E(\beta_\pi, \gamma_\pi, \beta_v, \gamma_v; \chi_\pi, \chi_v, x) = x \left[\frac{\beta^2_\pi}{1+\beta^2_\pi} + \frac{\beta^2_v}{1+\beta^2_v} \right] - \frac{1-x}{4} \left[Q^2_0(\beta_\pi, \gamma_\pi, \chi_\pi) \right.$$
$$+ \quad Q^2_0(\beta_v, \gamma_v, \chi_v) + 2 Q^2_2(\beta_\pi, \gamma_\pi, \chi_\pi) + 2 Q^2_2(\beta_v, \gamma_v, \chi_v)$$
$$+ \quad 2 Q_0(\beta_\pi, \gamma_\pi, \chi_\pi) Q_0(\beta_v, \gamma_v, \chi_v) + 4 Q_2(\beta_\pi, \gamma_\pi, \chi_\pi) Q_2(\beta_v, \gamma_v, \chi_v) \left. \right] \tag{10}$$

with

$$Q_0(\beta, \gamma, \chi) = \frac{1}{1+\beta^2} \left[2\beta \cos\gamma - \frac{2}{7} \beta^2 \chi \cos(2\gamma) \right] \tag{11}$$

$$Q_2(\beta, \gamma, \chi) = Q_{-2}(\beta, \gamma, \chi) = \frac{1}{1+\beta^2} \left[\sqrt{2}\beta \sin\gamma + \frac{1}{7} \beta^2 \chi \sin(2\gamma) \right]. \tag{12}$$

It should be noted that an alternative approach for discussing transitions from axial to triaxial shapes has been presented by Caprio and Iachello in this conference.

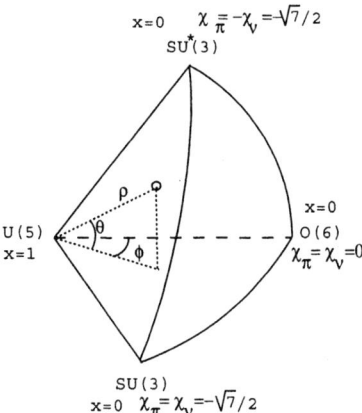

FIGURE 3. Schematic representation of the IBM-2 space with its dynamical symmetries

The transition from SU(3) to SU*(3)

As a natural extension of the Casten triangle for IBM-1, the representation of the IBM-2 space is presented in Fig. 3.

Any point in the IBM-2 space corresponding to a selection of the control parameters (x, χ_π, χ_ν) can be obtained with the following transformation to polar coordinates (see Fig. 3)

$$\rho = 1 - x \; ; \; \theta = -\frac{\pi}{3} \frac{\chi_\pi - \chi_\nu}{\sqrt{7}} \; ; \; \phi = -\frac{\pi}{3} \frac{\chi_\pi + \chi_\nu}{\sqrt{7}} \; . \tag{13}$$

In this way we have explored the transitions from axial to triaxial shapes in IBM-2. The general discussion will be presented elsewhere, here we just discuss the transition from SU(3) to SU*(3) [15]. This case corresponds to the selection $x = 0, \chi_\pi = -\sqrt{7}/2$ and χ_ν varying from $-\sqrt{7}/2$ to $\sqrt{7}/2$. The results are presented in Fig. 4 where it is clearly seen that in the transition $SU(3) \rightarrow SU^*(3)$ a second order phase transition develops at around $\chi_\nu = 0.4$. In this graph we have plotted a function proportional to $|z - z_{crit}|^{1/2}$, where z is the relevant control parameter.

CONCLUSIONS

In this paper we have revised in a mean field framework the shape phase transitions in IBM including a first study of the transition from axial to triaxial shapes. For this purpose we have used the IBM-2. We have shown that the transition from SU(3) to SU*(3) is second order. The full IBM-2 shape phase diagram will be presented in a forthcoming publication.

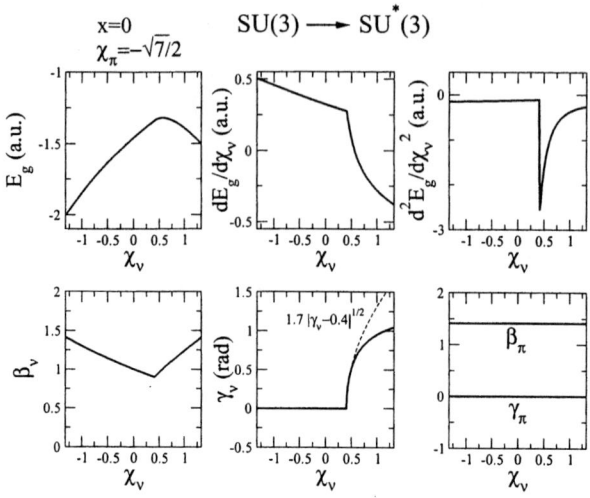

FIGURE 4. Transition from SU(3) to SU*(3): $x = 0$, $\chi_\pi = -\sqrt{7}/2$ and χ_v varies from $-\sqrt{7}/2$ to $\sqrt{7}/2$. In the panels are plotted in arbitrary units: the energy of the ground state, its first and second derivatives and the variation of the order parameters $\beta_\pi, \beta_v, \gamma_\pi(rad), \gamma_v$ (rad).

ACKNOWLEDGMENTS

This work has been done to honor Stuart Pittel and we acknowledge his continuous collaboration and friendship. This work was supported in part by the Spanish DGI under project numbers BFM2002-03315, BFM2000-1320-C02-02, and BFM2003-05316-C02-02.

REFERENCES

1. Iachello, F., and Arima, A., *The Interacting Boson Model*, Cambridge University Press, Cambridge, 1987.
2. Dieperink, A. E. L., Scholten, O., and Iachello, F. *Phys. Rev. Lett.* **44**, 1747 (1980).
3. Feng, D. H., Gilmore, R., and Deans, S. R., *Phys. Rev.* **C23**, 1254 (1981).
4. Frank, A., *Phys. Rev.* **C39**, 652 (1989).
5. Iachello, F., *Phys. Rev. Lett.* **85**, 3580 (2000); *ibid* **87**, 052502 (2001).
6. Iachello, F., *Phys. Rev. Lett.* **91**, 132502 (2003).
7. López-Moreno, E., and Castaños, O., *Phys. Rev. C* **54**, 2374 (1996).
8. Jolie, J., et. al., *Phys. Rev. Lett.* **89**, 182502 (2002).
9. Arias, J. M., Dukelsky, J., and García-Ramos, J. E., *Phys. Rev. Lett.* **91**, 162502 (2003).
10. Casten, R. F., and Warner, D. D., *Rev. Mod. Phys.* **60**, 389 (1988).
11. Ginocchio, J. N., and Kirson, M. W., *Nucl. Phys.* **A350**, 31 (1980).
12. Bohr, A., and Mottelson, B., *Phys. Scripta* **22**, 468 (1980).
13. Arias, J. M., et al., *Phys. Rev. C* **68**, 041302(R) (2003).
14. Ginocchio, J. N., and Leviatan, A., *Ann. Phys. (NY)* **216**, 152 (1992).
15. Dieperink, A. E. L., and Bijker, R., *Phys. Lett.* **116B**, 77 (1982).

Analytic X(5)–to–Rigid Rotor Transition in the Confined β-Soft Rotor Model

N. Pietralla[*†] and O.M. Gorbachenko[†]

[*]Department of Physics and Astronomy, State University of New York, Stony Brook, NY, U.S.A.
[†]Institut für Kernphysik, Universität zu Köln, 50937 Köln, Germany

Abstract. Iachello's X(5) solution for the Bohr-Hamiltonian for axially-symmetric prolate nuclei is generalized to the transition path between X(5) and the rigid rotor limit using infinite square-well potentials in the quadrupole deformation parameter β confined within boundaries at $\beta_M > \beta_m \geq 0$. Analytical solutions in terms of Bessel-functions of first and second kind are derived. They account well for the energies and E2 transition rates of $K = 0$ bands in transitional even-even nuclei with structural signature $2.9 < R_{4/2} = E_x(4_1^+)/E_x(2_1^+) < 3.25$. The relative 0_2^+ energies, $R_{0/2} = E_x(0_2^+)/E_x(2_1^+)$, for nuclei in the $A \approx 160$ mass region correlate to the $R_{4/2}$ values. The evolutionary trajectory of $R_{0/2}$ as a function of $R_{4/2}$ is predicted in a parameter-free way and the prediction agrees with the observations in this mass region within about 10%.

INTRODUCTION

One of the most interesting features of many-body quantum systems (*e.g.*, molecules, electronic quantum dots, Bose-Einstein condensates, atomic nuclei, and hadrons) is their ability to exhibit collective quantum effects such as spontaneously assuming deformed shapes. A prime example for deformed quantum systems are atomic nuclei. As the number of valence nucleons grows, spontaneous symmetry breaking to a deformed shape can occur. The mechanisms leading to nuclear deformation have attracted a great deal of attention both by experimentalists and theoreticians. Understanding these mechanisms microscopically has been one general theme of the contribution of Stuart Pittel to the development of nuclear physics *e.g.*, [1, 2]. Research interest on this topic has recently intensified and it is a great honor and pleasure for us to make the present contribution to this conference in honor of Stuart Pittel's 60*th* birthday.

Understanding of collective nuclear structure in shape-changing regions of the nuclear chart has, recently, gained from analytical solutions [3, 4, 5] of the Bohr-Hamiltonian's eigenvalue problem appropriate for the description of nuclei near the critical points of quadrupole-shape phase transitions, called E(5), X(5), and Y(5). These solutions serve as benchmarks for structures characterized by large fluctuations in at least one of the intrinsic quadrupole deformation parameters β and γ. Empirical examples for nuclei exhibiting properties close to the analytic solutions have first been found in the $A \approx 130$ and $A \approx 160$ mass regions [6, 7]. The *evolution* of quadrupole structure is, however, still left to numerical procedures.

Particularly appealing is the X(5) solution [4] for the phase transitional point from spherical vibrators to rigid rotors. Its predictions, namely, differ so significantly from

CP726, *Nuclear Physics, Large and Small: International Conference on Microscopic Studies of Collective Phenomena*, edited by R. Bijker, R. F. Casten, and A. Frank
© 2004 American Institute of Physics 0-7354-0207-8/04/$22.00

the predictions for pure vibrational or rotational structure that its experimental identification will not be obscured by non-collective perturbations outside of the model. Key signatures for X(5) structure are: an $R_{4/2}$ ratio of 2.90, pronounced centrifugal stretching along the ground band, a low-lying 0_2^+ state with $R_{0/2} = 5.65$, and strong $J \rightarrow J+2$ inter-band $E2$ transitions from the excited $K = 0$ band. Since the X(5) solution couples the excitation energy of the 0_2^+ state to the structure of the ground band, discussion of 0_2^+ state's energies became an important part of recent nuclear structure assignments, both for the structure of that state and the levels built on top of it and for the transitional character of a nucleus as a whole [7, 8, 9, 10, 11]. The nuclear shape phase transition is a topic of high current interest, e.g., [12, 13, 14, 15, 16, 17, 18, 19, 20, 21, 22]. While the recent discussions focused on the structural benchmarks, the *evolution* of 0_2^+ states [23] between them has not received much attention. A large amount of data exists on the spherical to prolate deformed shape transitional region at $A \approx 160$ on which we want to focus here.

CONFINED β-SOFT ROTOR

In analogy to the X(5) solution [4] we consider [24] the Bohr-Hamiltonian [25] with separable potentials $V(\beta, \gamma) = v(\beta) + u(\gamma)$ for axially symmetric prolate ($\gamma \approx 0°$) nuclei. We adopt here the same treatment of the γ degree of freedom as in Ref. [4] and focus again on the β degree of freedom in the limit of small fluctuations of γ about $\gamma \approx 0°$. This simplification is acceptable for $K = 0$ states that are discussed here.

The wave functions approximately separate into $\Psi(\beta, \gamma, \theta_i) = \xi_L(\beta)\eta_K(\gamma)\mathscr{D}^L_{M,K}(\theta_i)$, where $\mathscr{D}$ denotes the Wigner-functions of Euler-angles for the intrinsic system, and $\eta_K(\gamma)$ from Ref. [4]. We then seek solutions *between* X(5) and the rigid rotor for the "radial" differential equation (in the intrinsic quadrupole shape variables) for infinite square-well potentials $u(\beta)$ with boundaries at $\beta_M > \beta_m \geq 0$ [24]. In particular, we allow for $\beta_m \neq 0$, outside of the cases studied before [3, 4, 15]. The ratio $r_\beta = \beta_m/\beta_M$ parameterizes the stiffness of the potential and hence the structural evolution from the shape transitional point X(5) [$r_\beta = 0$, large fluctuations in β] to the rigid rotor [$r_\beta \rightarrow 1$ no fluctuations in β] in a unique way.

Using the new variables $z = \sqrt{E/(\hbar^2/2B)}\beta$ and $\tilde{\xi}[z] = \beta(z)^{3/2}\xi_L[\beta(z)]$ the differential equation for β transforms into the Bessel-equation [3, 4]. Solutions are the Bessel functions, $J_\nu(z)$ and $Y_\nu(z)$ of irrational order $\nu = \sqrt{L(L+1)/3+9/4}$, where $Y_\nu(z)$ is the Bessel function of second kind. The general solutions [24] are, thus, appropriate superpositions of the corresponding Bessel-J and Bessel-Y functions

$$\tilde{\xi}_\nu(z) \propto J_\nu(z) + \gamma_Y Y_\nu(z) . \tag{1}$$

The two boundary conditions

$$\tilde{\xi}_\nu(r_\beta z_M) = \tilde{\xi}_\nu(z_M) = 0 , \tag{2}$$

for $z_M = \sqrt{E/(\hbar^2/2B)}\beta_M$, determine the relative amplitude $\gamma_Y = -[J_\nu(r_\beta z_M)/Y_\nu(r_\beta z_M)]$ of the Bessel-Y function and serve as the quantization condition

$$Q_\nu^{r_\beta}(z_M) = J_\nu(z_M) Y_\nu(r_\beta z_M) - J_\nu(r_\beta z_M) Y_\nu(z_M) = 0 \ . \tag{3}$$

(B is the mass parameter of the Bohr-Hamiltonian.) For each value of $\nu(L)$ and r_β the appropriate z_M are obtained as the s-th zero, $z_{L,s}^{r_\beta}$, of the function $Q_\nu^{r_\beta}(z)$. The quantum number s counts the number of nodes of the wavefunction in β for $\beta > \beta_m$.

The normalized eigenfunctions of the differential equation in β are

$$\xi_{L,s}(\beta) = c_{L,s}\beta^{-3/2}\left[J_\nu(z_{L,s}^{r_\beta}\beta/\beta_M) + \gamma_Y Y_\nu(z_{L,s}^{r_\beta}\beta/\beta_M)\right] \tag{4}$$

with the eigenvalues

$$E_{L,s} = \frac{\hbar^2}{2B\beta_M^2}(z_{L,s}^{r_\beta})^2 \ . \tag{5}$$

Besides the choice of a finite value of r_β this solution is analogous to the X(5) solution of Iachello [4]. Considering potentials with no structure in β (being β-soft) over a range of values *confined* within the variable boundaries of an infinite square well, this approach represents a "Confined β-Soft" (CBS) rotor model [24]. We denote the 0_2^+ state with $s = 2$ as the (first) β-*excitation*.

Increasing the value of r_β shifts all levels to higher energies since the potential narrows. That energy shift increases with the s quantum number and with decreasing angular momentum. Consequently, the $R_{4/2}$ value increases with r_β until the rigid rotor limit with $R_{4/2} = 3.33$ is reached for $r_\beta \rightarrow 1$, where the gain in angular momentum fully originates in a gain of angular velocity with constant moments of inertia for all values of L. This geometrical description of nuclear level bands neglects, of course, non-collective single-particle degrees of freedom and can, therefore, work only until the back-bending region. For a given spin the energy shifts increase strongly with s as r_β increases. Therefore, the structural signature $R_{0/2}$ strongly increases with r_β and becomes infinite in the limit $r_\beta \rightarrow 1$.

It is certainly not obvious whether our choice of potentials can be justified microscopically as a relevant approximation to the realistic potentials for nuclei with $R_{4/2}$ values in the range of 2.9 – 3.3. However, the clear interpretability of analytic solutions supports easier comprehension of the structural mechanisms of the model (here, Bohr-Hamiltonian) and often justifies a sacrifice in predictive accuracy. Moreover, some "unrealistic" aspects of toy-potentials eventually turn out to have little impact on the solutions of interest. One example for this, is the assumption of an *infinite* square-well potential in the analytic E(5) critical point solution. Numerical studies of the corresponding wave functions for *finite* well depths show marginal deviations from the idealized situation for low-energy states [15], that do not justify the sacrifice in clarity when going from analytical to numerical solutions.

EVOLUTION OF 0_2^+ STATES IN THE $A \approx 160$ MASS REGION

Both structural signatures, $R_{4/2}$ and $R_{0/2}$, depend monotonically on r_β. Hence, the CBS rotor model yields a parameter-free relation between them for the evolution path from X(5) to the rigid rotor limit. This relation is confronted with data in Fig. 1. We consider the classical region of spherical-to-prolate deformed transitional nuclei, the light rare earth nuclei around neutron number $N = 90$. All nuclei with neutron numbers $N \in [90 \pm 6]$ and with at least one known excited 0^+ state are taken into account.

These $R_{0/2}$ data fall on a compact trajectory [23] when plotted as a function of $R_{4/2}$ until values of $R_{0/2} > 10$ are reached for $R_{4/2} > 3.25$. It can be concluded that the 0_2^+ states in these transitional nuclei with $2.9 < R_{4/2} \leq 3.2$ have a related collective structure because they correlate in a unique way to the (collective) structural signature $R_{4/2}$. This remarkable conclusion is model independent.

Second we note that the shape of the evolutionary path of $R_{0/2}$ data and within about 10% even their absolute values, as a function of $R_{4/2}$, are well predicted by the CBS rotor model in a parameter-free way. The β-excitation is comparably low-lying in transitional nuclei and increases strongly in energy when the rotor limit is approached. Therefore, β-excitations are difficult to observe in good rotors [27] where they are shifted to too high energies.

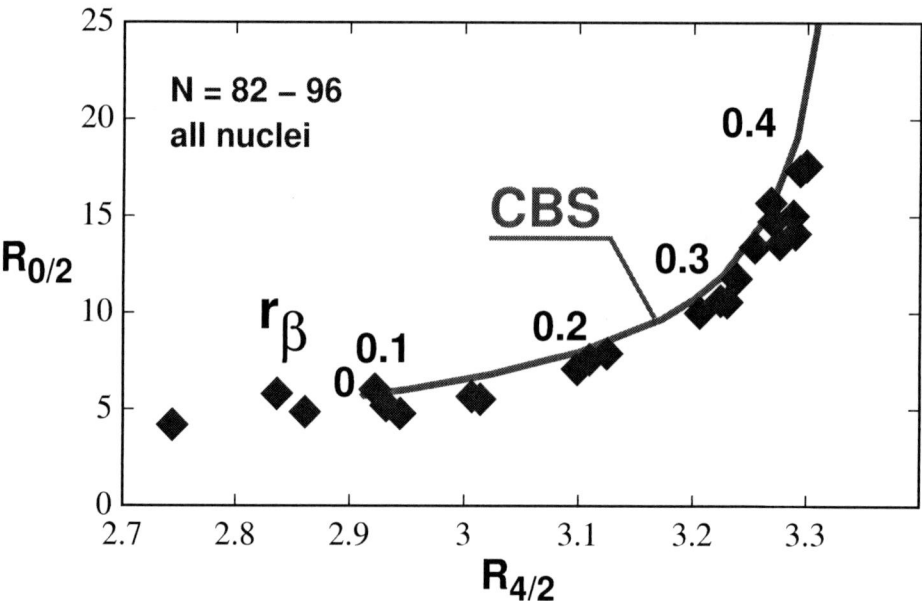

FIGURE 1. Correlation of $R_{0/2}$ and $R_{4/2}$ values [26] for all rare earth nuclei with the specified neutron numbers around $N = 90$ with known 0_2^+ state and $R_{4/2} > 2.7$. The solid curve corresponds to the parameter-free prediction of the CBS rotor model as a function of r_β.

E2 TRANSITIONS

For the calculation of $E2$ rates we use the $E2$ operator up to second order in the quadrupole deformation parameters with expansion coefficients treated as free parameters. The matrix elements are obtained from integration over the Euler angles and the intrinsic variables β and γ. Since we consider $K = 0$ states with $n_\gamma = 0$ [4], the $\Delta K = 0$ part of the $E2$ transition operator is relevant. It can be written as [24]

$$T_\mu^{\Delta K=0}(E2) = e_{\mathrm{eff}} \left[(\beta/\beta_M) + \chi(\beta/\beta_M)^2 \right] D_{\mu 0}^2 \qquad (6)$$

where constants from the L-independent integration over the γ variable (for states with $n_\gamma = 0$) have been absorbed in the effective charge e_{eff} and in the structural parameter χ, both to be adjusted to data.

It is interesting to compare energies and $E2$ rates for transitional nuclei to the predictions of the CBS rotor model. Since ^{152}Sm is well described by X(5) [CBS limit for $r_\beta = 0$] we consider ^{154}Sm with an $R_{4/2}$ value of 3.26. Figure 2 displays energies and $B(E2)$ values for ^{154}Sm and analytical predictions for the rigid rotor and for the CBS rotor model with $r_\beta = 0.35$ adjusted to the experimental $R_{4/2}$ ratio. Energies of excited $K = 0$ states are then parameter-free predictions. The good description of energies and the $E2$ rates results from the analytic solution of a unique, very simple Hamiltonian and a simple choice for the $E2$ operator (here $\chi = -0.5$ fixed from a fit to the neighboring nucleus ^{152}Sm). It does not involve any *ad-hoc* assumptions about band mixing [19].

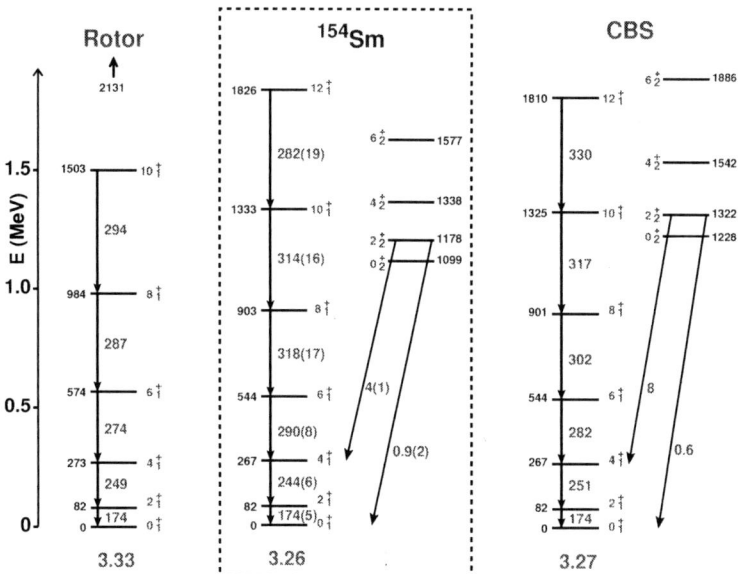

FIGURE 2. Comparison of data on $K = 0$ states of ^{154}Sm (middle) to the Rigid Rotor (left) and the CBS Rotor for $r_\beta = 0.35$ (left). Energies in keV are given next to the levels. Numbers at transitions indicate $B(E2)$ values in W.u. Model predictions are normalized to the experimental $B(E2; 2_1^+ \rightarrow 0_1^+)$ value.

CONCLUSION

We have studied the Bohr Hamiltonian for decoupled prolate axial symmetry ($\gamma \approx 0°$) with confined β-soft potentials. It has been solved analytically in terms of Bessel functions of first and second kind. This CBS rotor model interpolates between X(5) and the rigid rotor. It predicts the structural evolution for prolate transitional nuclei as a function of one parameter (r_β) and correlates the structural signatures $R_{0/2}$ and $R_{4/2}$. The shape of the empirical trajectory of $R_{0/2}$ as a function of $R_{4/2} \in [2.9, 3.2]$ for nuclei with $A \approx 160$ is well predicted by the model. The CBS rotor model well describes ground band energies as well as $E2$ rates for transitional and even near-rotational nuclei, *e.g.*, ^{154}Sm, analytically with two parameters (r_β, χ) and two scales ($\hbar^2/2B\beta_M^2, e_{\mathrm{eff}}$).

ACKNOWLEDGMENTS

We thank M.A. Caprio, R.F. Casten, C. Fransen, F. Iachello, P. Jung, D. Mücher, N.A. Smirnova, P. von Brentano, and N.V. Zamfir for discussions and various contributions. This work was supported by the DFG under grant No. Pi 393/1-2 and by the U.S. NSF under grant No. PHY-0245018.

REFERENCES

1. S. Pittel, P.D. Duval, B.R. Barrett, Ann. Phys. (NY) **144**, 168 (1982).
2. P. Federmann and S. Pittel, Phys. Rev. C **20**, 820 (1979).
3. F. Iachello, Phys. Rev. Lett. **85**, 3580 (2000).
4. F. Iachello, Phys. Rev. Lett. **87**, 052502 (2001).
5. F. Iachello, Phys. Rev. Lett. **91**, 132502 (2003).
6. R.F. Casten and N.V. Zamfir, Phys. Rev. Lett. **85**, 3584 (2000).
7. R.F. Casten and N.V. Zamfir, Phys. Rev. Lett. **87**, 052503 (2001).
8. R. Krücken *et al.*, Phys. Rev. Lett. **88**, 232501 (2002).
9. P.G. Bizzeti and A.M. Bizzeti-Sona, Phys. Rev. C **66**, 031301(R) (2002).
10. M.A. Caprio *et al.*, Phys. Rev. C **66**, 054310 (2002).
11. C. Fransen, N. Pietralla, A. Linnemann, V. Werner, and R. Bijker, Phys. Rev. C **69**, 014313 (2004).
12. J.M. Arias, Phys. Rev. C **63**, 034308 (2001).
13. J. Jolie, R.F. Casten, P. von Brentano, and V. Werner, Phys. Rev. Lett. **87**, 162501 (2001).
14. A. Frank, C.E. Alonso, and J.M. Arias, Phys. Rev. C **65**, 014301 (2002).
15. M.A. Caprio, Phys. Rev. C **65**, 031304 (2002).
16. N.V. Zamfir, P. von Brentano, R.F. Casten, and J. Jolie, Phys. Rev. C **66**, 021304(R) (2002).
17. V. Werner, P. von Brentano, R.F. Casten, and J. Jolie, Phys. Lett. **B527** 55 (2002).
18. J. Jolie, P. Cejnar, R.F. Casten, S. Heinze, A. Linnemann, and V. Werner, Phys. Rev. Lett. **89**, 182502 (2002).
19. R.M. Clark *et al.*, Phys. Rev. C **67**, 041302(R) (2003).
20. A. Leviatan and J.N. Ginocchio, Phys. Rev. Lett. **90**, 212501 (2003).
21. L. Fortunato and A. Vitturi, J.Phys.(London) **G29**, 1341 (2003).
22. R. Bijker, R.F. Casten, N.V. Zamfir, and E. A. McCutchan, Phys. Rev. C **68**, 064304 (2003).
23. W.T. Chou, Gh.Cata-Danil, N.V. Zamfir, R.F. Casten, N.Pietralla, Phys. Rev. C **64**, 057301 (2001).
24. N. Pietralla and O.M. Gorbachenko, Phys. Rev. C **70**, 011304(R) (2004).
25. A. Bohr, Mat. Fys. Medd. K. Dan. Vidensk Selsk. **26**, No. 14 (1952).
26. R.B. Firestone, V.S. Shirley, *Table of Isotopes* (Whiley & Sons, New York, 1996).
27. P.E. Garrett, J. Phys. G: Nucl. Part. Phys. **27**, R1 (2001).

Future Investigations in Nuclear Structure

J. Jolie

Institut für Kernphysik
Universität zu Köln, Zülpicher Strasse 77, D-50937 Köln, Germany

Abstract. We discuss in a more global context where nuclear structure studies might contribute to the development of our understanding of mesoscopic systems.

INTRODUCTION

The study of atomic nuclei and their dynamics at low excitation energy has been performed over several decades using more and more sophisticated experimental techniques. Nowadays the themes of nuclear structure research are changing. On one hand they are refocussing towards the study of low-spin properties and the associated complete spectroscopy. On the other hand nuclear structure research will be able to manipulate an extremely important degree of freedom, the neutron-to-proton ratio, with the advent of the new radioactive beam facilities.

THE ATOMIC NUCLEUS: AN UNIQUE MESOSCOPIC SYSTEM

Central in the study of the structure of atomic nuclei is the question how complex systems emerge from simple ingredients. In essence an atomic nucleus is based on the combination of two sets of indistinguishable fermions, the protons and neutrons. Here the Pauli principle will have an important influence, as it is only valid within one set and not for two fermions belonging to both sets. A second particularity deals with the highly complex short range force of Van der Waals type that acts between nucleons. In contrast to the Pauli principle the charge independence of the strong force makes this force equal between alike and dislike nucleons. Finally, we would like to stress also the importance of the weak force as it allows one kind of nucleon to become the other kind, a property totally absent of other kinds of mesoscopic systems.

When these nucleons are put together within the reach of the strong force, a nucleus forms with an enormous binding energy in comparison to other mesoscopic systems. Also at this moment the property of the weak interaction becomes an important ingredient, transforming one atomic nucleus into another with again an extra binding energy. The large binding energies released in radiation of all kinds is what makes the very complex atomic nucleus a particularly generous quantal system to study. In the formation of a new atomic nucleus, e.g. by fusion-evaporation reactions, neutron capture or beta decay, copiously radiation is emitted in order to get rid of the large

CP726, *Nuclear Physics, Large and Small: International Conference on*
Microscopic Studies of Collective Phenomena, edited by R. Bijker, R. F. Casten, and A. Frank
© 2004 American Institute of Physics 0-7354-0207-8/04/$22.00

excess of binding energies. Besides this many types of direct excitation mechanisms like transfer reactions, inelastic scattering of particles and photons or Coulomb excitation can be used without desintegrating the atomic nucleus. This is the wealth of nuclear structure. The price to pay is the need of considerably large accelerators and reactors needed to fuse atomic nuclei and the difficulty to manipulate the atomic nucleus under well defined external conditions. So, the atomic nucleus forms a unique two component mesoscopic system, which is hard to manipulate but genereous in the number of observables it emits.

Once the complex mesoscopic system is formed with its complex short range force dealing with the interaction of up to hundreds of nucleons, something marveleous happens. Effective (in-medium) forces generate simple collective motions that can be described in phenomenological theoretical models. The simple motions can be given with three double keywords: shell structure or the Pauli principle in action, Cooper pairing or the bosonification of a fermion system, and collective motion or the emergence of a quantum fluid out of a limited number of entities. These properties are related in some sense, but we do not know exactly how yet. One of the basic fields of research will therefore be how and why this happens. One could compare the importance of these questions with the times before and after the advent of the kinetic theory of gases when suddenly the description using the small microscopic entities explained the large macroscopic properties. However, as the atomic nucleus is itself a small finite quantal ensemble of strongly interacting and two-component particles, that are not really small compared to the nucleus, it also poses many new questions and therefor a lot of experimental input will be needed.

In comparison with other mesoscopic systems which are intensively studied in solid state physics, it should be remarked that each atomic nucleus is in some sense a fundamental object whose properties influence the composition of our world via the stellar nucleosynthesis.

MANIPULATION OF ALL DEGREES OF FREEDOM

Research in medium heavy nuclei over the last decades has concentrated on the study of both the spin and excitation energy dependence of nuclear structure. The spin degree of freedom was varied up to the point that superdeformation, chiral bands and magnetic rotation were observed. This degree of freedom allows a certain internal manipulation of the atomic nucleus that was not available before, and led to the development of new highly performant instruments. On the other hand the manipulation of the excitation energy at constant spin, equivalent to the classical temperature axis, allowed the exploration of another largely unknown degree of freedom, the study of yrare states. This domain showed the prevalence of collective motion up to multiphonon states, the appearance of new collective motions like mixed symmetry states, the importance of 2p-2h intruder states, and the appearance of beautiful dynamical symmetries and supersymmetries. At even higher excitation energies the onset of quantal chaos and pair breaking phase transitions becomes now within reach. This field relied on improved instrumentation and computing capacities. It still has a vary large potential of discovery.

In addition to these emerging fields of study, will be the possibilities offered by the upcoming and existing Radioactive Ion Beam (RIB) facilities, disclosing the third degree of freedom needed to study nuclear structure, the important neutron-to-proton ratio. Not only will we be able to study predictions of present day nuclear models, but moreover will the effects due to the underlying neutron-proton degree of freedom be throughoutly studied. Besides this new properties will be revealed like coupling of bound states with the continuum, dilute nuclear matter, clustering and new decay modes. With the availability of RIB, all essential degree of freedoms will so become available for manipulation. The price to pay here is an important infrastructure that can only be afforted on a generally supranational scale. However, the discovery potential is very large and new facilities, like RIA and the GSI extension, will be built.

QUESTIONS TO BE ANSWERED

Shell Structure at largely asymmetric N and Z Values

In contrast to other mesoscopic systems, the fact that one has a two-component system induces shell closures dependent on the number of neutrons N and the number of protons Z. A beautiful example is the Federman-Pittel mechanism [1] which due to the monopole or τ,τ,σ,σ interaction leads to shell quenching that was already observed in some exotic nuclei. Besides this the shell structure will drastically change for strongly asymmetrical nuclei, due to changing spin-orbit interactions and coupling with the continuum.

In order to studese this questions we will need to determine new magic numbers as a function of N and Z and therefore be able to produce the farest reachable atomic nuclei. Here mass measurements using traps and storage rings will be important. Further transfer, knock-out and fragmentation reactions as well as low-spin spectroscopy yielding B(E2) values, moments and excitation energies, will be crucial. *The outcome of these studies will not only contribute to nuclear physics but be vital for astrophysics.*

Pairing: bosonification of a fermionic system

Medium heavy and heavy nuclei can be very well described as an interacting boson system [2], due to the strong pairing between like nucleons occupying the same orbitals. However, because they are formed out of largely asymmetrical combinations of neutrons and protons the pairing is strongly influenced by the Pauli principle. This will not be the case for heavy N = Z nuclei, where neutron and protons occupy the same orbits and can form pairs with T = 0 and J = 1^+ (quasi-deuteron states) in addition to the common T = 1, J = 0^+ pairs. It would be of great interest to study what kind of pairing will be observed for heavy N=Z nuclei.

It was shown that random matrix elements in the shell model [3] and the interacting boson model (IBM) [4] generate in most cases a J = 0^+ ground state. This surprising finding happens in 76% (63%) of the cases in the shell model (IBM) while these states span only 10% (3.3 %) of the total Hilbert space. The fact that no even-even nucleus has been observed that does not possess a J = 0$^+$ ground state contradicts the conclusion that we are dealing with a chaotic system and is the strongest argument for pairing. Nevertheless, it would be extremely interesting to investigate if any nucleus can be found in nature that does not have a J = 0$^+$ ground state.

Finally, we know from the study of nuclei near stability, that the strong pairing force allows to bridge the gap between the shell model and the collective model using the IBM. In turn this boson model has shown to have an extremely rich structure exhibiting several dynamical symmetries and supersymmetries. It would then be of great interest to study whether these beautiful symmetries also occur in presently unknown nuclei.

In order to study pairing in atomic nuclei we will need complete low-spin spectroscopy, i.e. gamma-ray spectroscopy and transfer reactions, on stable and unstable nuclei that can be produced in large quantities. At RIB facilities the use of beta decay as a mean to populate low-spin states in a rather complete way will be an essential tool. *The outcome of these studies will be essential for nuclear physics but also for the study of the many other mesoscopic systems where pairing occurs.*

Quantum phase transitions at finite N

A new way to look at low-energy nuclear structure, has gained a lot of attention in the last five years. Instead of considering a particular atomic nucleus, one theoretically treats the nucleus as a finite-N system that can change its structure in a continuous way as a function of a limited number of control parameters. The system which was studied in detail is the Casten triangle parametrized by the simple hamiltonian:

$$\hat{H} = \eta \hat{n}_d - \frac{1-\eta}{N} \hat{Q}_\chi \cdot \hat{Q}_\chi, \quad \hat{Q}_\chi = (s^\dagger \tilde{d} + d^\dagger s)^{(2)} + \chi (d^\dagger \tilde{d})^{(2)} \quad (1)$$

where η and χ are the control parameters and N is the number of s and d bosons. This hamiltonian allows the study of quantum shape phase transitions in a system with finite N. This new subject in mesoscopic physics can then be analyzed with the help of the coherent state formalism that allows for an extrapolation towards the infinite N limit [5,6]. In this limit the IBM turns out to represent an particular example of Landau theory of second order phase transitions[7,8], showing three shape phases, spherical, oblate and prolate deformed, with first and second order transitions between them. In addition, Iachello has shown that at the phase transitions new critical point symmetries occur, albeit in a collective model framework [9].

Atomic nuclei cannot in a continuous way undergo the above mentioned phase transitions since they consist out of a discrete set of neutrons and protons. Therefore the experimental investigation of shape phase transitions has to rely on studies of nuclei near to the phase transitions or series of isotopes crossing the phase transition. Some excellent examples of such experimental studies, including those related to the critical point symmetries can be found in the works listed in [10].

In the future the question should be solved of how these phases evolve away from stability. Therefore it is of utmost importance to determine the experimental signatures of the phase transitions and to investigate the existence of up to now unknown phases. In this context the question whether triaxial nuclei or nuclear molecules can exist should be answered. To achieve these goals one needs to perform experiments on exotic nuclei produced in sufficient quantities for low-spin spectroscopy using gamma and conversion electron spectroscopy, as well as to determine masses. *The results of these studies will be important for the understanding of nuclear structure but also of mesoscopic systems with a finite number of constituents.*

Quantal Chaos versus integrable many-body systems

The atomic nucleus forms a unique laboratory to study chaos and regularity in quantal systems due to the ability to study complete sets of excited states in a given spin and energy window. For those states properties like spin and parity can be determined. The presence of nuclei exhibiting dynamical symmetries at low exitation energy as well as the knowledge that chaotic motion prevails at excitation energies around the neutron binding energies [11] (typically 6-10 MeV) are good examples of the richness of atomic nuclear excitations.

One of the main questions is then how collective motion washes out as a function of excitation energy and up to which excitation energy it can survive. One expects a phase transition, much like normal temperature driven phase transitions, when pairing breaks down, and then a transition towards a regime where only statistical descriptions by means of level densities and average properties will be applicable. The study of these can be performed near stability, but has a major impact on reaction theories needed to study exotic nuclei.

Another question related with quantal chaos and regularity can be studied at very low excitation energies and even at the groundstate: the question of integrability of quantum systems. The interacting boson model and the shell model are ideal realistic models to study this subject. Moreover they provide examples of nuclei that do allow the experimental determination of the characteristic observables. As a concrete example let us consider again hamiltonian (1). The hamiltonian exhibits not only three dynamical symmetries, first and second order phase transitions, but also two regions of increased regularity [12]. One is formed by the U(5) to O(6) leg of the triangle and is due to the presence of a conserved O(5) symmetry and can be solved in a SU(1,1)xSU(1,1) model [13]. The other one which we will call the Alhassid-Whelan arc of regularity is situated inside the triangle and goes from U(5) to SU(3). Its locus can be parametrised by the relation:

$$\chi = \frac{\sqrt{7}-1}{2}\eta - \frac{\sqrt{7}}{2}, \qquad (2)$$

and its nature is still not understood. Several nuclei could recently be identified as being situated close to this arc of regularity [14] (see Figure 1). They all exhibit a close lying second excited 0^+ and 2^+ state, which provides a clear experimental signature.

FIGURE 1. The Casten triangle with the Alhassid-Whelan arc of regularity. Indicated by symbols are the locations of isotopes in the Gd-Os region as determined in ref [15]. In bold are represented the isotopes close to the regular region for Gd (diamonds), Dy (cross), Hf (x) and Yb (squares).

143

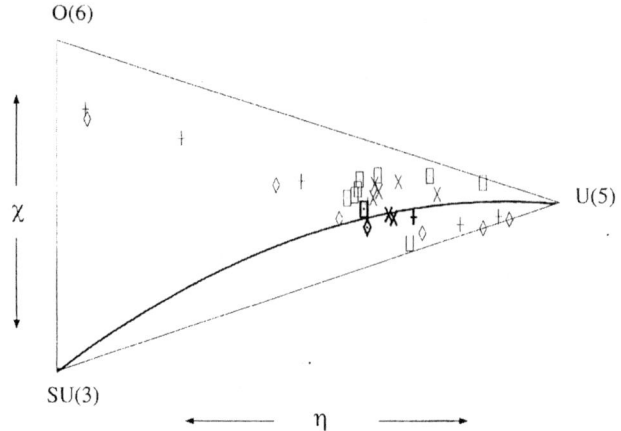

In the future we will need complete spectroscopy at all possible spins in order to study the chaos introduced by pair breaking, but we might also study the regularity versus chaotic behavior at low excitation energy relying on fewer properties, as indicated by the arc of regularity. *The results of these studies will be important for nuclear structure, but on a more general level for the whole of quantum mechanics.*

EXPERIMENTAL CHALLENGES

The challenges for the future lie to a large extent in the manipulation of the neutron-proton degree of freedom using RIB and in the determination of an as complete as possible set of excited states near stability. Three major challenges will have to be adressed here in the future.

The first concerns the use of highly performant detector arrays to study essentially low spin properties of stable and exotic nuclei. These spectrometers like Euroball and Gammasphere were developed for the study of high-spin physics where they have been very successful. Nowadays they should be converted towards low-spin applications. The challenge is here that low-spin physicists need the high spin technology and high-spin physicists need the low-spin methods and theoreticians. The most advanced in this direction is the Euroball spectrometer, which was dismantled in 2003 for this goal. The 15 Euroball Cluster detectors are now installed at the FRS of GSI in the RISING project [16] for RIB research. The RISING collaboration forms a nice illustration that the so-called I and J community can merge and pursue common goals in the future. Noteworthy, is also that other parts of the Euroball spectrometer were successfuly installed at the RITU spectrometer in Jyvaskula for stable beam experiments.

The second challenge concerns the development of new instrumentation. Especially at RIB facilities the beam is generally the target which leads to important Doppler effects which should be corrected for. Here the development of segmented Ge detectors for the MINIBALL array [17] has shown the feasibility of position sensitive gamma-ray detection. This opens the way to the technique of gamma-ray tracking as foreseen in the AGATA spectrometer which is now being constructed in an European effort [18].

The third challenge is the need of a sufficient number of experimental facilities. While frontier physics will be performed at the most advanced facilities, one should consider that these facilities which can reach nuclei very far away from stability will also allow to obtain more detailed information on atomic nuclei that are less exotic but can be produced in larger quantities. Here the physical mechanisms determining the structure of the more exotic nuclei can be studied in detail. In addition we will still need stable beam facilities to study many of the topics mentioned above and to answer new questions that will follow from the RIB research .

To conclude, the future of nuclear structure looks bright, but there is still a lot of work to be done.

ACKNOWLEDGMENTS

Many discussions with R.F. Casten, P. Cejnar, A. Dewald, J. Eberth, S. Heinze, N. Pietralla, P. Reiter and P. von Brentano have contributed to this work. This work was supported by the BMBF under grant 06-K-167.

REFERENCES

1. Federman, P., and Pittel, S., *Phys. Lett. B*, **65**, 385-389 (1977).
2. Iachello F., and Arima A, *The Interacting Boson Model*, Cambridge, England: Cambridge University Press 1987.
3. Johnson, C. W., Bertsch G.F. and Dean, D. J., *Phys. Rev. Lett.* **80**, 2749 (1998).
4. Bijker, R., Frank A., *Phys. Rev. Lett.* **84**, 420 (2000).
5. Iachello F., Zamfir N.V., Casten R.F., *Phys. Rev. Lett.,* **81,** 1191 (1998).
6. Cejnar P., Jolie J., *Phys. Rev. E* **61,** 6237 (2000).
7. Jolie J., Cejnar P., Casten R.F., Heinze S., Linnemann A., Werner V. *Phys. Rev. Lett.* **89**, 182502 (2002).
8. P. Cejnar, Heinze S., Jolie J., *Phys. Rev. C* **68**, 034326(2003).
9. Iachello F., *Phys. Rev. Lett.* **85**, 3580 (2000); **87**, 052502 (2001); **91**, 032502 (2003).
10. Casten, R.F., and Zamfir, N.V. *Phys. Rev. Lett.* **85**, 3584 (2000); **87**, 052503 (2001); Zamfir N.V., von Brentano P., Casten R.F., Jolie J., *Phys. Rev. C* **66** 021304R (2002); Jolie J. and Linnemann A., *Phys. Rev. C* **68** 031301(R) (2003); Tonev D. et al. *Phys. Rev. C* **69** 034334(2004).
11. Haq, R.U., Pandey, A. Bohigas O., *Phys. Rev. Lett.,* **48,** 1086 (1982).
12. Alhassid Y., Whelan N., *Phys. Rev C* **67** 816 (1991).
13. Pan F., Draayer J.P., *Nucl. Phys. A* **636**, 156 (1998).
14. Jolie J. , Casten R.F., Cejnar, P., Heinze, S., Mc.Cutchan, E. A:, Zamfir, N.V., *subm to Phys. Rev. Letters*.
15. Mc.Cutchan, E. A:, Zamfir, N.V., and Casten R.F., *Phys. Rev. C in press* (2004).
16. Wollersheim H.J. et al., *subm to Nucl. Instr. and Meth. A (2004)*.
17. Eberth J. et al. *Prog. in Part. And Nucl. Phys* 46, 389 (2001).
18. Simpson J., and Kruecken R., *Nucl. Phys. News Int.* **13** No 4 p.15 (2003)

Radioactive Ion Beams: Enlargement of Possibilities on Experimental Nuclear Physics. Three Examples Obtained with HRIBF at ORNL

María Esther Ortiz

Instituto de Física
Universidad Nacional Autónoma de México
Apartado Postal 20-364, 01000 México D.F., México

Basic mysteries of nature can be clarified by studying the nucleus of the atom. New tools have been developed all around the world, to expand the boundaries of capability to explore the nucleus experimentally, as for example the development of accelerators manipulating radioactive ion beams.

How nuclei are kept together and how elements were formed are some of the most fundamental questions of Nuclear Physics that can be undertaken with nuclear reactions performed with different beams. Radioactive beams opened the possibilities a great deal.

Some of the main areas that nuclear science pursues with the use of radioactive beams are Nuclear Astrophysics, Nuclear Structure, Reaction Physics, the discovery of new isotopes, properties of exotic nuclei, hot nuclear systems, the equation of state of dense matter and of course applications of nuclear science in medicine, materials engineering, archeology, security, space, etc.

Now due to the new instrumentation and the improved techniques at ORNL is possible to systematically study properties of nuclei over a wide range of N/Z. It has been of particular interest the loosely bound nuclei near the limits of nuclear existence, which are expected to exhibit modified shell structures, new collective modes, isotope abundances and exotic decays. The trio of studies mentioned in the title has to do with these issues.

Before discussing each paper, it is necessary and convenient to mention that radioactive ion beam intensities are several orders of magnitude lower than those reached with stable beams, so experimentation with them are several orders of magnitude more difficult and instrumentation more sophisticated. Intensities of the order of 10^5 ions/sec are the usual. The usefulness of the beam called beam quality is dependent on its intensity and composition, experts at all laboratories are working on developing new techniques to improve the quality of such beams, in particular higher production rates, faster transport, higher ionization efficiency in the ion source and lower beam emittance.

CP726, *Nuclear Physics, Large and Small: International Conference on Microscopic Studies of Collective Phenomena*, edited by R. Bijker, R. F. Casten, and A. Frank
© 2004 American Institute of Physics 0-7354-0207-8/04/$22.00

To understand several explosive events and to explain the abundance of elements in nature, precise measurements of particular cross sections are quite important. In the first example [1] they found, by measuring the excitation function ^{1}H(^{17}F, p)^{17}F with a radioactive ^{17}F beam, a 3* state in ^{18}Ne and its contribution to the ^{17}F(p, γ)^{18}Ne stellar reaction rate and how it contributes strongly to the abundances of ^{13}N, ^{14}O, ^{15}O and ^{17}F. The contributions to this rate of competing mechanisms at certain temperatures, are relevant and decide the regions where x-ray bursts and supernovae are very important and why at lower temperatures environments of novae direct capture contributions dominate this rate.

The second example [2] shows one of the most exotic and elusive decays: the simultaneous emission of two protons, searched for more than 40 years. This represents a powerful probe of the nuclear structure of very weakly bound systems. There are two different modes in which this decay can occur, by emission of a diproton (a 2He nucleus) which subsequently breaks up into two protons or as a direct three-body process, called colloquially democratic decay. This experiment was performed also with a 17F beam on a thick CH2 target and using inverse kinematics to populate an 18Ne state responsible of the two-proton emission to 16O. There are plans to go on this line of research, one is to decide the prevalence between the two modes diproton or the democratic decay in this particular reaction and secondly to work in the search of a dineutron decay.

One of the more common experimental searches in several laboratories around the world is the under- barrier enhancement of the fusion cross section if compared with usual predictions. The last experiment to mention here [3] was performed with a neutron-rich radioactive ^{132}Sn on ^{64}Ni to form a ^{196}Pt compound nucleus at sub-barrier energies where an evident enhancement of the fusion cross-section was observed. The large N/Z ratio seems to reduce the barrier height and the presence of a large number of nucleon transfer channels with large positive Q values, which act as doorway states, are supposed the main causes for this phenomenon. Sub-barrier fusion has been used to produce super heavy elements, and if projectile and target are closed shell nucleus the compound system have lower excitation energies and smaller fissility and therefore higher survival probability. Neutron-rich radioactive beams will help a lot in understanding this and other mechanisms.

Radioactive beams have been and will be a very important tool to perform essential Physics in this and similar laboratories all around the world.

REFERENCES

1. D. W. Bardayan et al., Phys. Rev. Lett. **83** (1) 45, (1999)
2. J. Gómez del Campo et al, Phys. Rev. Lett. **86** (1) 43, (2001)
3. J. F. Liang et al., Phys. Rev. Lett. **91**, 152701 (2003)

Proton-Neutron correlations in N=Z nuclei

C.J. (Kim) Lister

Physics Division, Argonne National laboratory, 9700 S. Cass Ave. Argonne IL 60439

Abstract. The issue of correlations between protons and neutrons has become a major focus of nuclear structure research along the proton dripline. The strength of these correlations determined when nuclear binding ends. The correlations can have isospin T=1, just like "normal" neutron-neutron and proton-proton pairs, or have more exotic structure in a deuteron-like T=0 configuration. I will discuss ways of studying these correlations and their relative importance. Odd-odd N=Z nuclei are good laboratories for this work.

The correlations between neutrons and protons that bind nuclei is a complicated issue that can only be properly addressed through large basis shell model calculations incorporating the many matrix elements that exist between all of the particles. In practice, beyond the fp-shell we can not attempt such calculations, so we resort to much more empirical methods which approximate to the full problem. Our most common path of simplification is to assume there is bulk binding, often described by a droplet model, modulated by quantum shell effects via the Strutinsky method, and with a very simplified pairing interaction to account for the substantial extra binding of even-even nuclei. The key truncations used to describe the pairing correlations are to assume that the neutron and proton correlations are completely independent, and that only pairs of like particles, coupled in time reversed orbits to T=1, J=0 are important. Other effects, like correlations between the neutrons and protons, and coupling to other spins, are assumed to be less significant and are absorbed into the overall mean-field binding energy. As the overall binding along the valley of stability is large, and quite well described by these models, these approximations seem well justified, especially for near-stable nuclei in the rare-earth and actinide regions, where the shapes are rigid, level densities are high, and valence particles numerous.

Although we have nearly perfected these useful empirical models, we have laid a minefield of obstacles against real, substantive progress in understanding effective pairing correlations in the nuclear mean field. We are certainly not in great shape when it comes to the extrapolative powers of these models which we need when exploring very far from stability...especially in exotic, near-dripline nuclei. The problems arises as we have arbitrarily, and not at all consistently, compartmentalized the residual interactions between particles into parts we call "mean-field binding", "shell corrections" and "pairing". This leaves a very shaky foundation on which to compare one model with another, extrapolate to new regions, or introduce new types of correlation. It is in these troubled waters Stuart Pittel has crafted much of his career, trying to find a way to properly incorporate the influences of correlations between

CP726, Nuclear Physics, Large and Small: International Conference on
Microscopic Studies of Collective Phenomena, edited by R. Bijker, R. F. Casten, and A. Frank
© 2004 American Institute of Physics 0-7354-0207-8/04/$22.00

neutrons and protons and build bridges between various theoretical approaches and between experimenters and theory.

At the driplines, these correlations need more special attention. The addition of relatively few protons quickly take us to the proton dripline. Here, in mid-mass nuclei between [56]Ni and [100]Sn the proton dripline lies close to the nuclei with N=Z. In these nuclei correlations between pairs of particles of all types cannot be ignored, as a few hundred keV of extra binding determines the difference between almost instant dissociation and life long enough for β-decay. The extra neutron-proton correlations are critical in nuclear survival, as can be seen from the jagged dripline which strongly favors even numbers of particles. It is through these nuclei that the nucleosynthesis rp-path passes, so the correlations have a very strong effect on nucleosynthesis, energetics of thermonuclear runaways, and reaction rates in X-ray bursts. This paper concerns reaching out to the proton dripline using stable beams and performing as detailed gamma-ray spectroscopy as possible to learn about the neutron-proton correlations.

We know much less about the location of the neutron dripline. Certainly, it lies much further away from stability than the proton dripline, as addition of neutrons does not add Coulomb repulsion. However, where it lies and the structure of nuclei which live there depend strongly on the decoupling of neutron and proton mean fields, and on the correlations experienced by weakly bound neutrons in the low-density neutron skin. That is for the future when next generation facilities like RIA, the Rare Isotope Accelerator, come online. However, learning how to more self-consistently deal with correlations will be very important in the neutron-rich world. Experimentally, we may never reach the neutron dripline in heavy nuclei, so reliable extrapolative methods will need to be developed.

A good starting point for discussing neutron-proton correlations is the schematic model of Chasman [1,2,3]. This is by no means the first attempt to include the effects of these correlations in calculating overall binding or in structure, but it is intuitive and allows us to compare various types of data with a model. A much more complete discussion of the history of neutron-proton pairing correlations has been made by Goodman [4,5], who has worked on this problem for many years. It is a very active field at the moment, with more than 20 papers being published in the last year. The Chasman model is a simple shell model, using a basis of Nilsson states. It allows many couplings of neutrons and protons to various spins, and in pairs and quartets. It carefully differentiates between diagonal matrix elements, the energy gained in making pairs, and off-diagonal matrix elements, the energy gained by the interactions and scattering between pairs. Most importantly, it is careful of number projection and maintains a track of how many pairs of each type are in the wave function.

A good starting exercise is consideration of the separation of the T=1 excited states in even-even N=Z nuclei from their T=0 ground states. In the Chasman model the ground states are highly correlated with each Nilsson level quartetted with exactly two protons and two neutrons. This leads to maximum binding and J=0. To make a T=1 state neutron and proton pairs must be broken and particles promoted. The cost in energy is twice the single particle energy plus the loss of diagonal correlations, which turn out to be four in number. Inserting reasonable numbers leads to a gap of ~5.5MeV. Unfortunately for the nuclei we are interested in, between Z=30 and 50,

152

these analog states are not experimentally known. However, we can calculate the Coulomb shifts in these N=Z systems quite well, and map states, or bands of states, from T_z=1 nuclei which lie nearer stability, and thus are better known. The Chasman prediction fits very well assuming this mapping is true. For example, in ^{72}Kr the lowest T=1 state is predicted to lie near 5.4MeV, exactly where the Coulomb shift mapping of ^{72}Br predicts them. Parenthetically, it is interesting to note that at very high spin, J > 25, the loci of T=0 and T=1 high spin states merge indicating that correlations have been largely destroyed, and what remaining correlations are left are equal in each channel. An interesting prediction is that there should be two of these T=1 states at very similar excitation energy. This T=1 to T=0 gap in even-even nuclei has been discussed by Janecke [6], Vogel [7], and Macchiavelli [8] in a slightly different way. In these approaches the key interest is to categorize the gap as arising form two distinct parts, one arising from the different isospin symmetries of the wavefunctions, the so-called "symmetry energy", which can be calculated in a liquid drop framework, and the rest due to blocking of pairing.

A natural progression is then to approach odd-odd nuclei, which are currently the key laboratories for experimenters in this field. In the Chasman model, two distinct predictions emerge. Firstly, in the limit of the T=0 and T=1 diagonal matrix elements being identical, the groundstate of N=Z odd-odd nuclei should be a degenerate pair of states, one with T=0 and J=1 and one with T=1 and J=0. Any splitting of this doublet marks a clear experimental verification that the nuclei are away from this symmetry. Secondly, the level densities in N=Z odd-odd nuclei should be much lower than all neighboring N>Z odd-odd nuclei, due entirely to the need to break up the highly correlated quartets in the N=Z systems. Until recently, very little was known about these nuclei, so the issue was a bit abstract. Several key experiments have been performed on ^{62}Ga [9,10], ^{66}As [11,12], ^{70}Br [13,14] and ^{74}Rb [15,16] that help clarify the situation. These are all very difficult studies and need arrays like Gammasphere, Euroball, or Gasp, plus channel selective auxiliary detector systems to pick these exotic reaction products, which typically are less than 1 part in 1,000 of the reaction products (usually 100's of μ-barns). The experimental observations are clear: the nuclei all have T=1 ground states, clearly characterized by their fast superallowed Fermi β-decays. However, it now emerges that there are NO low-lying doublets. The T=1 state always lies above 800keV in excitation in all cases. In fact, the T=0 state has not been seen in most of these nuclei, but can be excluded by inference, in so far that if a low-lying T=0 state existed then its population and decay would have been observed. Thus, on the face of it, in these nuclei, it seems difficult to form T=0, J=1 collective pairs. The experimental level density situation is also clear. In all the N=Z odd-odd nuclei, there are NO quasi-particle bandheads found below 1 MeV, exactly in accordance with the Chasman prediction, and is in drastic contrast to the N=Z+2 nuclei which have many low-lying bandheads. This situation is shown in figure 1.

It is natural to ask where the T=0 J=1 pairs have gone? They form the groundstate of many light N=Z odd-odd nuclei, most significantly in the deuteron, this simplest neutron-proton system where the T=0 state is 2.2 MeV below the T=1 level. One consideration is the influence of spin orbit splitting. In heavier nuclei the coupling is j.j not l.s, so forming objects with T=0, S=1 and L=0 becomes increasingly difficult. Two possibilities exist; having the two particles in the same j-orbit coupled to T=0,

S=1, or population of spin-orbit partners j=L+1/2 and j=L-1/2. We can calculate the coefficient of fractional parentage (cfp) which tells us the likelihood of these couplings. Geometrically, the "j²" situation, for example two particles in a "stretched" configuration is not favored, as when the spin vectors are parallel, then so too are the orbital angular momentum vectors, so they cannot easily couple to L=0 and the probability of this arrangement falls rapidly with L. The spin-orbit partner solution is more favored by cfp's as when the spin vectors are parallel, the orbital vectors cancel, so T=0,S=1 is relatively L-independent. However, in heavier nuclei, on average, the L-values are higher and the

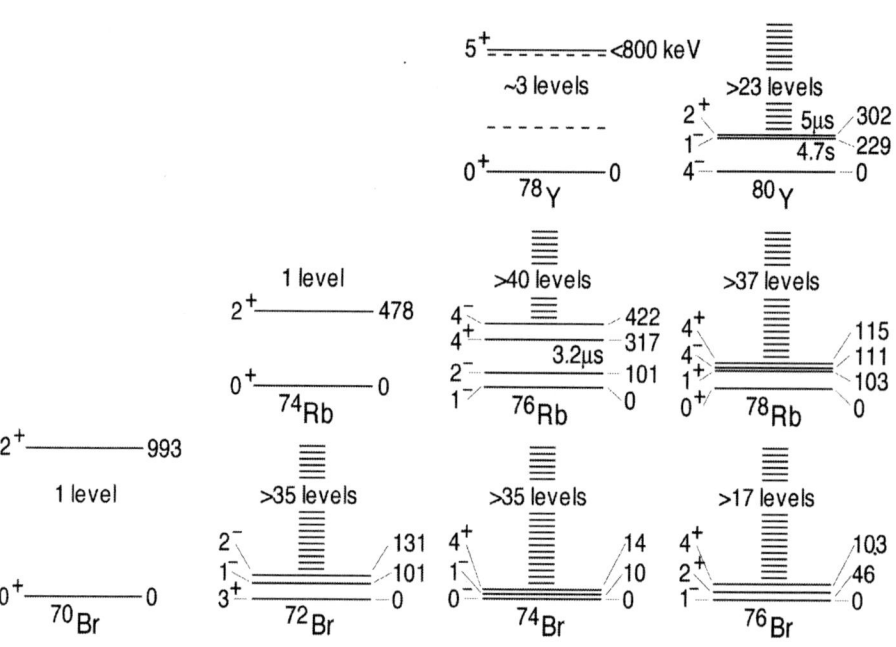

FIGURE 1. A compilation of data on odd-odd nuclei with N ~ Z in the mass 70 region, for bromine, rubidium and yttrium isotopes. Note the dramatic change in level density between N=Z and N=Z+2 nuclei.

splitting of the spin-orbit partners gets bigger. Thus, the energy required to promote one of the particles to the spin-orbit partner level high above the Fermi surface becomes large and thus this coupling is also seldom favored. It should be noted though that these arguments do not apply smoothly with increasing particle number, and when low L-states are near the Fermi surface, then deuteron-like T=0, S=1 object may re-emerge. In the A=60-70 mass region that we are discussing this can happen when the $p_{1/2}$ and $p_{3/2}$ orbits are near the Fermi surface. Macchiavelli [8] and Goodman [5] have both pointed out that the T=1 pairing field is also not monotonic with Fermi-level, becoming weaker when level density is low, like near shell closures. Consequently, the interplay between T=1 and T=0 fields is expected to have a complicated mass dependence.

In conclusion, in the last few years considerable progress has been made in understanding the collective correlations of neutrons and protons. At first it was thought that the high-spin domain was a good place to look for "smoking guns" concerning a T=0 pair field, as it is less susceptible to the Coriolis force [17-19]. However, the experimental evidence [20] has shown that in the mass 60-90 region at least, the nuclear shapes are too unstable and unpredictable to allow the nuclei to be used as "pairing laboratories". Thus, much of the recent work has been focused on complete spectroscopy of low lying levels in odd-odd N=Z nuclei. We have not reached the end of this story, but with much better new data coming along and the hope of full Monte Carlo shell model calculations above ^{56}Ni, we have reached a promising line of attack. The key experiments that we must do are two-nucleon transfer reactions that actually count the numbers of correlated pairs in these nuclei. Unfortunately, these experiments probably will have to await RIA. In the shorter term, it will be very nice to reach T= -1 nuclei and observe their β-decay back to T=1 and T=0 states in odd-odd N=Z nuclei. β-decay studies of this kind are very selective in populating the states of most interest and the experiments are feasible now.

ACKNOWLEDGMENTS

This work was supported in part by the U.S. Department of Energy, Office of Nuclear Physics, under contract W-31-ENG-109-ENG38

REFERENCES

1. R.R. Chasman, Phys. Lett. **B524** (2002) 81
2. R. R. Chasman, Phys. Lett. **B533** (2003) 581
3. R.R. Chasman., Phys. Lett. **B577** (2003) 47
4. A Goodman, Adv. Nucl. Phys. **11** (1979) 263
5. A. Goodman, Phys. Scr. **T88** (2000) 170; Phys. Rev. **C63** (2001) 04325
6. J. Janecke, Nucl. Phys. **A73** (1965) 97
7. P. Vogel., Nucl. Phys. **A662** (2000) 148
8. A. O. Macchiavelli *et al.*, Phys. Rev. **C61** (2000) 041303
9. S. M. Vincent *et al.*, Phys. Lett. **437B** (1998) 264
10. D. Rudolph *et al.*, Phys. Rev. **C69** (2004) 034309
11. R Grzywacz *et al.*, Nucl. Phys. **A682** (2001) 41c
12. G. DeAngelis *et al.*, Private Communication (2003)
13. D. Jenkins *et al.*, Phys. Rev. **C65** (2002) 064307
14. G. DeAngelis *et al.*, Eur. Phys J **A12** (2001) 51
15. D. Rudolph *et al.*, Phys. Rev. Lett. **76** (1976) 376
16. C.D. O'Leary *et al.*, Phys Rev **C67** (2003) 021301
17. W. Satula and R. Wyss, Phys. Rev. Lett. **86** (2001) 4488; **87** (2001) 052504
18. S. Frauendorf and J.A. Sheikh, Phys. Rev. **C59** (1999) 1400; Phys. Scr. **T88** (2000) 162
19. J. Dobaczewski, J Dudek and R Wyss, Phys. Rev. **C67** (2003) 034308
20. S. M. Fischer, C.J. Lister and D. Balamuth, Phys. Rev. **C67** (2003) 064318

Structural Evolution in the (N,Z,I^π) Coordinate Frame for $A{\sim}100$

P.H. Regan[1,2], C. Wheldon[1,3], A.D. Yamamoto[1,2], C.Y. Wu[4], S. Ashley[1],
P. Mumby-Croft[1], D. Seaborne[1], A.O. Macchiavelli[5], D. Cline[4],
J.F. Smith[6], R.S. Chakrawarthy[6], M. Cromaz[5], P. Fallon[5], S.J. Freeman[6],
W. Gelletly[1], A. Görgen[5], A. Hayes[4], H. Hua[4], S.D. Langdown[1,2],
I.Y. Lee[5], C.J. Pearson[1], Zs. Podolyák[1], R. Teng[4], J.J.Valiente-Dobón[1,7]

[1]Department of Physics, University of Surrey, Guildford, GU2 7XH, UK
[2]Wright Nuclear Structure Laboratory, Yale University, 272 Whitney Avenue, New Haven CT USA
[3]SF7, HMI, Glienicker-straße 100, D-14109 Berlin, Germany
[4]Nuclear Structure Research Laboratory, Department of Physics, University of Rochester, Rochester, NY 14627 USA
[5]Nuclear Science Division, Lawrence Berkeley National Laboratory, Berkeley, CA 94720, USA
[6]Department of Physics and Astronomy, The University of Manchester, Manchester M13 9PL, UK
[7]Department of Physics, University of Guelph, Guelph N1H 2W1 Ontario Canada

Abstract. Multinucleon transfer reactions between a thin ^{100}Mo target and a ^{136}Xe beam at an energy of 700 MeV have been used to investigate the evolution of structure with increasing angular momentum in a range of nuclei with $A{\sim}100$. The use of kinematically complete experiments coupled with isomer tagging allow the selection of specific exit channels. The evolution from quasi-vibrational states at low spins to quasi-rotational structures following the population of a low-Ω, 'rotationally aligned' $h_{11/2}$ neutron configuration is demonstrated to be a standard feature of the region using the so-called E-Gamma Over Spin (*E-GOS*) empirical prescription. This treatment is extended to odd-A nuclei.

INTRODUCTION

The question of shape change and structural evolution for near-stable and neutron-rich nuclei around the $Z{=}40$ sub-shell closure has been a major theme in nuclear structure physics for more than 25 years [1,2,3]. The dramatic change in the excitation energy of the first 2^+ states in the $N{=}60$ isotones, ^{98}Sr and ^{100}Zr compared to the neighbouring $N{=}58$ isotones has been explained in terms of (i) the monopole proton-neutron interaction between spin-orbit partner orbitals ($\pi g_{9/2}$ and $\nu g_{7/2}$) [1]; (ii) the role of the prolate deformation favouring intruder $h_{11/2}$ neutron orbital [2]; and (iii) the presence of a deformed shell closure for $N{=}60$ [3]. Figure 1 shows the energy systematics for the $N{=}58$ and 60 even-Z isotopes, highlighting this dramatic effect. Also plotted are the analogous quantities for the decoupled $h_{11/2}$ neutron structure in the odd-A, $N{=}59$ isotonic chain. The aim of the present work is to use heavy-ion binary reactions to populate the medium-to-high spin states of these near-stable nuclei to investigate the role of specific orbitals in determining how the nuclear structure

CP726, Nuclear Physics, Large and Small: International Conference on
Microscopic Studies of Collective Phenomena, edited by R. Bijker, R. F. Casten, and A. Frank
© 2004 American Institute of Physics 0-7354-0207-8/04/$22.00

properties evolve as a function of angular momentum in this region. An off-shoot of the work also provides an insight into the generation of angular momentum in multi-nucleon transfer reactions which is important because of its potential for the study of other predicted nuclear shape phenomena such as superdeformation at high spins in these nuclei [4].

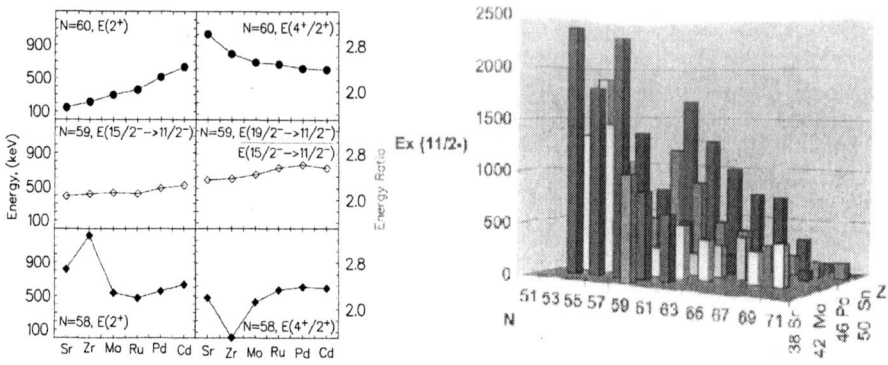

FIGURE 1. (Left) Energy systematics of the even-Z nuclei between Sr and Cd. (Right) Excitation energies for the lowest lying $I^{\pi}=11/2^{-}$ state in the odd-A, even-Z nuclei between Sr and Sn.

EXPERIMENTAL DETAILS, ANALYSIS AND RESULTS

The data reported in the current work were obtained using a heavy-ion binary reaction between a self-supporting 420 $\mu g/cm^2$ ^{100}Mo target and a 700 MeV beam of ^{136}Xe ions, provided by the 88" cyclotron at the Lawrence Berkeley National Laboratory. The binary fragments were detected using the position sensitive gas-filled detector, CHICO [5], which provided the positional information required to allow the Doppler correction to be made for both beam- and target-like fragments (BLF and TLFs respectively). The reaction γ rays were detected in the GAMMASPHERE detector array [6], comprising 102 Compton suppressed hyper-pure germanium detectors. The beam had a natural pulsing period of 64 ns with the time resolution of each pulse being between 2 and 3 ns FWHM. The experimental master gate was set such that at least three 'prompt' γ rays had to be detected within approximately 50 ns of two co-planar binary fragments being detected in CHICO. Delayed γ rays de-exciting isomeric states in fragments which were stopped in the walls of CHICO could also be detected up to 1 μs after the initial master gate. Typical beam currents for the four-day experiment were 1-2 pnA, giving rise to a total of 9×10^8 Compton-suppressed γ-ray triples and higher-fold coincidences being written to tape for subsequent off-line analysis. In this experiment, the hevimet collimators were removed from the BGO suppression shields, enabling a measurement of the total γ-ray multiplicity for each event. Further details of this experiment can be found in references [7,8].

One of the advantages of using this 'thin-target' set-up is that the events are essentially kinematically complete, which allows a clear separation between quasi-elastic (*i.e.* cold transfer and Coulomb excitation) and deep-inelastic events [9]. Figure 2 shows the effect of gating on different angles of the reaction products as measured in CHICO. By selecting only fragments measured at the most backward angles, Coulomb excitation events of the target can be selected. Moving this gating condition forward by only 10^0 into the region expected to cover the grazing angle [9] at which the target-like fragments should be scattered in this reaction (approximately 50^0), a clear difference can be observed, with many more multi-nucleon transfer reactions channels identified. Figure 3 shows the scattered-particle angular distributions and γ-ray total-fold distributions of the target-like fragments gated by specific pairs of transitions in a range of final reaction products

The Coulomb excitation component is evident in the angular distributions for the ^{100}Mo residues up to spins of $8\hbar$. At spins of $10\hbar$ and above the reaction residue distributions are centred around the grazing angle. This can be understood in terms of the change in structure in ^{100}Mo at $10\hbar$ associated with the population of the maximally aligned $(h_{11/2})^2$ neutron configuration [10]. Also perhaps worthy of note is the rather narrow fold and angular distributions associated with the 2n transfer channel to ^{102}Mo. Both the fold and particle-angle distributions gated by the low-spin members of the yrast sequence of this nucleus have much narrower, well-defined distributions compared to the other, multi-nucleon transfer products. Most likely, this reflects to the 'cold' nature of this particular transfer process. The other final products might be expected to be populated following neutron evaporation after the initial multi-nucleon transfer, whereas the ^{102}Mo product is predominantly made by a direct, 'cold' transfer, resulting in narrower angular and fold distributions, compared to the other, more complex reactions channels.

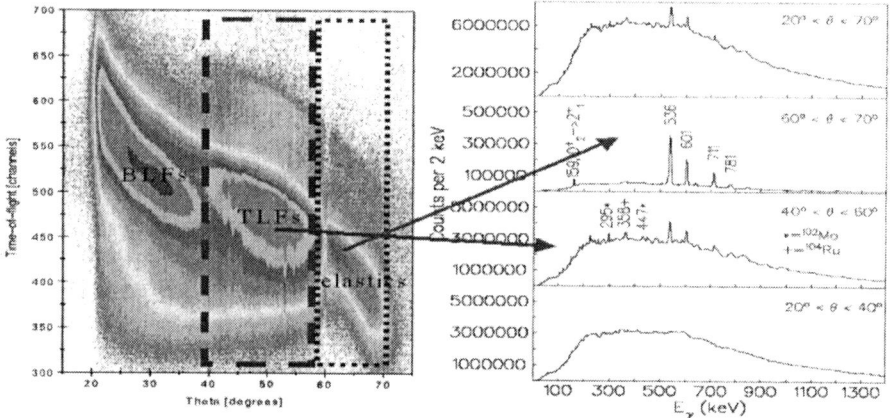

FIGURE 2. (Left) Time-of-flight difference between the two fragments measured in CHICO versus laboratory angle. (Right) Total projections of the γ-ray spectra, Doppler corrected assuming target-like fragments, gated by different particle detection angles in CHICO. Note the clear separation of Coulomb excitation events for more backward angles, with a preponderance of nuclear interaction events (e.g., cold transfer and deep-inelastic products) centred about the grazing angle at 50^0.

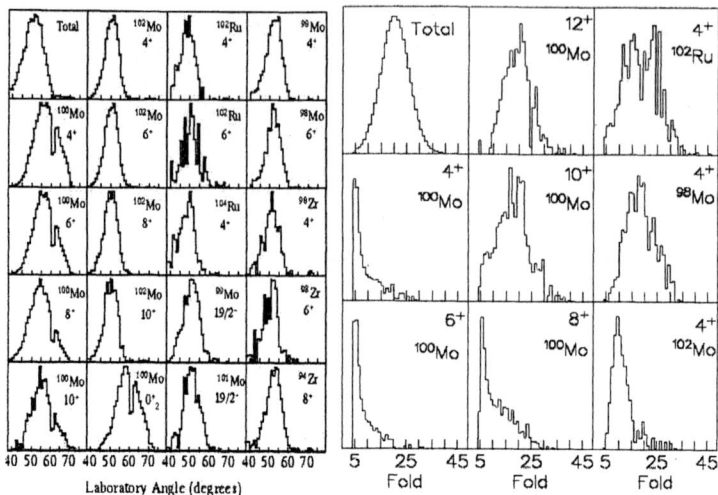

FIGURE 3. (Left) Double-γ-gated projections of charged-particle angular distributions and (right) measured γ-ray folds for different spins in target-like fragments in the ^{136}Xe+^{100}Mo reaction at a ^{136}Xe beam energy of 700 MeV.

Figure 4 shows examples of the spectra for quasi-rotational states obtained in the current work. The decoupled structure in ^{101}Mo was obtained by gating on the delayed transitions below the $I^{\pi}=11/2^{-}$ isomeric state.

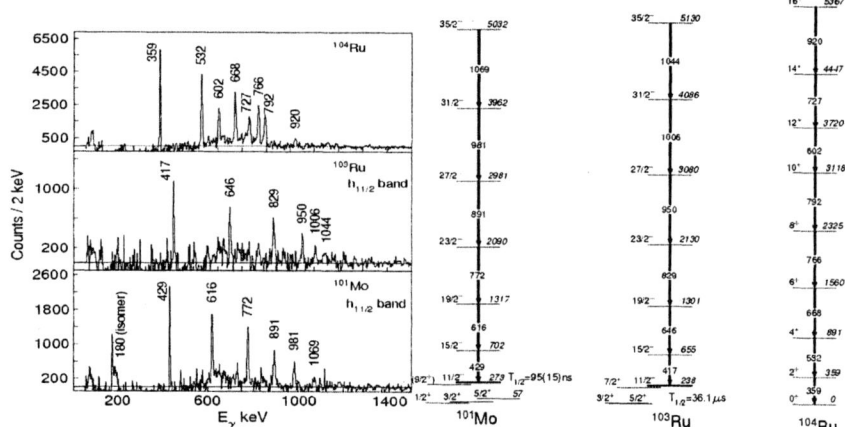

FIGURE 4. Partial decay schemes and double-gated γ-ray spectra (*i.e.* triple and higher coincidences) showing the $h_{11/2}$ decoupled bands in ^{101}Mo and ^{103}Ru and the yrast sequence in ^{104}Ru.

E-GOS Plots for Even- and Odd-A Structures

The recently reported 'E-Gamma Over Spin' or *E-GOS* prescription (see refs. [11,12]) provides an intuitive mechanism for viewing changes in the mechanism for angular momentum generation in even-even transitional nuclei by taking the quotient of the γ-ray transition energy between yrast states which differ in spins by $2\hbar$ and dividing this by the spin value of the higher-lying level. This empirical treatment highlights the transition between quasi-vibrational sequences giving way to statically deformed rotational sequences associated with the population of a pair of low-Ω $h_{11/2}$ neutron orbitals maximally coupled to an angular momentum of $10\hbar$ [11].

In trying to extend the *E-GOS* picture to odd-*A* nuclei, a simple, weak-coupling limit has been previously assumed (*e.g.*, see ref. [12]), with the *E-GOS* ratio obtained by dividing the transition energy by the spin of the state minus the band-head spin, *i.e.*, $R(I \rightarrow I-2) = \dfrac{E_\gamma(I \rightarrow I-2)}{I-j}$. While this seemed to adequately describe the vibrator-to-rotor transition associated with the $h_{11/2}$ neutron alignment in the positive parity bands in the odd-A systems [12,13], this analysis also suggested that the $h_{11/2}$ negative parity decoupled structures were quasi-vibrational. This was surprising since it is the population of these specific 'down-sloping', prolate-favouring orbitals which have been suggested to be the underlying cause of the vibrator-to-rotor phase change in this region [10].

Wheldon [14] has suggested the following extension of the *E-GOS* prescription which renormalizes the rotational energies at the bandhead (for bands with non-zero bandhead spins, j). This is necessary for odd-*A* nuclei and $j \neq 0$ bands to correctly identify the vibrator→rotor transition by preventing the *E-GOS* ratio from becoming unphysically large at low spins, simply due to the higher transition energies compared to an equivalent $j=0$ γ-ray energy at $(I-j)\hbar$. For odd-*A* cases, a constant dependent on the bandhead spin must be subtracted. This depends on the *E-GOS* ratio of the first stretched *E2* decay in the band. If the bandhead spin is given by j then by substituting $I-j$ in place of I, one obtains,

$$R(I) = \frac{E_\gamma}{I} = \frac{\hbar^2}{2\Im}\frac{(4I-2)}{(I)} \xrightarrow{I \to I-j} \frac{\hbar^2}{2\Im}\frac{[4(I-j)-2]}{(I-j)} = \frac{E_\gamma - \left(4j\dfrac{\hbar^2}{2\Im}\right)}{I-j} = R(I-j)$$

$$\text{This can be simplified by,} \left(4\frac{\hbar^2}{2\Im}\right) \approx \frac{E_\gamma(I=j+2 \to j)}{I(=j+2)} = R(I=K+2) = R_{j+2}$$

$$\therefore R_{E-GOS} = \frac{E_\gamma - jR_{j+2}}{I-j} \quad \text{which reduces to } \frac{E_\gamma}{I} \text{ for the (even-even) } j = 0 \text{ case.}$$

This ratio tends towards a constant value for good rotors and has a hyperbolic shape for vibrators. The difference between these two extremes is shown in Fig.5.

Future work in this region includes plans to study the higher-spin regime around [100]Mo using a heavier ion-induced (Pb, Bi or U) binary reaction. This should provide sufficient input spin to populate the angular momentum range where a superdeformed ($\beta_2 \sim 0.4$) configuration in this nucleus is predicted to approach the yrast line [4,7]. Rolling-mode estimates suggest that for beam energies approximately 40-50% above the Coulomb barrier, studies of spins approaching $40\hbar$ might be possible.

161

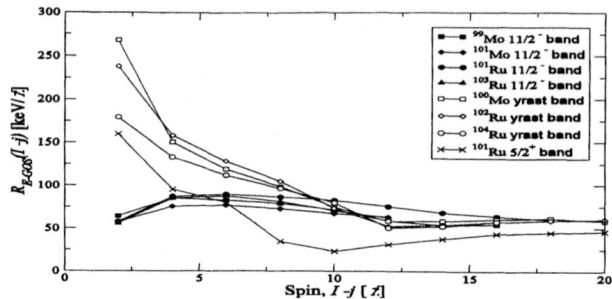

FIGURE 5: *E-GOS* plots for the yrast sequences in even-even *N*=58 and 60 isotones and the $h_{11/2}$ decoupled bands in ^{99,101}Mo and ^{101,103}Ru [7,8,13]. A change from vibrator to rotor is evident above 10$\hbar$ for the even-even isotopes. A bandhead spin dependent constant has been subtracted for the odd-A cases as described in the text.

ACKNOWLEDGMENTS

This work is supported by EPSRC (UK), the US Department of Energy under grants DE-FG02-91-ER-40609 and DE-AC03-76SF00098 and by the National Science Foundation. PHR acknowledges support from the Yale University Flint and Science Development funds and discussions with R.F.Casten, N.V.Zamfir, C.W. Beausang and J.Y. Zhang with regards to the *E-GOS* empirical formalism.

REFERENCES

1. Federman, P. and Pittel, S., *Phys. Lett.* **77B**, 28 (1978); *Phys. Rev.* **C20**, 820 (1979); Federman, P. Pittel, S. and Campos, R., *Phys. Lett.* **82B**, 9 (1979); Federman, P., Pittel, S. and Etchegoyen, A., *Phys. Lett.* **140B**, 269 (1984); Lhersonneau, G., et al., *Phys. Rev.* **C49**, 1370 (1994)
2. Urban, W. *et al.*, *Nucl. Phys.* **A689**, 605 (2001)
3. Hotchkis, M., *et al.*, *Phys. Rev. Lett.* **64**, 3123 (1990); *Nucl. Phys.* **A530**, 111 (1991)
4. Skalski, J., Mizutori, S., and Nazarewicz, W., *Nucl. Phys.* **A617** , 282 (1997)
5. Simon, M.W., *et al.*, *Nucl. Inst. Meth. Phys. Res.* **A452**, 205 (2000)
6. Lee, I.Y., *Nucl. Phys.* **A520**, 641c (1990)
7. Regan, P.H., *et al.*, Phys. Rev. **C68**, 044313 (2003)
8. Regan, P.H., *et al.*, "Studies Around A~100 Using Binary Reactions" in *International Conference on the Labyrinth of Nuclear Structure* edited by C.Kalfas, AIP Conference Proceedings **701**, New York: American Institute of Physics. 2004, pp 329-333
9. Bock, R., *et al.*, Nukleonika **22**, 529 (1977)
10. Regan, P.H. *et al.*, Phys. Rev. **C55**, 2305 (1997)
11. Regan, P.H., *et al.*, Phys. Rev. Lett. **90**, 152502 (2003)
12. Regan, P.H., *et al.*, "The Highs and Lows of the A=100 Region: Vibration-to-Rotation Evolution in Mo and Ru Isotopes" in *Frontiers of Nuclear Structure* edited by P.Fallon and R.Clark, AIP Conferences Proceedings **656** , New York, American Institute of Physics, 2003, pp422-428
13. Yamamoto, A.D., et al., *Phys. Rev.* **C66**, 024302 (2002)
14. Wheldon, C., unpublished

Exactly Solvable Pairing Plus Mean Field Model

Feng Pan*[†], V. G. Gueorguiev[1][†] and J. P. Draayer[†]

*Department of Physics, Liaoning Normal University, Dalian, 116029, P. R. China
[†]Department of Physics and Astronomy, Louisiana State University, Baton Rouge, LA 70803

Abstract. A mean-field plus extended pairing interaction Hamiltonian that includes multi-pair scattering terms is proposed for describing well-deformed nuclei. Eigenvalues of the model are easily obtained. The investigation shows that even through the extended pairing includes many-body terms, the one- and two-body interactions continue to dominate the dynamics for relatively small values of the pairing strength. As the strength of the pairing interaction grows, however, the higher many-body interaction terms grow in importance. A numerical study of even-odd mass differences in the $^{154-171}$Yb isotopes demonstrates the applicability of the theory.

Pairing is a important interaction that is widely used in nuclear physics and other fields. And it is one that Stuart Pittel, whose 60th birthday we celebrate, along with his collaborators have added so much to by articulating its microscopic underpinnings as well as macroscopic outpinnings [1]. Through this contribution, my colleagues and I wish to step into that world and offer what we believe is a heretofore unknown solution of an extended pairing model — so, Stu, in what follows we present to you and your colleagues an 'algebraic solution' to an extended version of the pairing problem that has been so central to your work in nuclear physics [2].

The Bardeen-Cooper-Schrieffer (BCS) [3] and Hartree-Fock-Bogolyubov (HFB) [4] methods for finding approximate solutions are well known. However, the limitations of BCS methods, when applied in nuclear physics, are also well known. First of all, not only is the number of nucleons in a nucleus typically small, the number of valence particles ($n \sim 10$) which dominate the behavior of low-lying states is far too few to support the underlying assumptions of the approximations, that is, particle number fluctuations are non-negligible. As a result, particle number-nonconservation effects can lead to serious problems such as spurious states, nonorthogonal solutions, and so on. Furthermore, an essential feature of pairing correlations are differences between neighboring even and odd mass nuclei, which are driven mainly by Pauli blocking effects, and it is difficult to treat even-odd differences with either the BCS or HFB theories because different quasi-

---

[1] On leave from Institute of Nuclear Research and Nuclear Energy, Bulgarian Academy of Sciences, Sofia 1784, Bulgaria.

CP726, *Nuclear Physics, Large and Small: International Conference on Microscopic Studies of Collective Phenomena*, edited by R. Bijker, R. F. Casten, and A. Frank
© 2004 American Institute of Physics 0-7354-0207-8/04/$22.00

particle bases must be introduced for different blocked levels. Another problem with approximate treatments of the pairing Hamiltonian is related to the fact that both the BCS and the HFB approximations break down for an important class of physical situations. A remedy that uses particle number projection techniques complicates the algorithms considerably and does not help to achieve a better description of the higher-excited part of the spectrum of the pairing Hamiltonian. The importance of having exact solutions of the pairing Hamiltonian has driven a great deal of work in recent years. In particular, building on Richardson's early work [5, 6, 7] and extensions to it based on the Bethe ansatz, several authors have introduced novel approaches [8, 9, 10, 11, 12, 13, 14]. For the algebraic approaches based on the Bethe ansatz, the solutions are provided by a set of highly non-linear Bethe Ansatz Equations (BAEs). Though these applications demonstrate that the pairing problem is exactly solvable, solutions are not easily obtained and normally require extensive numerical work, especially when the number of levels and valence pairs are large. This limits the applicability of the methodology to relatively small systems; for example, it cannot be applied to large systems such as well-deformed nuclei.

The standard pairing Hamiltonian for well-deformed nuclei is given by

$$\hat{H} = \sum_{j=1}^{p} \epsilon_j n_j - G \sum_{i,j=1}^{p} a_i^+ a_j, \tag{1}$$

where p is the total number of Nilsson levels, $G > 0$ is the pairing strength, ϵ_j is single-particle energies taken from the Nilsson model, $n_j = c_{j\uparrow}^\dagger c_{j\uparrow} + c_{j\downarrow}^\dagger c_{j\downarrow}$ is the fermion number operator for the j-th Nilsson level, and $a_i^+ = c_{i\uparrow}^\dagger c_{i\downarrow}^\dagger$ $(a_i = (a_i^+)^\dagger = c_{i\downarrow} c_{i\uparrow})$ are pair creation (annihilation) operators. The up and down arrows in these expressions denote time-reversed states. Since each level can only be occupied by one pair due to the Pauli Principle, the Hamiltonian (1) is also equivalent to a finite site hard-core Bose-Hubbard model with infinite range one-pair hopping and infinite on-site repulsion. Specifically, the operators a_i^+, a_i, and $n_i^a = n_i/2$ satisfy the following hard-core boson algebra:

$$(a_i^+)^2 = 0, \quad [a_i, a_j^+] = \delta_{ij}(1 - 2n_i^a), \quad [a_i^+, a_j^+] = [a_i, a_j] = 0. \tag{2}$$

As an extension of the usual approach (1), we construct the following new (extended) pairing Hamiltonian:

$$\hat{H} = \sum_{j=1}^{p} \epsilon_j n_j - G \sum_{i,j=1}^{p} a_i^+ a_j$$

$$-G \left(\sum_{\mu=2}^{\infty} \frac{1}{(\mu!)^2} \sum_{i_1 \neq i_2 \neq \cdots \neq i_{2\mu}} a_{i_1}^+ a_{i_2}^+ \cdots a_{i_\mu}^+ a_{i_{\mu+1}} a_{i_{\mu+2}} \cdots a_{i_{2\mu}} \right), \tag{3}$$

where no pair of indices among the $\{i_1, i_2, \cdots, i_{2\mu}\}$ are the same for any μ. Besides the usual mean-field and the standard pairing interaction (1), this form includes many-pair hopping terms that allow nucleon pairs to simultaneously scatter (hop) between and among different levels. With this extension, we will show that the model is exactly solvable.

Because of infinite on-site repulsion, the sum in (3) truncates for $\mu \leq [p/2]$, where $[x]$ is the integer part of x. It is also easy to see that each term of the form $a_i^+ \cdots a_j^+$ that enters into eigenstates of (3) should come with different indices $i \neq \cdots \neq j$. Let $|j_1, \cdots, j_m\rangle$ be the pairing vacuum state that satisfies

$$a_i|j_1, \cdots, j_m\rangle = 0 \qquad (4)$$

for $1 \leq i \leq p$, where $j_1, j_2, \cdots, j_m$ indicates those m levels that are occupied by single nucleons. Any singly-occupied state is blocked by the Pauli principle.

The algebraic ansatz introduced in [15] suggests that the k-pair eigenstates of (3) can be written as

$$|k; \zeta; j_1, \cdots, j_m\rangle = \sum_{1 \leq i_1 < i_2 < \cdots < i_k \leq p} C_{i_1 i_2 \cdots i_k}^{(\zeta)} a_{i_1}^+ a_{i_2}^+ \cdots a_{i_k}^+ |j_1, \cdots, j_m\rangle, \qquad (5)$$

where $C_{i_1 i_2 \cdots i_k}^{(\zeta)}$ is an expansion coefficient that must be determined, and the strict ordering to the indices $i_1, i_2, \cdots, i_k$ is a reminder that double occupation is not allowed. It is always assumed that the level indices $j_1, j_2, \cdots, j_m$ should be excluded from the summation in (5). Since the formalism for even-odd systems is similar, in what follows we focus on the even-even seniority zero case.

The expansion coefficient $C_{i_1 i_2 \cdots i_k}^{(\zeta)}$ can be expressed very simply as

$$C_{i_1 i_2 \cdots i_k}^{(\zeta)} = \frac{1}{1 - x^{(\zeta)} \sum_{\mu=1}^{k} \epsilon_{i_\mu}}, \qquad (6)$$

where, similar to the results given in the Bethe ansatz approach, $x^{(\zeta)}$ is a c-number that is to be determined. To prove that the algebraic ansatz given in (5) and (6) are consistent, one may directly apply Hamiltonian (3) on the k-pair state (5). Using the hard-core boson algebraic relation given by (2) and a procedure that is similar to that used in Ref. [10] for finding exact solution of a Heisenberg algebra Hamiltonian, one can determine that for the mean-field part of the Hamiltonian (3)

$$\sum_j \epsilon_j n_j |k; \zeta; 0\rangle = \frac{2}{x^{(\zeta)}} \left(|k; \zeta; 0\rangle - \sum_{1 \leq i_1 < i_2 < \cdots < i_k \leq p} a_{i_1}^+ a_{i_2}^+ \cdots a_{i_k}^+ |0\rangle \right), \qquad (7)$$

and for the re-arranged extended pairing part of the Hamiltonian (3)

$$\left(\sum_i a_i^+ a_i + \sum_{\mu=1}^{\infty} \frac{1}{(\mu!)^2} \sum_{i_1 \neq i_2 \neq \cdots \neq i_{2\mu}} a_{i_1}^+ a_{i_2}^+ \cdots a_{i_\mu}^+ a_{i_{\mu+1}} a_{i_{\mu+2}} \cdots a_{i_{2\mu}} \right) |k; \zeta; 0\rangle =$$

$$\left(\sum_{1 \leq i_1 < i_2 < \cdots < i_k \leq p} C_{i_1 i_2 \cdots i_k}^{(\zeta)} \right) \sum_{1 \leq i_1 < i_2 < \cdots < i_k \leq p} a_{i_1}^+ a_{i_2}^+ \cdots a_{i_k}^+ |0\rangle + (k-1)|k; \zeta; 0\rangle \qquad (8)$$

By combining Eqs. (7) and (8), the k-pair excitation energies of (3) are given by:

$$E_k^{(\zeta)} = \frac{2}{x^{(\zeta)}} - G(k-1), \tag{9}$$

where the undetermined variable $x^{(\zeta)}$ is given by

$$\frac{2}{x^{(\zeta)}} + \sum_{1 \leq i_1 < i_2 < \cdots < i_k \leq p} \frac{G}{\left(1 - x^{(\zeta)} \sum_{\mu=1}^{k} \epsilon_{i_\mu}\right)} = 0. \tag{10}$$

The additional quantum number ζ now can be understood as the ζ-th solution of (10). Similar results for even-odd systems can also be derived by using this approach except that the index j of the level occupied by the single nucleon should be excluded from the summation in (5) and the single-particle energy term ϵ_j contributing to the eigenenergy from the first term of (3) should be included. Extensions to many broken-pair cases are straightforward.

Comparing Eqs. (9) and (10) to exact solutions of the Heisenberg algebraic Hamiltonian with a one-body interaction [10], one can consider the operator product $a_{i_1}^+ a_{i_2}^+ \cdots a_{i_k}^+$ in (5) as a 'grand' boson. The corresponding 'single-particle energy' of this 'grand' boson is $E_{i_1 i_2 \ldots i_k} = \sum_{\mu=1}^{k} 2\epsilon_{i_\mu}$, since (10) and the eigenstates (5) are similar to those for a multi-boson system with a one-body interaction as shown in [10], even though the Hamiltonians are totally different. It should be noted that even through all of the eigenstates (5) with distinct roots given by (10) are orthogonal [10], they are not normalized (5) but can be made so using standard procedures.

Eigenenergies of the standard pairing model can be expressed in terms of k variables that satisfy k coupled nonlinear equations. Such a system is very difficult to solve numerically, especially when the number of pairs k and number of levels p are large. However, in contrast to a BAE solution of the standard pairing model, for the extended pairing model there is but a single variable $x^{(\zeta)}$. It should be noted that the solution (10) requires the single-particle energies $\sum_{\mu=1}^{k} \epsilon_{i_\mu}$ to be all different. Fortunately, this is the case when the single-particle energies are generated from typical mean fields such as, for example, the Nilsson potential. When this holds, (10) has $\frac{p!}{(p-k)!k!}$ distinct roots, which could be a large number, for example, for an entire deformed major shell.

If (10) is rewritten in terms of a new variable $z^{(\zeta)} = 2/(Gx^{(\zeta)})$ and the dimensionless energy of the 'grand' boson $\tilde{E}_{i_1 i_2 \ldots i_k} = \sum_{\mu=1}^{k} \frac{2\epsilon_{i_\mu}}{G}$, (10) takes the form:

$$1 = \sum_{1 \leq i_1 < i_2 < \cdots < i_k \leq p} \frac{1}{\left(\tilde{E}_{i_1 i_2 \ldots i_k} - z^{(\zeta)}\right)}. \tag{11}$$

Since there is only a single variable $z^{(\zeta)}$ in (11), the zero points of the function can be determined graphically, in a manner that is similar to the one-pair solution of the TDA and RPA approximations with separable potentials[4]. From expression (11), it is clear that any solution $z^{(\zeta)}$ of (11) is located between two nearby values of the dimensionless 'grand' boson energy $\tilde{E}_{i_1 i_2 \ldots i_k} = \sum_{\mu=1}^{k} \frac{2\epsilon_{i_\mu}}{G}$ and the smallest solution $z^{(\zeta)}$ would be smaller than $\min(\{\tilde{E}_{i_1 i_2 \ldots i_k}\})$.

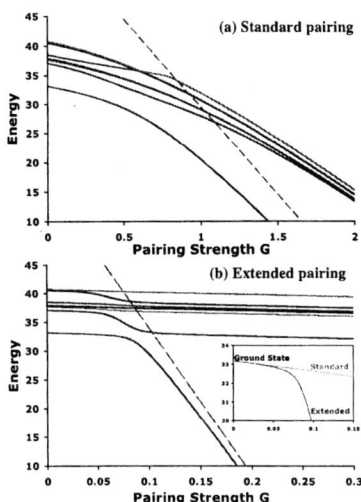

FIGURE 1. (a) Spectral structure of the standard pairing interaction given by Eq.(1), and (b) spectral structure of the extended pairing interaction given by Eq.(3), as functions of the pairing interaction strength G for $k = 5$ pairs for a system with $p = 10$ levels and single-particle energies $\epsilon_1 = 1.179$, $\epsilon_2 = 2.65$, $\epsilon_3 = 3.162$, $\epsilon_4 = 4.588$, $\epsilon_5 = 5.006$, $\epsilon_6 = 6.969$, $\epsilon_7 = 7.262$, $\epsilon_8 = 8.687$, $\epsilon_9 = 9.899$, $\epsilon_{10} = 10.201$, where the single-particle energies and G are given in arbitrary units. The straight dashed line is the expectation value of the Hamiltonian in the pure pairing ($\epsilon_i = 0$) ground state.

It is important to understand the differences between the extended pairing introduced by (3) and the standard pairing given in (1). For this purpose, we consider a simple example in which there are $p = 10$ levels and the single-particle energies are given by $\epsilon_i = i + \chi_i$ for $i = 1, 2, \cdots, 10$, where χ_i are random numbers within the interval $(0, 1)$ to avoid accidental degeneracy required for exact solvability, and the pairing strength G is allowed to vary from 0.01 to 0.10. Fig. 1 shows the lowest few energies of the standard and extended pairing models. From this graph it is very clear that there are essential differences in the spectral structure of these two models. As shown by Inset (b), the extended pairing model very rapidly develops a paired ground-state configuration which is strongly dependent on the pairing strength G. In this case the transition from mean-field eigenstates to pairing eigenstates is very sharp and fast, while the standard pairing model, Inset (a), exhibits a much slower and smoother transition. The quantitative difference in the two spectra, with the extended pairing case showing a much stronger dependence on G than for standard pairing, is a very clear distinguishing characteristic and can be used to explore cases where the extended pairing concept might be more relevant and appropriate than the standard pairing model.

It is important to know whether the dynamics is still dominated by the one- and two-body interactions, and, if not, under what conditions the higher order terms

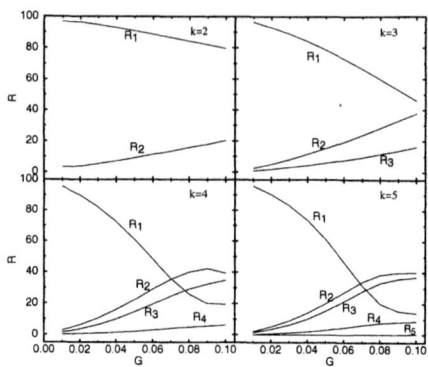

FIGURE 2. Ratios $R_\mu(\%)$ with $\mu = 1, 2, \cdots, 5$ as function of the pairing interaction strength G for $k = 2, \cdots, 5$ for a system with $p = 10$ levels. The parameters that were used are the same as those shown in Fig. 1.

can be treated perturbatively. To explore this, we calculated as a function of G the expectation value of each higher order term $\langle V_\mu \rangle$ defined by

$$V_1 = \sum_{i,j} a_i^+ a_j, \quad V_\mu = \frac{1}{(\mu!)^2} \sum_{i_1 \neq i_2 \neq \cdots \neq i_{2\mu}} a_{i_1}^+ a_{i_2}^+ \cdots a_{i_\mu}^+ a_{i_{\mu+1}} a_{i_{\mu+2}} \cdots a_{i_{2\mu}} \quad (12)$$

with $\mu = 2, 3, \cdots$, for k-pair ground states. Then, we calculate the ratio $R_\mu = \langle V_\mu \rangle / \langle V_{\text{total}} \rangle$, where $\langle V_{\text{total}} \rangle$ is the sum of all terms given in (12). The results up to the half-filled case are shown in Fig. 2. It can be seen that the two-body pairing interaction (V_1) dominates the dynamics of the system as long as the interaction strength G is small. With increasing interaction strength, the system is driven mainly by V_2, less by V_3 (but it may be comparable with V_2), and much less by the higher order terms. As one would expect, increasing the number of pairs k drives the critical point where V_2 becomes dominant towards smaller G values. The situation and the graphs beyond the half-filled case are qualitatively similar because of the particle-hole symmetry. The critical point where V_2 becomes dominant is actually at higher G values for $p - k$ pairs than for the $k < p/2$ pairs. Even though higher order terms appear beyond the half-filled case, these terms are always less important than V_1 for small G and V_2 for big values of G. It is important to note that for large values of G when the dynamics is dominated by the pairing interaction, and thus independent of G, the V_2 term dominates followed in importance by the V_3 term.

As an example of an application of the theory to well-deformed nuclei, we fit even-odd mass differences of the $^{154-171}$Yb isotopes. The single-particle energies of each nucleus were calculated using the Nilsson deformed shell model with experimentally determined deformation parameters [16]. Fig. 3 shows the theoretical fit in comparison with the corresponding experimental values[17]. Except for small deviations for $^{157-161}$Yb, the experimental results are well reproduced. The deviations

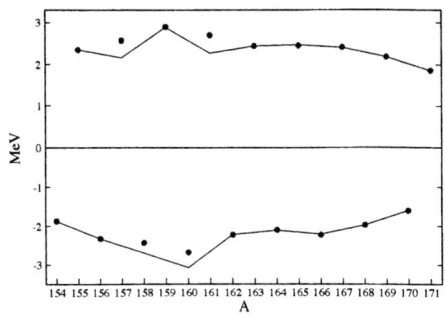

FIGURE 3. Even-odd mass difference $P(A) = E(A) + E(A-2) - 2E(A-1)$ for $^{154-171}$Yb, where $E(A)$ is the total binding energy, and the dots correspond to the experimental numbers. The theoretical values for even-odd mass $P(A)$ are connected by the lines.

TABLE 1. Pairing interaction strength $|G|$ (keV) used in Fig. 3 for the $^{154-171}$Yb isotopes.

	^{154}Yb	^{156}Yb	^{158}Yb	^{160}Yb	^{162}Yb	^{164}Yb	^{166}Yb	^{168}Yb	^{170}Yb
k	1	2	3	4	5	6	7	8	9
G	245	41.1	6.5	3.02	1.11145	0.4675	0.2337	0.1376	0.094816

	^{155}Yb	^{157}Yb	^{159}Yb	^{161}Yb	^{163}Yb	^{165}Yb	^{167}Yb	^{169}Yb	^{171}Yb
k	1	2	3	4	5	6	7	8	9
G	270.7	42.0	10.9	1.0	1.471	0.649	0.3477	0.2185	0.16113

for $^{157-161}$Yb can be traced to the fact that there are singular points in Eq. (10) when $x^{(\varsigma)}$ approaches the value $1/\sum_{\mu=1}^{k} \epsilon_{i_\mu}$. So although the results are quite good, we do not get a perfect fit for these nuclei with the Nilsson single-particle-energies provided by [17]. One could also consider an empirical expression for the pairing strength G as a function of the number of valence pairs, but that goes beyond the purpose of the present paper and such enhancements will therefore be considered elsewhere.

The corresponding G values are given in Table 1, from which one can see that the pairing interaction strength decreases with increasing number of pairs k from 245keV for 1 pair to 0.0948keV for 9 pairs, while the single-particle energy gaps are always about a few hundreds keV. This situation is characteristic of the extended pairing model. However, since the strengths of the many-pair terms in (3) are the same as the one-pair term, the model has little flexibility. The results suggest that the extended pairing model may be applicable to well-deformed nuclei and in other physical systems where pairing plays an important role.

ACKNOWLEDGMENTS

Support from the U.S. National Science Foundation (0140300), the Natural Science Foundation of China (10175031), and the Education Department of Liaoning Province (202122024) is acknowledged.

REFERENCES

1. P. Federman and S. Pittel, Phys. Rev. C **20**, 820 (1979).
2. Feng Pan, V. G. Gueorguiev, and J. P. Draayer, Phys. Rev. Lett. **92**, 112503 (2004).
3. J. Bardeen, L. N. Cooper, and J. R. Schrieffer, Phys. Rev. **108**, 1175 (1957).
4. P. Ring and P. Schuck, *The Nuclear Many-Body Problem* (Springer Verleg, 1980, Berlin).
5. R. W. Richardson, Phys. Lett. **3**, 277 (1963).
6. R. W. Richardson, Phys. Lett. **5**, 82 (1963).
7. R. W. Richardson and N. Sherman, Nucl. Phys. **52**, 221 (1964).
8. Feng Pan, J. P. Draayer, and W. E. Ormand, Phys. Lett. **B422**, 1 (1998).
9. Feng Pan and J. P. Draayer, Phys. Lett. **B442**, 7 (1998).
10. Feng Pan and J. P. Draayer, Ann. Phys. (NY) **271**, 120 (1999).
11. Feng Pan, J. P. Draayer, and Lu Guo, J. Phys. A: Math. Gen. **33**, 1597 (2000).
12. J. Dukelsky, C. Esebbag, and P. Schuck, Phys. Rev. Lett. **87**, 066403 (2001).
13. J. Dukelsky, C. Esebbag, and S. Pittel, Phys. Rev. Lett. **88**, 062501 (2002).
14. H. -Q. Zhou, J. Links, R. H. McKenzie, and M. D. Gould, Phys. Rev. B **65**, 060502(R) (2002).
15. Feng Pan and J. P. Draayer, J. Phys. A: Math. Gen. **33**, 9095 (2000).
16. J. R. Nix and K. L. Kratz, Atomic Data Nucl. Data Tables **66**, 131 (1997).
17. G. Audi et al., Nucl. Phys. **A624**, 1 (1997).

Nuclear Astrophysics and Nuclear Structure

Ani Aprahamian

Institute for Structure and Nuclear Astrophysics
University of Notre Dame, Notre Dame, IN 46556 USA

Abstract. We explore the impact of nuclear structure on nucleosythesis processes via the nuclear mass.

It was my pleasure to be at Hacienda Cocoyoc for **StuFiesta** honoring the 60th Birthday of Stuart Pittel. I met Stuart at Brookhaven National Laboratory when I was just starting my graduate studies in various aspects of Nuclear Structure. Today, my research interests have drifted to nuclear astrophysics looking at the impact of nuclear structure on stellar processes. One of the most important goals of nuclear astrophysics is the attempt to understand nucleosynthesis processes that take place in the cosmos by the simulation of various astrophysical scenarios. These scenarios are strongly dependent on nuclear structure which sets the time scale for the stellar processes from giga-years of stellar evolution to milli-seconds of stellar explosions. In each case, they leave signatures in stellar luminosities, elemental and/or isotopic abundances, and neutrino fluxes from distant supernovae.

One of the most basic nuclear structure effects is the nuclear mass. There have been a number of excellent reviews recently on Nuclear Masses [1] reporting on the status of present day mass models. We also heard from Jorge Hirsch at StuFiesta that one cans set bounds on the chaotic motion of nuclei and possibly allow the prediction of nuclear masses with higher accuracy. Masses or mass-differences in the form of reaction Q-values come in exponentially into reaction rates. Mass differences determine the energy generation in nucleosynthesis processes as well as the decay channels which are open for compound states and determine reaction branchings that affect the reaction path. The latter part naturally guides the abundance patterns that result from a specific nucleosynthesis scenario. An example of this is the waiting point phenomenon. Whether we speak about a thermonuclear runaway ignited in the high temperature and density conditions of an accretion disk that may be a part of the rp-process (rapid sequential proton capture reactions) or the high neutron densities and temperatures that characterize the r-process (rapid sequential neutron capture reactions) thought to occur on the shockfronts of core collapse supernovae.

In the rp-process, the p-capture reactions are thought to drive the nucleosynthesis path towards the proton drip-line passing through the N=Z nuclei beyond ^{56}Ni. The Q-values for p-capture decrease significantly and sometimes become negative as N=Z nuclei are approached. This causes an enrichment in the N=Z nuclei as an equilibrium is established between (p,γ)-(γ,p) reactions. The N=Z nucleus becomes a "waiting

CP726, Nuclear Physics, Large and Small: International Conference on
Microscopic Studies of Collective Phenomena, edited by R. Bijker, R. F. Casten, and A. Frank
© 2004 American Institute of Physics 0-7354-0207-8/04/$22.00

point" and the reaction flow is delayed by the beta-decay half-life of the N=Z nucleus unless 2p-capture reactions become feasible enabling the nucleosynthesis to jump over the waiting-point. One of the waiting points for the rp-process occurs at N=Z=34, ^{68}Se. The mass of ^{68}Se is particularly important in determining whether further p-capture is likely to occur within the rp-process. The Q value for proton capture on ^{68}Se leading to ^{69}Br that can be predicted from various mass models differs from a positive +70keV [2] to a maximum of -730keV [3] with the Audi extrapolations lying somewhere in between. There have been several attempts to search for ^{69}Br by fragmentation studies at GANIL and MSU. No ^{69}Br was observed but an upper limit of 150ns was set for the half-life indicating that ^{69}Br may be p-unbound. The only way that the rp-process can proceed is via 2p-capture or β-decay of ^{68}Se. The mass of ^{68}Se was measured [4] via the β–decay endpoint. ^{68}Se was produced by the ^{12}C(^{58}Ni,2n) ^{68}Se reaction and subsequently implanted onto a moving tape system using the Fragment Mass Analyzer at the ATLAS facility of Argonne National Laboratory. A mass excess value of (-54189+/-240) keV was determined from the b-endpoint measurement of Q_{EC}=4710(200) keV. Evaluations of proton separation energies based on the measured mass were used in a one zone type I x-ray burst model. It is concluded, that 2p-capture reactions are not very likely to play a significant role at N=Z=34 isotope and that ^{68}Se remains a waiting point for rp-process nucleosynthesis. While we have shown that the likelihood of jumping over the waiting point via 2p-capture reactions is low in this case, there is still the possibility that a long lived isomeric state in ^{68}Se can itself capture protons decreasing the reaction Q value by the excitation energy of the isomer and still provide a path for the rp-process. If the isomeric state is at approximately 1.5 MeV in excitation, the Q value changes and p-capture on ^{68}Se becomes feasible. To date, there is no information on the existence of any isomeric states in the spectrum of ^{68}Se.

I conclude by emphasizing that in addition to describing global properties of the nucleus via nuclear structure, we can also see the signatures of nuclear structure transformed into time scales of stellar events, abundance distributions of various scenarios, as well as the energy production from both steady and explosive nucleosynthesis processes.

Acknowledgments

This work was supported by the National Science Foundation under contract PHY-0140324 and the Joint Institute of Nuclear Astrophysics under contract PHY02-16783.

REFERENCES

1.Lunney,D. Pearson,J.M., Thibault,C, *Reviews of Modern Physics 75, (2003)* 1021
2. Aprahamian,A. Gadala-Maria,A. and Cuka,N., *Rev. Mex. Fis. 42 (1996)* 1
3. Brown,B.A. et al., *Phys. Rev. C 65 (2002)* 045802
4. Woehr,A. et al., *Nucl. Phys. A (2004)* in press.

An Alternate Mechanism for E0 Transitions in Transitional and Deformed Nuclei

R.F. Casten[1,2], P. von Brentano[2], V. Werner[2], C. Scholl[2],
E.A. McCutchan[1], R. Krücken[3] and J. Jolie[2]

[1]*Wright Nuclear Structure Laboratory, Yale University, New Haven Connecticut 06520-8124, USA*
[2]*Institut für Kernphysik, Universität zu Köln, Köln, GERMANY*
[3]*Physik Department E12, Technische Universität München, 85748 Garching, GERMANY*

Abstract. It is shown that the simple IBA–1, which acts in a single space, robustly predicts a sharp rise in $\rho^2(E0; 0_2^+ \to 0_1^+)$ values in spherical–deformed transition regions, in agreement with the data. These predictions are effectively parameter–free and provide an alternative to the usual coexistence model for these transitions.

INTRODUCTION

E0 transitions, though much less studied than E2 deexcitations, are a sensitive measure of nuclei shapes and radii. At the same time their interpretation has long been beset by ambiguity. Here we focus mostly on $0_2^+ \to 0_1^+$ ground state E0 transitions but similar results would apply to higher spin states in the same bands.

The standard approach to E0 transitions is the shape coexistence model, widely discussed by Heyde and others [1], in which the mixing of spherical and deformed shape coexisting configurations is suggested to lead to large E0 transitions, especially in transitional regions such as those near A = 100 and 150. In the IBA model, this shape coexistence, or two–space, approach is codified in the well–known Duval–Barrett (D–B) formalism [2] in which the more collective configurations are imagined to arise from 2–particle–2–hole excitations from the valence space. Often, these 2p - 2h excitations are considered to occur across a shell (Z = 82) or subshell (Z = 40, 64) gap. Thus, ignoring the differences in the single particle orbits occupied in the two shells, if the spherical configuration built in the normal valence space, has N s and d bosons, the more collective one is built in an (N + 2) boson space. E0 transitions are forbidden between the two spaces but can occur if there is mixing between them. The D–B formalism then has a Hamiltonian of the generic form H = H_N + H_{N+2} + H_{mix}. This has been used successfully by many authors to interpret regions of shape coexistence. In particular, Sambataro and Molnar [3] have used it in the Mo isotopes to correctly predict the sharp rise in $\rho^2(E0; 0_2^+ \to 0_1^+)$ values in the spherical–deformed shape transition region near A = 100.

The downside of this success is the large number of parameters, as in the case of ref. [3] where each of the three Hamiltonians has a minimum of two or three parameters. The purpose of this paper is to present an alternative approach that we have recently

CP726, *Nuclear Physics, Large and Small: International Conference on Microscopic Studies of Collective Phenomena*, edited by R. Bijker, R. F. Casten, and A. Frank
© 2004 American Institute of Physics 0-7354-0207-8/04/$22.00

discussed [4], which uses a single space IBA–1 model. It *robustly* predicts the same rise in $\rho^2(E0)$ strength in transitional regions and is therefore effectively *parameter free*. Moreover, we will show that, in fact, many of the predictions of large E0 strengths in the two–space D–B formalism in fact arise from a single space which is identical to the one we use. Finally, we will show that the origins of the E0 strengths in our approach are not mixing of spherical and deformed configurations but rather result from mixing of n_d values, and that the E0 strength is directly related, not to shape mixing, but to the fluctuations in n_d, that is, to the β–softness.

E0 TRANSITIONS IN A SINGLE SPACE IBA–1 MODEL

The study of E0 transitions in the IBA–1 model date back at least to the classic foundation paper by Scholten *et al.* [5] who showed that the rise in E0 transitions in Sm could be accounted for in the IBA–1. The E0 operator can be written as [1, 5]

$$T(E0) = \alpha N + \beta (d^\dagger \tilde{d})^{(0)} \tag{1}$$

Matrix elements for E0 transitions vanish for the first term, and hence, given initial and final states with wave functions of the form $\psi = \sum_i \alpha_i n_{d_i}$ we obtain, simply

$$\rho(E0; 0_2^+ \to 0_1^+) = \sum_i \alpha_i^1 \alpha_i^2 n_{d_i} \tag{2}$$

where α_i^1 and α_i^2 are the components of the initial and final state with d–boson number n_{d_i}. Anticipating the discussion below, we note that $\rho(E0)$ is non–zero only if the wave functions have amplitudes in common for the same n_d value(s). Since the 0_1^+ and 0_2^+ states in U(5) have $n_d = 0, 2$, respectively, this implies that finite $\rho(E0)$ values only if there is n_d–mixing.

We have calculated $\rho^2(E0)$ values throughout the symmetry triangle [6] of the IBA–1. We use the simple Hamiltonian

$$H = \varepsilon n_d + \kappa Q \cdot Q = a(1 - \zeta) n_d - \frac{\zeta}{4N_B} Q \cdot Q \tag{3}$$

where ζ ranges from 0 (spherical) to 1 (deformed) and χ from 0 to $-\sqrt{7}/2$ (-1.32). We present calculations for a boson number typical of transitional nuclei, N = 10. The results are shown in Fig. 1 (left).

The most conspicuous feature is the sharp rise at the point of the shape transition. Nuclei on the spherical side have small E0 transitions, those just after the transition to a deformed structure have large $\rho(E0)$ values and the rise in between is sharp. Note that the rise occurs *independent of* χ: that is, it cannot be avoided in going from spherical to deformed nuclei in the IBA–1. The prediction of the rise is therefore robust and effectively parameter free. The only parameter dependence is in the specific spherical–deformed trajectory taken, which depends on other data (*e.g.*, levels and E2 transitions) characterizing the specific nuclei involved. The falloff in $\rho^2(E0)$ for deformed nuclei and small χ values (*i.e.*, nuclei approaching O(6)) is due to an interchange of the 0_2^+ and

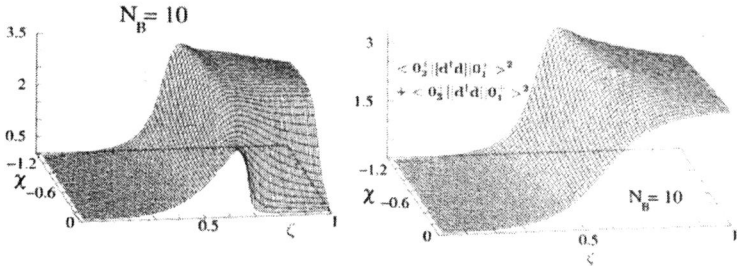

FIGURE 1. Contour plots of $\rho^2(E0)$ values in the IBA–1 against ζ and χ. Left: $0_2^+ \to 0_1^+$ transitions. Right: Sums of $\rho^2(E0)$ values for the $0_2^+ \to 0_1^+$ and $0_3^+ \to 0_2^+$ transitions. N = 10. Based on ref. [4].

0_3^+ states. If the $\rho^2(E0)$ values from each of these are summed, one obtains the plot on the right in Fig. 1, which is almost flat for all deformed nuclei.

Figure 2 shows a comparison of these predictions with the data for several isotopic chains (data taken from the compilation in ref. [1]). The calculations in the figure are highly schematic and therefore one does not expect exact agreement. The calculations all have $\chi = -1.32$ rather than fitting χ to each nucleus. The ζ value is obtained simply by fitting the $R_{4/2}$ value for each nucleus. And the boson number is kept constant. Thus, there can only be a single calculated trajectory. Nevertheless, it agrees remarkably well with the data, showing in particular the large and rapid rise as deformation sets in.

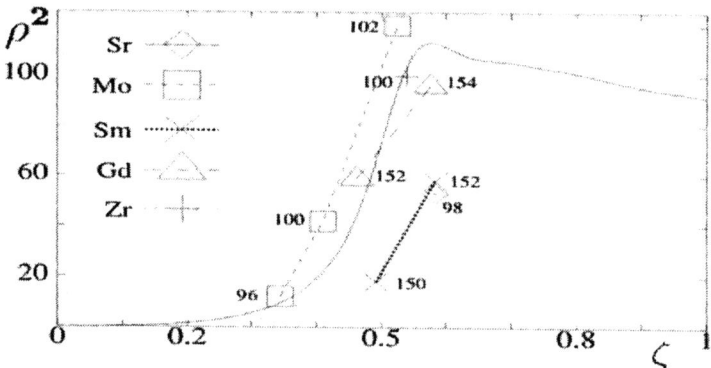

FIGURE 2. Comparison of IBA calculations of $\rho^2(E0; 0_2^+ \to 0_1^+)$ values with the data compiled from ref. [1]. The zeta values were obtained by fitting the $R_{4/2}$ value for each nucleus. For simplicity χ was kept constant at -1.32 and the boson number was fixed at N = 10. Based on ref. [4].

As noted earlier, for the Mo isotopes, calculations based on ref. [3], shown in ref. [7], also reproduce the empirical results. This raises the question of what is the actual origin of the large E0 strengths, mixing of coexisting spherical and deformed states, or mixing of n_d values in the single–space IBA–1. It turns out that the question is moot. The sharp rise in E0 strength to large values ($\sim$100) in the transition region in the D–B calculations [3] themselves in fact arises from only one of the two spaces. To see this, we show in Fig. 3 the probability for the N + 2 boson space in the ground state in the calculations of ref. [3]. There is strong mixing for ^{98,100}Mo, but essentially *none* for ^{96,102}Mo. ^{96}Mo is dominated by the more spherical configuration, which gives a small E0 strength since it is close to U(5) while ^{102}Mo is dominated by the deformed configuration with a more collective structure and hence (see Fig. 1) has a large E0 strength. It is *only* for the moderate E0 strengths ($\sim$20-40) early in the transitional region where mixing could play a role. Thus, although heretofore unrealized, the origin of the largest E0 strengths in the D–B formalism is in fact the single space IBA–1.

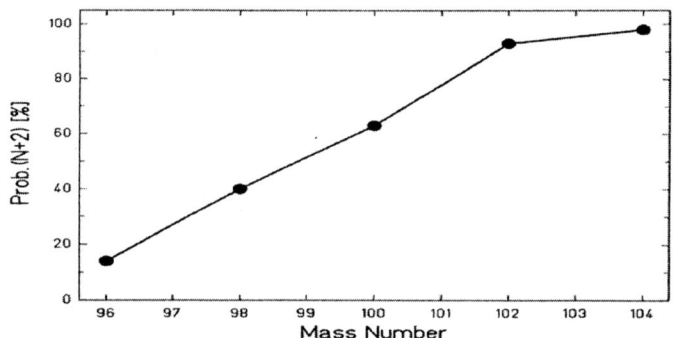

FIGURE 3. Probability for the N + 2 configuration for the ground state of Mo isotopes. Based on ref. [3].

Since the single space IBA–1 will always give an increase in E0 strength across a transition region, the question of whether mixing of coexisting states also contributes can be answered only by consideration of other data that might point to such mixing. In the case of Mo, such data does show the need for such mixing in the intermediate nuclei ^{98,100}Mo.

Finally, it is informative to understand the origin of the E0 strengths in the single space IBA–1. From the 0_2^+ and 0_1^+ wave functions we can calculate the contribution to $\rho^2(E0)$ from each n_d value. Typical results are shown in Fig. 4, for ζ values before, at, and well after the shape transition (all calculations are for $\chi = -1.32$). Before the transition region, only small n_d values occur. By eq. 2, these contributions give relatively small values. Hence $\rho^2(E0)$ is small. As the transition region is traversed, and for deformed nuclei, the n_d mixing is greatly enhanced and ultimately shows both positive and negative contributions. Thus the resulting $\rho^2(E0)$ values are larger, due to the presence in eq. 2 of large n_d values, and they also result as a consequence of the *specific d–boson coherence* inherently characteristic of the IBA.

We close by noting a fascinating aspect of this d–boson structure. It is easy to show that

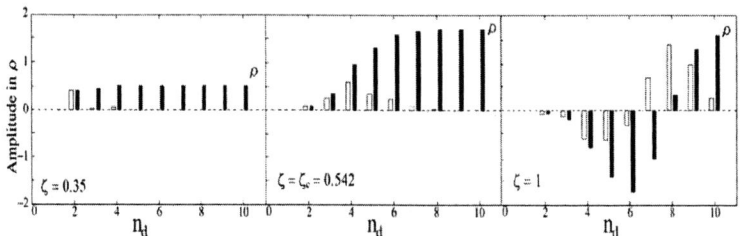

FIGURE 4. Contributions of different n_d values and the total $\rho(E0)$ value up to that n_d value for $0_2^+ \rightarrow$ 0_1^+ transitions, for nuclei before, at, and after the spherical to deformed shape transitions. Based on ref. [4].

$$\sum_i \rho^2(E0; 0_i^+ \rightarrow 0_1^+) = <n_d^2> - <n_d>^2 \qquad (4)$$

Most of the strength comes from the 0_2^+ state (see Fig. 1). Hence the E0 strength arises, *not* from *mixing* of different deformations but from the *fluctuations* in n_d, that is, the variance in n_d, or the n_d–softness. Since n_d is directly related, in the coherent state formalism [8], to quadrupole deformation, this suggests that the E0 strength comes from the *softness* in deformation (β–softness).

ACKNOWLEDGMENTS

We are grateful to T. Otsuka and F. Iachello for useful discussions, especially of the n_d content of the wave functions. Work supported by USDOE grant number DE-FG02-91ER-40609, by contract number Br 799/11–1/, and by BMBF grant number 06MT190.

REFERENCES

1. J.L. Wood *et al.*, Nucl. Phys. **A651**, 323 (1999); see also K. Heyde *et al.*, Phys. Reports **102**, 291 (1983)
2. P.D. Duval and B.R. Barrett, Phys. Lett **100B**, 223 (1981)
3. M. Sambataro and G. Molnar, Nucl. Phys. **A376**, 201 (1982)
4. P. von Brentano *et al.*, submitted to Phys. Rev. Lett.
5. O. Scholten, F. Iachello and A. Arima, Ann. Phys. (N.Y.), **115**, 235 (1978)
6. R.F. Casten in *Interacting Bose-Fermi Systems in Nuclei*, ed. F. Iachello, Plenum, p. 1 (1981)
7. R.J. Estep *et al.*, Phys. Rev. **C35**, 1485 (1987)
8. A.E.L. Dieperink and O. Scholten, Nucl. Phys. **A346**, 125 (1981)

Pentaquark spectroscopy: exotic Θ baryons

R. Bijker*, M.M. Giannini† and E. Santopinto†

*Instituto de Ciencias Nucleares, Universidad Nacional Autónoma de México, A.P. 70-543, 04510
México, D.F., México
†Dipartimento di Fisica dell'Università di Genova, I.N.F.N., Sezione di Genova, via Dodecaneso
33, 16164 Genova, Italy

Abstract. We propose a collective stringlike model of $q^4\bar{q}$ pentaquarks with the geometry of an equilateral tetrahedron in which the four quarks are located at the four corners and the antiquark in its center. The nonplanar equilibrium configuration is a consequence of the permutation symmetry of the four quarks. In an application to the spectrum of exotic Θ baryons, we find that the ground state pentaquark has angular momentum and parity $J^P = 1/2^-$ and a small magnetic moment of $0.382\ \mu_N$. The decay width is suppressed by the spatial overlap with the decay products.

INTRODUCTION

The building blocks of atomic nuclei, the nucleons, are composite extended objects, as is evident from their anomalous magnetic moment, their excitation spectrum and their charge distribution. Examples of excited states of the nucleon are the $\Delta(1232)$ and N(1440) Roper resonances. To first approximation, the internal structure of the nucleon at low energy can be ascribed to three bound constituent quarks q^3.

The discovery of the Θ(1540) with positive strangeness $S = +1$ by the LEPS Collaboration [1] as the first example of an exotic baryon (with quantum numbers that can not be obtained with q^3 configurations) has motivated an enormous amount of experimental and theoretical studies. The width of this state is observed to be very small < 20 MeV (or perhaps as small as a few MeV's). More recently, the NA49 Collaboration [2] reported evidence for the existence of another exotic baryon Ξ(1862) with strangeness $S = -2$. The Θ^+ and Ξ^{--} resonances are interpreted as $q^4\bar{q}$ pentaquarks belonging to a flavor antidecuplet with quark structure $uudd\bar{s}$ and $ddss\bar{u}$, respectively. In addition, there is now the first evidence [3] for a heavy pentaquark $\Theta_c(3099)$ in which the antistrange quark in the Θ^+ is replaced by an anticharm quark. The spin and parity of these states have not yet been determined experimentally. For a review of the experimental status we refer to [4].

Theoretical interpretations range from chiral soliton models [5] which provided the motivation for the experimental searches, correlated quark (or cluster) models [6], and various constituent quark models [7, 8]. A review of the theoretical literature of pentaquark models can be found in [9].

In this contribution, we introduce a collective stringlike model of $q^4\bar{q}$ pentaquarks in which the four quarks are located at the four corners of an equilateral tetrahedron and the antiquark in its center. This nonplanar equilibrium configuration is a consequence of the permutation symmetry of the four quarks [10]. As an application, we discuss the

CP726, Nuclear Physics, Large and Small: International Conference on
Microscopic Studies of Collective Phenomena, edited by R. Bijker, R. F. Casten, and A. Frank
© 2004 American Institute of Physics 0-7354-0207-8/04/$22.00

spectrum of exotic Θ baryons, as well as the parity and magnetic moments of the ground state decuplet baryons. The decay width is suppressed by the spatial overlap with the decay products.

PENTAQUARK STATES

We consider pentaquarks to be built of five constituent parts whose internal degrees of freedom are taken to be the three light flavors u, d, s with spin $s = 1/2$ and three colors r, g, b. The corresponding algebraic structure consists of the usual spin-flavor and color algebras $SU_{sf}(6) \otimes SU_c(3)$. The full decomposition of the spin-flavor states into spin and flavor states can be found in [8]

$$SU_{sf}(6) \supset SU_f(3) \otimes SU_s(2) \supset SU_I(2) \otimes U_Y(1) \otimes SU_s(2) \,. \tag{1}$$

The allowed flavor multiplets are singlet, octet, decuplet, antidecuplet, 27-plet and 35-plets. The first three have the same values of the isospin I and hypercharge Y as q^3 systems. However, the antidecuplets, the 27-plets and 35-plets contain exotic states which cannot be obtained from three-quark configurations. These states are more easily identified experimentally due to the uniqueness of their quantum numbers. The recently observed Θ^+ and Ξ^{--} resonances are interpreted as pentaquarks belonging to a flavor antidecuplet with isospin $I = 0$ and $I = 3/2$, respectively. In Fig. 1 the exotic states are indicated by a •: the Θ^+ is an isosinglet $I = 0$ with hypercharge $Y = 2$ (strangeness $S = +1$), and the cascade pentaquarks $\Xi_{3/2}$ have hypercharge $Y = -1$ (strangeness $S = -2$) and isospin $I = 3/2$.

In the construction of the classification scheme we are guided by two conditions: the pentaquark wave function should be a color singlet and it should be antisymmetric under any permutation of the four quarks [8]. The permutation symmetry of the four-quark system is given by S_4 which is isomorphic to the tetrahedral group $\mathcal{T}_d$. We use

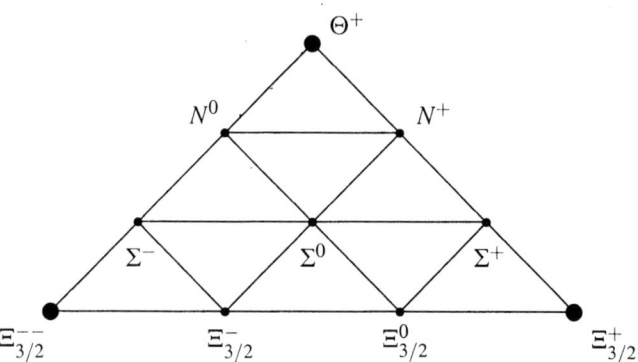

FIGURE 1. $SU(3)$ antidecuplet. The isospin-hypercharge multiplets are $(I,Y) = (0,2)$, $(\frac{1}{2},1)$, $(1,0)$ and $(\frac{3}{2},-1)$. Exotic states are indicated with •.

the labels of the latter to classify the states by their symmetry character: symmetric A_1, antisymmetric A_2 or mixed symmetric E, F_2 or F_1.

The relative motion of the five constituent parts is described in terms of the Jacobi coordinates

$$
\begin{aligned}
\vec{\rho}_1 &= \tfrac{1}{\sqrt{2}}(\vec{r}_1 - \vec{r}_2)\,, & \vec{\rho}_3 &= \tfrac{1}{\sqrt{12}}(\vec{r}_1 + \vec{r}_2 + \vec{r}_3 - 3\vec{r}_4)\,, \\
\vec{\rho}_2 &= \tfrac{1}{\sqrt{6}}(\vec{r}_1 + \vec{r}_2 - 2\vec{r}_3)\,, & \vec{\rho}_4 &= \tfrac{1}{\sqrt{20}}(\vec{r}_1 + \vec{r}_2 + \vec{r}_3 + \vec{r}_4 - 4\vec{r}_5)\,,
\end{aligned}
\tag{2}
$$

where $\vec{r}_i$ ($i = 1,..,4$) denote the coordinate of the i-th quark, and $\vec{r}_5$ that of the antiquark. The last Jacobi coordinate is symmetric under the interchange of the quark coordinates, and hence transforms as A_1 under $\mathscr{T}_d$ ($\sim S_4$), whereas the first three transform as three components of F_2.

The total pentaquark wave function is the product of the spin, flavor, color and orbital wave functions $\psi = \psi^s \psi^f \psi^c \psi^o$. Since the color part of the pentaquark wave function is a singlet and that of the antiquark an anti-triplet, the color wave function of the four-quark configuration is a triplet with F_1 symmetry. The total q^4 wave function is antisymmetric (A_2), hence the orbital-spin-flavor part has to have F_2 symmetry

$$
\psi = \left[\psi^c_{F_1} \times \psi^{osf}_{F_2} \right]_{A_2}.
\tag{3}
$$

Here the square brackets $[\cdots]$ denote the tensor coupling under the tetrahedral group $\mathscr{T}_d$.

STRINGLIKE MODEL

In this section, we discuss a stringlike model for pentaquarks, which is a generalization of a collective stringlike model developed for q^3 baryons [11]. We introduce a dipole boson with $L^P = 1^-$ for each independent relative coordinate, and an auxiliary scalar boson with $L^P = 0^+$, which leads to a compact spectrum-generating algebra of $U(13)$ for the radial excitations. As a consequence of the invariance of the interations under the permutation symmetry of the four quarks, the most favorable geometric configuration is an equilateral tetrahedron in which the four quarks are located at the four corners and the antiquark in its center [10]. This configuration is in agreement with that of [12] in which arguments based on the flux-tube model were used to suggest a nonplanar structure for the $\Theta(1540)$ pentaquark to explain its narrow width. In the flux-tube model, the strong color field between a pair of a quark and an antiquark forms a flux tube which confines them. For the pentaquark there would be four such flux tubes connecting the quarks with the antiquark.

Mass spectrum of Θ baryons

Hadronic spectra are characterized by the occurrence of linear Regge trajectories with almost identical slopes for baryons and mesons. Such a behavior is also expected on basis of soft QCD strings in which the strings elongate as they rotate. In the same spirit as in algebraic models of stringlike q^3 baryons [11], we use the mass-squared operator

$$
M^2 = M_0^2 + M_{\mathrm{vib}}^2 + M_{\mathrm{rot}}^2 + M_{\mathrm{sf}}^2.
\tag{4}
$$

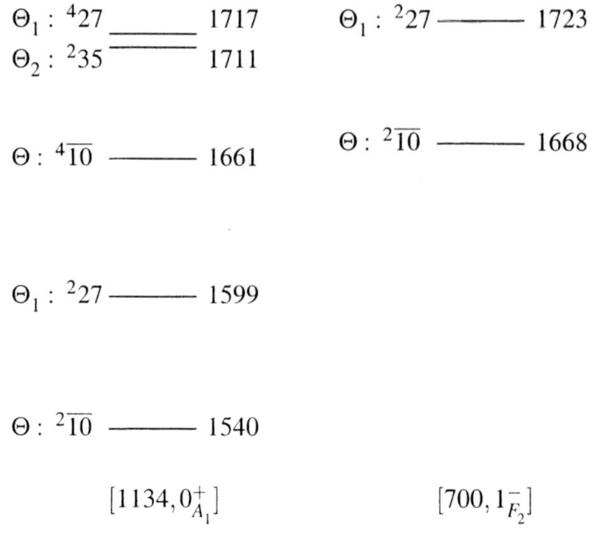

$\Theta_1 : {}^4 27$ _____ 1717 $\Theta_1 : {}^2 27$ ——— 1723

$\Theta_2 : {}^2 35$ ‾‾‾‾‾‾ 1711

$\Theta : {}^4 \overline{10}$ ——— 1661 $\Theta : {}^2 \overline{10}$ ——— 1668

$\Theta_1 : {}^2 27$ ——— 1599

$\Theta : {}^2 \overline{10}$ ——— 1540

$[1134, 0^+_{A_1}]$ $[700, 1^-_{F_2}]$

FIGURE 2. Spectrum of Θ pentaquarks. Masses are given in MeV.

The vibrational term M^2_{vib} describes the vibrational spectrum corresponding to the normal modes of of a tetrahedral $q^4\bar{q}$ configuration

$$M^2_{\text{vib}} = \varepsilon_1 \, v_1 + \varepsilon_2 \left(v_{2a} + v_{2b}\right) + \varepsilon_3 \left(v_{3a} + v_{3b} + v_{3c}\right) + \varepsilon_4 \left(v_{4a} + v_{4b} + v_{4c}\right) . \qquad (5)$$

The rotational energies are given by a term linear in the orbital angular momentum L which is responsable for the linear Regge trajectories in baryon and meson spectra

$$M^2_{\text{rot}} = \alpha L . \qquad (6)$$

The spin-flavor part is expressed in a Gürsey-Radicati form, i.e. in terms of Casimir invariants of the spin-flavor groups of Eq. (1)

$$M^2_{\text{sf}} = a C_{2SU_{\text{sf}}(6)} + b C_{2SU_{\text{f}}(3)} + c C_{2SU_{\text{s}}(2)} + d C_{1U_Y(1)} + e C^2_{1U_Y(1)} + f C_{2SU_{\text{I}}(2)} . \qquad (7)$$

The coefficients α, a, b, c, d, e and f are taken from a previous study of the nonstrange and strange baryon resonances [11], and the constant M^2_0 is determined by identifying the ground state exotic pentaquark with the recently observed $\Theta(1540)$ resonance. Since the lowest orbital states with $L^P = 0^+$ and 1^- are interpreted as rotational states, for these excitations there is no contribution from the vibrational terms ε_1, ε_2, ε_3 and ε_4. The results for the lowest Θ pentaquarks (with strangeness $+1$) are shown in Fig. 2.

The lowest pentaquark belongs to the flavor antidecuplet with spin $s = 1/2$ and isospin $I = 0$, in agreement with the available experimental information which indicates that the $\Theta(1540)$ is an isosinglet. In the present calculation, the ground state pentaquark belongs to the $[42111]_{F_2}$ spin-flavor multiplet, indicated in Fig. 2 by its dimension 1134, and an orbital excitation 0^+ with A_1 symmetry. Therefore, the ground state has angular momentum and parity $J^P = 1/2^-$, in agreement with recent work on QCD sum rules [13] and lattice QCD [14], but contrary to the chiral soliton model [5], various cluster models [6] and a lattice calculation [15] that predict a ground state with positive parity. The first excited state at 1599 MeV is an isospin triplet Θ_1 state with strangeness $S = +1$ of the 27-plet with the same value of angular momentum and parity $J^P = 1/2^-$. The lowest pentaquark state with positive parity occurs at 1668 MeV and belongs to the $[51111]_{A_1}$ spin-flavor multiplet (with dimension 700) and an orbital excitation 1^- with F_2 symmetry. In the absence of a spin-orbit coupling, in this case we have a doublet with angular momentum and parity $J^P = 1/2^+, 3/2^+$.

There is some preliminary evidence from the CLAS Collaboration for the existence of two peaks in the nK^+ invariant mass distribution at 1523 and 1573 MeV [16]. The mass difference between these two peaks is very close to the mass difference in the stringlike model between the ground state pentaquark at 1540 MeV (fitted) and the first excited state Θ_1 at 1599 MeV.

Magnetic moments

The magnetic moment of a multiquark system is given by the sum of the magnetic moments of its constituent parts

$$\vec{\mu} = \vec{\mu}_{\text{spin}} + \vec{\mu}_{\text{orb}} = \sum_i \mu_i (2\vec{s}_i + \vec{\ell}_i) , \tag{8}$$

where the quark magnetic moments μ_u, μ_d and μ_s are determined from the proton, neutron and Λ magnetic moments and satisfy $\mu_q = -\mu_{\bar{q}}$.

The $SU_{\text{sf}}(6)$ wave function of the ground state pentaquark has the general structure

$$\psi_{A_2} = \left[\psi_{F_1}^{\text{c}} \times \left[\psi_{A_1}^{\text{o}} \times \psi_{F_2}^{\text{sf}} \right]_{F_2} \right]_{A_2} . \tag{9}$$

Since the ground state orbital wave function has $L^P = 0^+$, the magnetic moment only depends on the spin part. For the Θ^+ exotic state we obtain

$$\mu_{\Theta^+} = (2\mu_u + 2\mu_d + \mu_s)/3 = 0.382 \, \mu_N , \tag{10}$$

in agreement with the result obtained [17] for the MIT bag model. These results for the magnetic moments are independent of the orbital wave functions, and are valid for any quark model in which the eigenstates have good $SU_{\text{sf}}(6)$ spin-flavor symmetry.

SUMMARY AND CONCLUSIONS

In this contribution, we have discussed a stringlike model of pentaquarks, in which the four quarks are located at the corners of an equilateral tetrahedron with the antiquark in its center. Geometrically this is the most stable equilibrium configuration. The ground state pentaquark belongs to the flavor antidecuplet, has angular momentum and parity $J^P = 1/2^-$ and, in comparison with the proton, has a small magnetic moment. The width is expected to be narrow due to a large suppression in the spatial overlap between the pentaquark and its decay products [12].

The first report of the discovery of the pentaquark has triggered an enormous amount of experimental and theoretical studies of the properties of exotic baryons. Nevertheless, there still exist many doubts and questions about the existence of this state, since in addition to various confirmations there are also several experiments in which no signal has been observed. Hence, it is of the utmost importance to understand the origin between these apparently contradictory results, and to have irrefutable proof for or against its existence [4]. If confirmed, the measurement of the quantum numbers of the $\Theta(1540)$, especially the angular momentum and parity, and the excited pentaquark states, may help to distinguish between different models and to gain more insight into the relevant degrees of freedom and the underlying dynamics that determines the properties of exotic baryons.

ACKNOWLEDGEMENTS

It is a great pleasure to dedicate this article to the 60th anniversary of Stuart Pittel in acknowledgement of many years of collaboration and friendship. This work is supported in part by a grant from CONACyT, México.

REFERENCES

1. LEPS Collaboration: T. Nakano et al., Phys. Rev. Lett. **91**, 012002 (2003).
2. NA49 Collaboration, C. Alt et al., Phys. Rev. Lett. **92**, 042003 (2004).
3. H1 Collaboration: A. Aktas et al., Phys. Lett. B **588**, 17 (2004).
4. Q. Zhao and F.E. Close, hep-ph/0404075; M. Karliner and H.J. Lipkin, hep-ph/0405002.
5. D. Diakonov et al., Z. Phys. A **359**, 305 (1997); J. Ellis et al., hep-ph/0401127.
6. R. Jaffe and F. Wilczek, Phys. Rev. Lett. **91**, 232003 (2003); M. Karliner and H.J. Lipkin, Phys. Lett. B **575**, 249 (2003); E. Shuryak and I. Zahed, hep-ph/0310270.
7. Fl. Stancu, Phys. Rev. D **58**, 111501 (1998); C. Helminen and D.O. Riska, Nucl. Phys. A **699**, 624 (2002); A. Hosaka, Phys. Lett. B **571**, 55 (2003); C.E. Carlson et al., Phys. Lett. B **573**, 101 (2003); *ibid* **579**, 52 (2004); L.Ya. Glozman, Phys. Lett. B **575**, 18 (2003); Fl. Stancu and D.O. Riska, Phys. Lett. B **575**, 242 (2003); J.J. Dudek and F.E. Close, Phys. Lett. B **583**, 278 (2004).
8. R. Bijker, M.M. Giannini and E. Santopinto, hep-ph/0310281, hep-ph/0312380 and hep-ph/0403029.
9. B.K. Jennings and K. Maltman, hep-ph/0308286.
10. R. Bijker, M.M. Giannini and E. Santopinto, work in progress, to be published.
11. R. Bijker, F. Iachello and A. Leviatan, Ann. Phys. (N.Y.) **236**, 69 (1994), *ibid.* **284**, 89 (2000).
12. Xing-Chang Song and Shi-Lin Zhu, hep-ph/0403093.
13. S.L. Zhu, Phys. Rev. Lett. **91**, 232002 (2003); J. Sugiyama et al., Phys. Lett. B **581**, 167 (2004).
14. F. Csikor et al., J. High Energy Phys. **0311** (2003) 070; S. Sasaki, hep-lat/0310014.
15. T.-W. Chiu and T.-H. Hsieh, hep-ph/0403020.
16. M. Battaglieri, http://www.tp2.ruhr-uni-bochum.de/talks/trento04/battaglieri.pdf .
17. Y.-R. Liu, P.-Z. Huang, W.-Z. Deng, X.-L. Chen and S.-L. Zhu, hep-ph/0312074.
18. S. Coleman and S.L. Glashow, Phys. Rev. Lett. **6**, 423 (1961).
19. S.-T. Hong and G.E. Brown, Nucl. Phys. A **580**, 408 (1994).
20. H.-C. Kim and M. Praszałowicz, Phys. Lett. B **585**, 99 (2004).

Evolution of Nuclear Observables in the Spherical-Deformed Phase Transition and the Interacting Boson Model

N. V. Zamfir, E.A. McCutchan, and R. F. Casten

Yale University, Wright Nuclear Structure Laboratory, New Haven CT, 06520-8124, USA

Abstract. We discuss the empirical evolution of structure in the spherical-deformed phase transition region in the context of the behavior of basic structural signatures in the IBA symmetry triangle. The main signatures of a phase/shape transition are a sharp increase in $R_{4/2} \equiv E(4_1^+)/E(2_1^+)$ and minima in $E(0_2^+)$ and $E(2_\gamma^+)$.

INTRODUCTION

Understanding the evolution of nuclear structure from spherical shapes, near closed shells, to deformed shapes, toward the middle of shells, is one of the most important challenges in nuclear physics. Recently, there has been substantial progress in this direction with the discovery [1, 2, 3] that finite nuclei can exhibit behavior closely resembling that of phase transitions observed in macroscopic systems. Theoretical interpretation [4] of the observed behavior has led to new insights into phase/shape transition regions, to the development [5, 6] of a new class of symmetries to describe nuclei at the phase transitional point, and to the discovery of nuclei manifesting the properties of these symmetries [7].

Microscopically, the sudden onset of deformation is intimately related to the Federman-Pittel mechanism [8] involving the interaction between spin-orbit partner orbitals.

In this article, dedicated to Stu Pittel on the occasion of his 60-th birthday, we analyze the behavior of nuclei in transitional regions, with a focus on the structure in the vicinity of the phase transition. We do this by studying the expected and observed behavior of key observables across the spherical-deformed transition region.

NATURE OF STRUCTURAL EVOLUTION IN THE IBA

In discussing the evolution of nuclear shape, we will use the Interacting Boson Model (IBA) [9]. It is well known [10, 11] that the IBA, in the large boson number limit, exhibits a phase transition in the evolution from U(5) to the SU(3)−O(6) leg of the triangle. In Fig. 1 the shaded area represents the region of phase/shape transition and coexistence [4].

CP726, Nuclear Physics, Large and Small: International Conference on
Microscopic Studies of Collective Phenomena, edited by R. Bijker, R. F. Casten, and A. Frank
© 2004 American Institute of Physics 0-7354-0207-8/04/$22.00

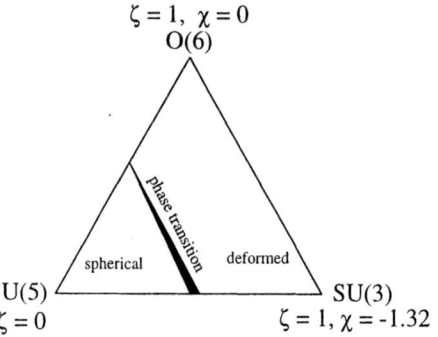

FIGURE 1. The symmetry triangle of the IBA with the three dynamical symmetries and the phase/shape transition region (shaded area).

We base the following discussion on the simple IBA Hamiltonian:

$$H(\zeta) = c[(1-\zeta)\hat{n}_d - \frac{\zeta}{4N_B}\hat{Q}^\chi \cdot \hat{Q}^\chi] \tag{1}$$

where N_B is the total number of bosons, $\hat{n}_d = d^\dagger \cdot \tilde{d}$ and $\hat{Q}^\chi = (s^\dagger \tilde{d} + d^\dagger s) + \chi(d^\dagger \tilde{d})^{(2)}$. This Hamiltonian contains 2 parameters, ζ and χ (c is only a scaling factor) and describes the entire IBA symmetry triangle. The dynamical symmetries correspond to limiting values of these 2 parameters: $\zeta=0$, any χ for U(5), $\zeta=1$, $\chi = -\sqrt{7}/2$ for SU(3), and $\zeta=1$, $\chi = 0$ for O(6). The parameter ζ is related to quadrupole deformation and χ to the γ-degree of freedom - axially symmetric for $\chi = -\sqrt{7}/2$ and γ-flat for $\chi=0$. Variations of ζ from 0 to 1 and of χ from $-\sqrt{7}/2$ to 0 span the symmetry triangle of Fig. 1. The phase/shape transition takes place for the control parameter $\zeta \sim 0.5$-0.6, the exact values being slightly dependent on the χ value and on the boson number [12].

The IBA description of the empirical evolution along different isotopic chains can be done only by a detailed fit. In Fig. 2 are shown the parameters ξ and χ obtained for Z=62-76 isotopic chains with N=86-104 [13, 14, 15].

The parameter ξ is increasing, in general, with N, as expected when the nuclei evolve from spherical to deformed shape. However, for N>90 different isotopic chains present different evolution of ξ and in particular for Yb, Hf, W, and Os, ξ is relatively constant. The different trends is even more pronounced for χ. As we will see in the next section, the different evolution with N of the IBA parameters is reflected in a different evolution of nuclear observables along the transitional region.

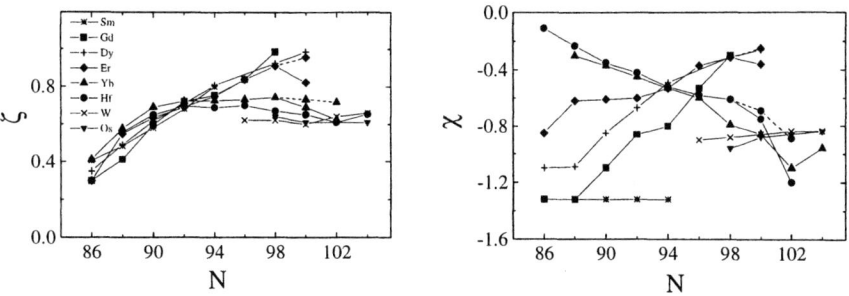

FIGURE 2. The IBA parameters ξ and χ for the Sm-Os isotopes [13, 14, 15] as a function of N.

THE EVOLUTION OF NUCLEAR OBSERVABLES

In this section we will explore how three key structural observables behave for different trajectories in the IBA triangle. We will compare the IBA description to the empirical behavior of nuclei in the rare-earth region (Z=62-76, N=86-104).

Three characteristic observables that provide insight into nuclear structure are the energy ratio $R_{4/2} \equiv E(4_1^+)/E(2_1^+)$, the energy of the first excited 0^+ state (0_2^+) and the energy of the (quasi) γ-band, that is, the 2_γ^+ state.

We first consider $R_{4/2}$, shown in Fig. 3a. This ratio evolves with increasing neutron number from $\sim$2.0, characteristic for spherical nuclei just above the closed shell N=82, to $\sim$3.33, characteristic for an axially symmetric rotor, towards the middle of the shell. The details of the evolution are different for various isotopic chains: In Sm-Dy the evolution is quite dramatic, exhibiting a sharp rise at N=88-90, while, with increasing proton number, the evolution is more and more gradual. However, the difference between the behavior of, say, the Sm and Os isotopes is not that Sm exhibits a rapid phase transition and that deformation develops in Os as a result of a gradual accumulation of collectivity. It is impossible to go from a spherical to a deformed equilibrium configuration *without* passing through a phase transition. The difference in behavior is due to the fact that the rapidity of structural change in units of N is not equivalent to the rapidity in units of the control parameter ζ. The gradualness of the Os spherical-deformed transition as N varies reflects the fact that the values of the control parameter ζ describing successive even-even Os nuclei change more slowly with N than those for Sm where $\zeta \sim$ N [14].

A complete survey of the behavior of this ratio in the IBA shows that the sharpness of the change in $R_{4/2}$ depends on the particular (ζ, χ) trajectory. Figure 3(bottom) gives a 3D illustration of this ratio in the symmetry triangle. In the U(5) - SU(3) transition, i.e. $\zeta = 0 \rightarrow 1$ with $\chi = -\sqrt{7}/2$, the $R_{4/2}$ ratio presents a sharp increase in the critical point region of the spherical to deformed transition ($\zeta \sim$ 0.5-0.6). For any other trajectory, the evolution of the ratio from 2.0 [U(5)] to 3.33 [SU(3)] is less sharp.

The empirical $R_{4/2}$ ratios are plotted as a function of the control parameter ζ in Fig. 3b. In terms of this parameter, the sharpness of the transition is, indeed, similar in all

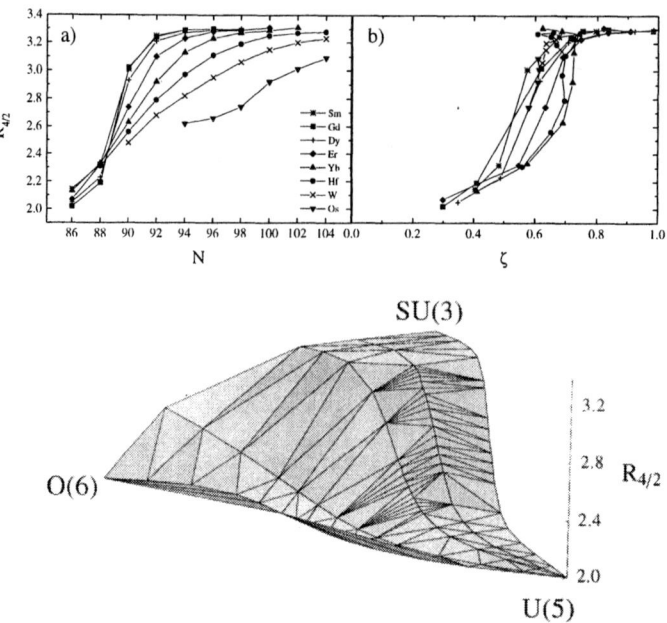

FIGURE 3. Top: a) Empirical $R_{4/2}$ values for the Sm-Os isotopic chains with $82<N\leq104$ showing the different rapidity of structural change with N across the phase transitional region. b) The same data as in part a) but as a function of the IBA parameter ζ corresponding to the best fit to each isotope [13]. Bottom: 3D plot for the evolution of the $R_{4/2}$ ratio in the symmetry triangle for $N_B=10$.

isotopic chains, increasing rapidly in the critical region.

The evolution of two other basic observables in nuclear spectra, $E(0_2^+)$ and $E(2_\gamma^+)$, is related to the change of their character from multiphonon states in the U(5) symmetry to intrinsic excitations in SU(3). This is illustrated in Fig. 4(top) where the change of their character along the phase transition region is reflected in a "valley" along this line: large values in U(5), minimum across phase transition region and growing again in the SU(3) limit. The minimum is more pronounced for 0_2^+ than for 2_γ^+. For the latter the minimum is attenuated due to the fact that close to the phase transition region, there is an interchange of positions between 2_γ^+ and $2_{0_2^+}^+$ states. As can be seen in Fig. 4(middle), for ζ small, 2_γ^+ is the second excited 2^+ state but, as ζ approaches the critical phase/shape transition region ($\zeta = 0.40$-0.45) the third 2^+ state become the band-head of the (quasi)-gamma band. A more pronounced minimum in the evolution of $E(0_2^+)$ compared with a flatter evolution of $E(2_\gamma^+)$ is also clearly seen in the empirical evolution of individual isotopic chains (Fig. 4 bottom).

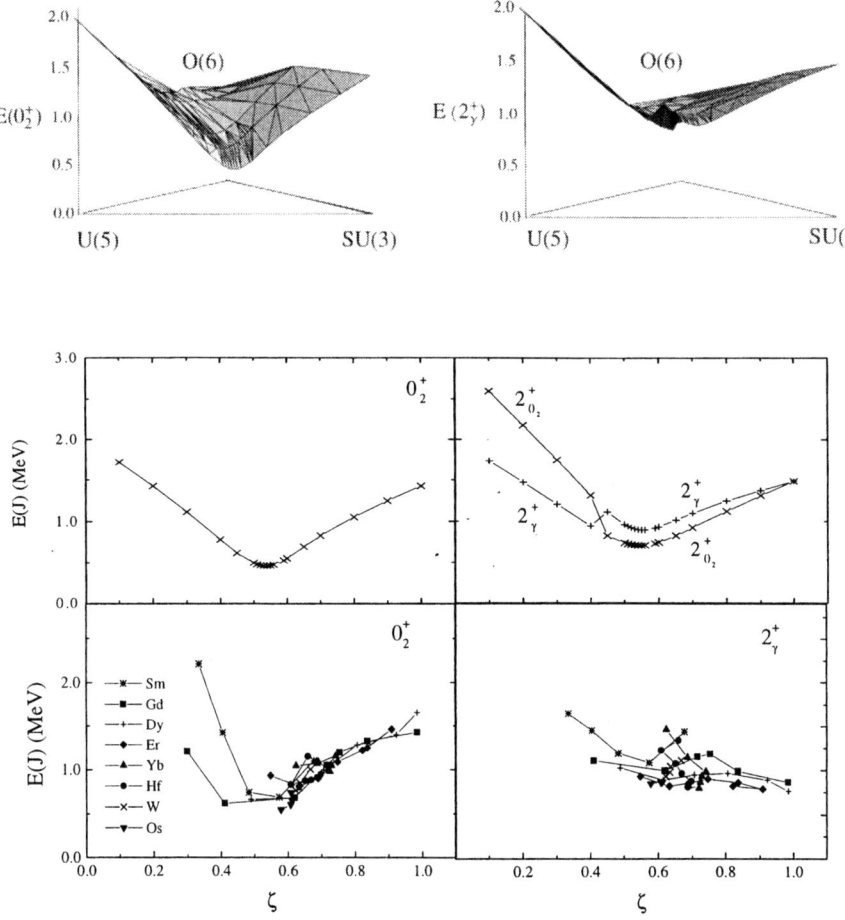

FIGURE 4. Top: 3D plots for the evolution of the calculated (c=1 MeV and N_B=10 in eq. 1) $E(0_2^+)$ and $E(2_\gamma^+)$ values in the symmetry triangle. Middle: Evolution of calculated $E(0_2^+)$, $E(2_{0_2}^+)$, and $E(2_\gamma^+)$ for $\chi = -\sqrt{7}/2$. Bottom: Empirical energies of the 0_2^+ and 2_γ^+ states for Sm-Os isotopic chains with $82<N\leq104$ as a function of the IBA parameter ζ corresponding to the best fit to each isotope [13].

CONCLUSION

We have studied the evolution of nuclear structure in phase transitional regions in terms of the behavior of characteristic observables. The main signatures of the phase/shape transition are a sharp increase in $R_{4/2} \equiv E(4_1^+)/E(2_1^+)$ and a minimum in the evolution of the energy of the first excited 0^+ state (0_2^+) and the energy of the 2_γ^+ state.

ACKNOWLEDGEMENTS

Useful discussions with Franco Iachello and collaboration in the early stage of this work with Gustavo Fernandes are acknowledged. Work supported by the US DOE under Grant number DE-FG02-91ER-40609.

REFERENCES

1. R.F. Casten, N.V. Zamfir, and D.S. Brenner, Phys. Rev. Lett. **71**, 227 (1993).
2. A. Wolf, R.F. Casten, N.V. Zamfir, and D.S. Brenner, Phys. Rev. **C49**, 802 (1994).
3. R.F. Casten, D. Kusnezov, and N.V. Zamfir, Phys. Rev. Lett. **82**, 5000 (1999)
4. F. Iachello, N.V. Zamfir, and R.F. Casten, Phys. Rev. Lett. **81**, 1191 (1998).
5. F. Iachello, Phys. Rev. Lett. **85**, 3580 (2000)and Phys. Rev. Lett. **87**, 052502 (2001).
6. F. Iachello, in "Mapping the Triangle" AIP Proc. 638, eds. A. Aprahamian, J. Cizewski, S. Pittel, and N.V. Zamfir., 2002.
7. R.F. Casten and N.V. Zamfir, Phys. Rev. Lett. **85**, 3584 (2000) and **87**, 052503 (2002).
8. P. Federman and S. Pittel, Phys. Lett. **69B**, 385 (1977).
9. F. Iachello and A. Arima, *The Interacting Boson Model* (Cambridge University Press, Cambridge, England, 1987).
10. A.E.L. Dieperink, O. Scholten, and F. Iachello, Phys. Rev. Lett **44**, 1747 (1980); Nucl. Phys. A346, 125 (1980).
11. D.H. Feng, R. Gilmore, and S.R. Deans *Phys. Rev.* **C 23**, 1254 (1981).
12. N.V. Zamfir and G.E. Fernandes, in Proceedings of the Eleventh International Symposium on Capture Gamma Ray Spectroscopy and Related Topics, Prohonice near Prague, Czech Republic, September 2-6, 2002, World Scientific Singapore 2003, eds J. Kvasil, P. Cejnar, and M. Krticka.
13. E.A. McCutchan, N.V. Zamfir, and R.F. Casten, Phys. Rev. C - in press.
14. O. Scholten, F. Iachello, and A. Arima, Ann. Phys. (N.Y.) **115**, 325 (1978).
15. E.A. McCutchan and N.V. Zamfir, to be published.

Shell model representation
with antibound states

T. Vertse[*†], R. Id Betan[†**], R. J. Liotta[†] and N. Sandulescu[†‡]

[*]ATOMKI, P.O. Box 51, H-4001, Debrecen, Hungary
[†]Royal Institute of Technology, Alba Nova University Center, SE-10691 Stockholm, Sweden
[**]Departamento de Fisica, FCEIA, UNR, Av. Pellegrini 250, 2000 Rosario, Argentina
[‡]Institute of Physics and Nuclear Engineering, P. O. Box MG-6, Bucharest-Magurele,Romania

INTRODUCTION

Present experimental facilities allow us to measure drip line nuclei, which might decay by particle emission. For the theoretical description of these nuclei the continuum part of the nuclear spectrum plays crucial role. One way for treating the continuum and the resonances embedded into it is using the completeness relation introduced by Berggren[1] already in 1968. By bending the real energy continuum to a complex path he included complex energy resonances, i.e. pole solutions of the scattering matrix $S(E)$ into the completeness relation. The use of this Berggren basis in shell model calculations has started only recently[2, 3, 4, 5].

The first attempt for generalizing the Berggren representation by including antibound poles into the basis and use it in numerical calculation was made already in 1989 by Vertse et al.[6] with partial success due to the neglect of the contour contribution. Antibound states were included later into an expansion of the resolvent by Berggren and Lind [7] however, without any numerical calculation.

The first use of antibound states in a completeness relation and without serious truncation and the application of this to shell model calculation is our recent work[8]. Our method can be considered as a generalization of the original Berggren formalism. A most recent attempt for using antibound states numerically in a generalized contour deformation method for expanding N-N interactions was made by Hagen et al.[9].

As an application of the new method, the generalized basis is used for constructing two-neutron shell model wave functions in a somewhat simplified description of the well known halo nucleus ^{11}Li.

SINGLE PARTICLE REPRESENTATIONS

Let us review the different completeness relations on the example of a spherically symmetric single particle Hamiltonian: $h = t + v(r)$, in which the potential $v(r) = v_N(r) + v_{so}(r) + v_C(r)$ is a sum of finite range nuclear potential $v_F(r) = v_N(r) + v_{so}(r)$ $v_F(r > r_{max}) = 0$ and for protons an infinite range Coulomb

CP726, Nuclear Physics, Large and Small: International Conference on
Microscopic Studies of Collective Phenomena, edited by R. Bijker, R. F. Casten, and A. Frank
© 2004 American Institute of Physics 0-7354-0207-8/04/$22.00

term $v_C(r) = \frac{Ze^2}{r}$ $r > R_C$. The radial part $u_{lj}(r,k)$ of the wave function: $\psi_{ljm}(\vec{r})$ satisfies the radial equation:

$$u''_{lj}(r,k) = [\frac{l(l+1)}{r^2} + V(r) - k^2]u_{lj}(r,k).$$

Its energy dependence is shown explicitly through the wave number k ($k^2 = \frac{2\mu}{\hbar^2}E$). The radial wave function should be regular at $r = 0$, i.e. $u_{lj}(0,k) = 0$ and beyond r_{max} and for $E > 0$ it should match

$$u_{lj}(r,k) = x_{lj}(k)O_l(kr,\eta) + y_{lj}(k)I_l(kr,\eta)$$

to the linear combinations of the incoming, outgoing solutions: $I_l = G_l - iF_l$, $O_l = G_l + iF_l$ of the Coulomb differential equation. For $l = 0$ neutrons ($\eta = 0$) the asymptotic solutions are:

$$O_0(\rho,0) = e^{i\rho} \quad I_0(\rho,0) = e^{-i\rho}.$$

The partial wave S-matrix element is the ratio of the Jost functions $x_{lj}(k)$ and $y_{lj}(k)$:

$$S_{lj}(k) = -\frac{x_{lj}(k)}{y_{lj}(k)}.$$

For $E < 0$ k is imaginary $k = i\gamma$ and for $\gamma > 0$ we have square integrable solution only at discrete $k_b = i\gamma_b$ values, where $y_{lj}(k_b) = 0$ i.e. where the S-matrix has poles. These solutions are the bound states with discrete negative energy values $E_b < 0$. Bound states satisfy purely outgoing boundary condition:

$$u_{nlj}(r,k_b) = NO_l(k_b r,\eta) \quad O_0(k_b r,0) = e^{ik_b r} = e^{-\gamma_b r}.$$

The *standard completeness relation* is composed of bound states and the real k continuum of the scattering states:

$$\delta(r-r') = \sum_{n=b} u_{nlj}(r,k_n)\, u_{nlj}(r',k_n)$$
$$+ \frac{2}{\pi}\int_0^\infty dk u_{lj}(r,k)u_{lj}(r',k)^* .$$

In 1968 Berggren[1] changed the real k continuum into a complex path L^+ and he included other type of complex poles of $S_{lj}(k)$, the decaying resonances into the basis.

$$\delta(r-r') = \sum_{n=b,d} u_{nlj}(r,k_n)\, u_{nlj}(r',k_n)$$
$$+ \frac{2}{\pi}\int_{\mathsf{L}^+} dk u_{lj}(r,k)u_{lj}(r',k) .$$

Therefore the elements of the *Berggren completeness relation* are

- i. complex k continuum i.e. scattering states along the path, $k \in \mathsf{L}^+$,
- ii. bound states and
- iii. decaying (proper) Gamow resonances with discrete complex energy values: $E_n \sim k_n^2$ ($k_n = \kappa_n - i\gamma_n$) $\kappa_n > \gamma_n$. Since they are also poles of $S_{lj}(k_n)$, at large r they also satisfy outgoing wave asymptotics:

$$u_{nlj}(r,k) = NO_l(k_n r,\eta) \quad O_0(k_n r,0) = e^{ik_n r} = e^{\gamma_n r}e^{i\kappa r} .$$

Since neither the resonances nor the scattering states are square integrable functions, Berggren introduced[1] a generalized scalar product (complex metric) with a bilinear basis set in which in bra position the mirror image state $\langle \tilde{n} |$ have to be used instead of the original state: $\langle n |$. Moreover special regularization methods have to be used for calculating the often divergent radial integrals. The generalization of the scalar product however, does not change the spin-angular part of the matrix elements. The standard completeness can be considered as a special case of the Berggren completeness with a path $+$ coinciding with the positive real k-axis.

It is a very natural idea to deform the Berggren contour further and include the rest of the poles of $S_{lj}(k)$ (the virtual resonances with $\kappa_n < \gamma_n$ and the antibound states with $k_a = -i\gamma_a$ with $\gamma_a > 0$) into the completeness relation. The antibound states have exponentially growing asymptotics, since for $r > r_{max}$:

$$u_{nlj}(r, k_a) = NO_l(k_a r, \eta) \qquad O_0(k_a r, 0) = e^{\gamma_a r} .$$

The *generalized Berggren representation* which includes antibound states is the following:

$$\delta(r - r') = \sum_{n=b,d,a} u_{nlj}(r, k_n) u_{nlj}(r', k_n)$$
$$+ \frac{2}{\pi} \int_{+} dk u_{lj}(r, k) u_{lj}(r', k) .$$

The generalized Berggren-basis can have the following elements:

- i. bound states,
- ii. antibound states with $k_a = -i\gamma_a$ $\gamma_a > 0$ and negative real energy $E_a < 0$,
- iii. Gamow resonant states with complex discrete energy: $E_n \sim k_n^2$ $(k_n = \kappa_n - i\gamma_n)$ without restriction i.e. including virtual resonances with $\kappa_n < \gamma_n$,
- iv. complex k continuum of scattering states from a contour $+$ sinking now deeper into the complex k region.

TWO PARTICLE SYSTEMS IN GENERALIZED BERGGREN REPRESENTATION

Let us use the generalized Berggren-basis in a simple shell model problem, in which we study two neutrons outside a closed neutron core. The core is represented by the fully occupied states of a WS potential. The two-neutron basis states with angular momentum λ are formed as tensorial products of the ordered single-particle basis states belonging to the representation defined by the path $+$ (standard, Berggren, generalized). For simplicity we use a separable interaction between the neutrons. The matrix elements of the separable interaction have the simple form:

$$< \widetilde{kl}; \lambda | V | ij; \lambda > = -G_\lambda D_\lambda(kl) D_\lambda(ij) ,$$

where G_λ is the strength of the interaction and in the one-body matrix element

$$D_\lambda(ik) = < \tilde{i} | f(r) Y_{\lambda 0}(\hat{r}) | k > = R(ik) M_{ik}^\lambda ,$$

radial and the angular integrals are separated. In the radial integral

$$R(ij) = \int_0^{r_{max}} u_i(r,k_i) f(r) u_j(r,k_j) dr ,$$

we use a WS shape for $f(r)$. The dispersion relation for a two-particle system with this force is:

$$-\frac{1}{G_\lambda} = \sum_{i \leq j} \frac{R(ij)^2 |M_{ij}^\lambda|^2}{\omega_\alpha - \varepsilon_i - \varepsilon_j} ,$$

where ω_α values are the correlated energies. Due to Berggren metric we have $R(ij)^2$ and not $|R(ij)|^2$ as in the standard Hilbert metric. The complex wave function amplitudes for state $E = \omega_\alpha$ are:

$$X_{ik}^\alpha = N \frac{D_\lambda(ik)}{\omega_\alpha - \varepsilon_i - \varepsilon_k}$$

and the coefficient N is fixed by the normalization condition: $\sum_{i \leq k} (X_{ik}^\alpha)^2 = 1$.

Let us calculate only 0^+ two-neutron states. Then $j_1 = j_2 = j$ and $l_1 = l_2 = l$ and we have only $(lj)^2$ configurations in the two-particle basis. The partial wave content of a correlated bound state α,

$$A_{lj}^\alpha = \sum_{i(lj),k(lj)} (X_{ik}^\alpha)^2$$

depends only on ω_α but not on the shape of the complex contour. However the fragmentation of the correlated state α will depend on the contour and we can analyze how A_{lj}^α composed of pole-pole $(nlj)^2$ and continuum (contour-contour and pole-contour) components.

Let us show how our new method can be used for the halo nucleus: ^{11}Li.

HALO NUCLEUS ^{11}LI

This nucleus has been studied extensively both in $^9Li+n+n$ 3-body picture and recently in the $\alpha+t+n+n$ multi cluster model by Varga, et al.[10]. Naturally our simple model can not compete with these microscopical descriptions and it serves only to demonstrate how the new method works in a physical application and gives meaningful result. We consider the ^{11}Li as a system of ^{9}Li passive core and two valence neutrons moving around the core. The passive core is the $^9Li_{gs}$ $J^\pi = 3/2^-$ with $\nu(0s_{1/2})^2(0p_{3/2})^4 \otimes \pi(0s_{1/2})^2 0p_{3/2}$ configuration. The ground state of the ^{10}Li is unbound. For the structure of it we assume that a parity inversion takes place (as in the ^{11}Be) and we have a virtual $1s_{1/2}$ state coupled to the passive core: $[(\pi 0p_{3/2}) \otimes (\nu 1s_{1/2})]_{1^-}$ or $[(\pi 0p_{3/2}) \otimes (\nu 1s_{1/2})]_{2^-}$.

The core is simulated by a WS potential ($r_0 = 1.27$ fm and $a = 0.67$ fm) with parity dependent V_0. For even l values with $V_0 = 50$ MeV and $V_{so} = 16.5$ MeV and for odd l with $V_0 = 36.9$ MeV and $V_{so} = 12.624$ MeV. This later gives the $0p_{1/2}$ resonance at $(0.195, -0.047)$ MeV in agreement with the experiment[11]: $E_{p_{1/2}} \sim 0.21 \pm 0.05$

MeV. The $1s_{1/2}$ antibound state and the $0d_{5/2}$ resonance are at $E_a = -0.05$ MeV and at $E_d = (2.731, -0.545)$ MeV, respectively. These three unbound single particle states form the discrete part of the generalized Berggren basis. The complex scattering states along the (optionally different) contours for $l < 3$ and $j < 7/2$ up to $E_{max} = 10$ MeV form the continuum part of the basis. Two-body correlations between the two valence neutrons should result a bound ground state with $J^\pi = 3/2^-$ in ^{11}Li at the measured energy: $E_{gs} = -0.295 \pm 0.035$ MeV[12]. We assume a $[\psi_{gs}(^9Li)^{3/2^-} \otimes \psi_{2v}^{0^+}]^{3/2^-}$ ground state configuration and the G strength of our separable residual interaction is adjusted to reproduce the ground state energy $E_{gs} = -0.295$ MeV (of the $\psi_{2v}^{0^+}$ subsystem). This adjustment can be done even with the use of the "standard" basis, since the use of the Berggren basis or generalized Berggren basis does not change the l contribution of the state. With $G = G_0 = 0.00194$ MeV we get $\omega_{0_1^+} = -0.295$ MeV and the l compositions of the ground state wave function are the following: $A_{s_{1/2}}^{0_1^+} = 0.44$, $A_{l=1}^{0_1^+} = 0.48$ and $A_{l=2}^{0_1^+} = 0.08$ in good agreement with other findings[13]. With the standard basis the wave function is spread over many components. The few largest of them correspond to $(cp_{1/2})^2$ configurations with $\varepsilon_k + \varepsilon_i \sim 2\Re(E_{0p_{1/2}}) = 0.4$ MeV and the $(cs_{1/2})^2$ lying close to threshold (i. e. $\varepsilon_k + \varepsilon_i \sim 2E_a$).

If we use the generalized Berggren contour for the A_{lj} values do not change but their fragmentation do change and there are large contributions from the antibound pole-pole term and also from the $l = 0$ contour around the pole. The $(1s_{1/2})^2$ pole-pole configuration has a large $X_{(1s_{1/2})^2}^2$ component because its unperturbed energy $E_a + E_a = -0.100$ MeV is close to the correlated one: $E_{gs} = \omega_{0_1^+} = -0.295$ MeV.

If we keep G fixed at $G = G_0$ and leave out the $l = 0$ contour around the pole, the energy of the state shifts to -1.567 MeV and the $A_{s_{1/2}}^{0_1^+}$ increases to 0.77. If we keep $G = G_0$ and leave out the antibound pole but keep the $l = 0$ contour around the pole, the energy of the state changes to -2.691 MeV and the $A_{s_{1/2}}^{0_1^+}$ increases again to 0.77. If we neglect either the antibound pole or the $l = 0$ contour around the pole and readjust the value of G in order to get the state at -0.295 MeV then $A_{s_{1/2}}^{0_1^+}$ increases even more up to 0.98 . This shows that the contributions of the antibound pole and the scattering states along the $l = 0$ complex path add up with very strong destructive interference and this reduces the $l = 0$ content: $A_{s_{1/2}}^{0_1^+}$ considerably, even below the $l = 1$ content: $A_{l=1}^{0_1^+} = 0.48$.

We calculated higher-lying 0^+ excitations in ^{11}Li. We found that the first excited state (i. e. the state 0_2^+) is a resonance which appears at the complex energy $(0.202, -0.137)$ MeV. The corresponding wave function dominated by the p-states. We can follow the trajectory of the resonance as G increases from G=0, to the ground state strengths G_0. By increasing G starting form $G = 0$ the resonance first becomes narrower and moves down in energy. Here the state is dominated by the $(0p_{1/2})^2$ component and it is a physically meaningful resonance. For $G > 0.0005$ MeV however, the resonance starts widening and $l = 1$ pole-contour configurations become important. At G_0 there is a strong mixing with the contour configurations. Increasing G farther the $(0p_{1/2})^2$ pole configuration looses its importance and the resonance disappears.

SUMMARY

The generalized Berggren representation treats antibound states on an equal footing with bound states and resonances. The advantage of this approach is that the effects of the antibound poles and the complex continuum can be studied separately. In order to show the power of the method we applied it for shell model problem with two valence neutrons for drip line nuclei.

In the example of ^{11}Li nucleus we generated the bound ground state from the unbound basis elements and observed that the antibound state is very important in the building up the halo. However the huge contribution of the antibound pole is partly canceled by that of the complex continuum. We predicted a two-particle resonance in ^{11}Li but for making more reliable predictions probably a more sophisticated model has to be used.

ACKNOWLEDGMENTS

This work has been supported by FOMEC and Fundación Antorcha (Argentina), by the Hungarian OTKA fund No. T37991 and No. T46791 and by the Swedish Foundation for International Cooperation in Research and Higher Education (STINT).

REFERENCES

1. T. Berggren, Nucl. Phys. **A109**, 265 (1968).
2. R. Id Betan, R. J. Liotta, N. Sandulescu, and T. Vertse, Phys. Rev. Lett. **89**, 042501 (2002); Phys. Rev. **C67**, 014322 (2003).
3. N. Michel, W. Nazarewicz, M Ploszajczak, and K. Bennaceur, Phys. Rev. Lett. **89**, 042502 (2002).
4. Id Betan, R., Liotta, R. J., Sandulescu, N., and Vertse, T., *Phys. Rev. C.*, **67**, 014322 (2003).
5. N. Michel, W. Nazarewicz, M Ploszajczak, and J. Okolowicz, Phys. Rev. **C 67**, 054311 (2003).
6. Vertse, T., Curutchet, P., Liotta, R. J., and Bang, J., *Acta Phys. Hung.*, **65**, 305 (1989).
7. Berggren, T., Lind, P., *Phys. Rev. C.*, **47**, 768 (1993).
8. Id Betan, R., Liotta, R. J., Sandulescu, N., and Vertse, T., *Phys. Letters B*, **584**, 48–57 (2004).
9. Hagen, G., Hjorth-Jensen, M., Vaagen, J. S., submitted to *Phys. Rev. C*
10. Varga, K., Suzuki, Y., Lovas, R. G., and Bang, J., *Phys. Rev. C*, **66**, 041302R (2002).
11. Zinser, M., et al. *Phys. Rev. Lett.*, **75**, 1719 (1995).
12. Young, B. M., et al. *Phys. Rev. Lett.*, **71**, 4124 (1993).
13. Simon, H., et al. *Phys. Rev. Lett.*, **83**, 496 (1999).

Symmetry conserving mean-field theory and applications for halo phenomena at the neutron drip line

P. Ring

Physikdepartment T30, Technische Universität München, D-85748 Garching, Germany

Abstract. Symmetry projected Hartree-Fock and Hartree-Fock-Bogoliubov equations have been introduced recently in non-relativistic many-body calculations. This method provides a very useful frame to include projection methods in covariant density functional theory. We derive the number projected Relativistic Hartree-Bogoliubov equations and we discuss first applications for the treatment of halo-phenomena in light nuclei near the neutron drip line.

INTRODUCTION

Mean-field models with effective forces have been very successful in describing the gross features of many-body systems determined by quantum mechanics. It is know that mean-field approaches are appropriate for systems with a very large number of particles. Nonetheless they have also been very useful for the description the of finite quantum systems, for instance mesoscopic systems as the the atomic nucleus. The ground-state properties of atomic nuclei have been well described within Hartree-Fock (HF) and Hartree-Fock-Bogoliubov (HFB) mean-field approaches based on various effective density dependent energy functionals [1]. However, the applications of the mean-field approximation to a finite system suffers from a fundamental problem in regions of phase transitions: it leads to sudden sharp changes of the characteristic observables, when the corresponding order parameter goes to a phase transition, a property, which is characteristic for an infinite system. The fact that phase transitions in finite system are smoothed out in finite systems is not reproduced properly in the mean field approximation, because fluctiations are not taken into account in a appropriate way. Therefore energy functional theories bases on product states with particles moving independently in an averaged field cannot be exact for finite systems in regions of phase transitions. In such cases a sharp phase transition is an artefact of the mean-field approach and is not observed in experimental data.

There are various methods in the literature to consider quantal fluctuations on top of the mean-field solution for the finite system. In cases where the phase transition is connected with a broken symmetry of the original Hamiltonian projection methods have turned out to be very powerful method do deal with this problems [1]. In this context it is essential that the projection has to be carried out before the variaton, because otherwise the abrupt changes in underlying mean field cannot be smoothed by an additional projection after the variation. The variation after projection is usually connected with

CP726, *Nuclear Physics, Large and Small: International Conference on*
Microscopic Studies of Collective Phenomena, edited by R. Bijker, R. F. Casten, and A. Frank
© 2004 American Institute of Physics 0-7354-0207-8/04/$22.00

a considerable numerical effort. Therefore it has been applied in the literature mostly in simple models like the BCS-approximation [2] or with additional approximations as for instance the cranking approximation in case of angular momentum projection [3] or the Lipkin-Nogami approach [4] in superfluid systems: One way to carry ont an exact projection before the variation is the gradient method, where the minima in the projected energy surface are following a path in the direction of the steepest decent [5, 6, 7]. In the realistic cases of heavy nuclei, where the number of variational parameters reaches astronomical values, these methods are connected with a tremendous numerical effort. In fact they have been applied in the literature so far only to the case of particle number violation in superfluid systems [7]. An additional drawback of gradient methods lies in the fact that they can be applied only for the global or local minima in the energy surface. General variational problems connected with stationary stationary solutions such as saddly points cannot be treated by gradient methods.

Recently a new method has been proposed by Sheikh and Ring [8] to carry out a variation before projection. It proves to be considerably simpler than the gradient method, because is is based on the usual technique to derive non-linear differential equations for the solution of the variational problem. Without projection these equations are the well known Hartree-Fock and Hartree-Fock-Bogoliubov equations. Very powerful techniques have been develop over the years to solve such equations. In Refs. [8, 9] projected Hartree-Fock and Hartree-Fock-Bogoliubov equations have been developed, which are nearly identical to the unprojeded equations. The conventional codes to solve the unprojected problem can be extended without much effort to include an exact projection before the variation. In particular these methods can also be used for cases where the variational solution of the variational problem does not correspond to an minimum in the energy surface.

We therefore discuss in this contribution the application of the Sheikh-Ring method to covariant density functional theories. These theories are based on selfconsistent Dirac-Hartree equations or Dirac-Hartree-Bogoliubov equations [10] and it is know that the solutions of these equations correspond to saddle points in the energy surface. Because of the no-sea approximation for each of these solutions there exists an infinite number of directions on the energy surface, where the energy decreases. These directions are connected with the admixture of negative energy solutions usually neglected in the the no-sea approximation. Therefore gradient methods cannot be applied for the solution of the projected covariant density functionals. This does not apply, however for the Sheikh-Ring method, which is based on non-linear differential equations and not on gradients.

COVARIANT DENSITY FUNCTIONAL THEORY FOR SUPERFLUID NUCLEI

The starting point of relativistic mean field theory is usually a Lagrangian containing the nucleon spinors ψ_i ($i = 1 \ldots A$), various classical meson fields $\phi_m = \sigma, \omega^\mu, \vec{\rho}^\mu$ and the electro-magnetic field A^μ. However, it can also be derived from a energy density functional of the form

$$E_{RMF}[\psi, \bar{\psi}, \phi_m] = \int \psi_i^+ h_0 \psi_i d^3r \pm \int \left[\frac{1}{2} (\nabla \phi_m)^2 + U(\phi_m) \right] d^3r$$

$$+ \int \left[\bar{\psi}_i \Gamma_m \psi_i \phi_m \right] d^3r. \tag{1}$$

$\hat{h}_0 = \alpha \cdot \mathbf{p} + \beta m$ is the free Dirac Hamiltonian. The upper sign holds for scalar and the lower sign for vector fields ϕ_m. The vertices Γ_m characterize the Dirac structure of the various mesons ($\Gamma_\sigma = g_\sigma$, $\Gamma_\omega^\mu = g_\omega \gamma^\mu$, etc.) and $U(\phi_m)$ describes a density dependence through a non-linear self-interaction of the mesons [11]. By using the definition of the relativistic single-nucleon density matrix

$$\hat{\rho}(\mathbf{r}, \mathbf{r}', t) = \sum_{i=1}^{A} |\psi_i(\mathbf{r}, t)\rangle \langle \psi_i(\mathbf{r}', t)|, \tag{2}$$

the total energy can also be written as a functional of the density matrix $\hat{\rho}$ and of the meson fields ϕ_m

$$E_{RMF}[\hat{\rho}, \phi_m] = \mathrm{Tr} \left[(\hat{h}_0 + \beta \Gamma_m \phi_m) \hat{\rho} \right] \pm \int \left[\frac{1}{2} (\nabla \phi_m)^2 + U(\phi_m) \right] d^3r \tag{3}$$

The trace operation involves a sum over the Dirac indices and an integral in coordinate space.

According to density functional theory the single-particle Hamiltonian $\hat{h}$ is obtained as the functional derivative of the energy with respect to the single-particle density matrix $\hat{\rho}$

$$\hat{h}_D = \frac{\delta E}{\delta \hat{\rho}} = \hat{h}_0 + \beta \Gamma_m \phi_m \tag{4}$$

Pairing correlations are essential for a correct description of structure phenomena in spherical open-shell nuclei and in deformed nuclei. HFB theory [1] provides a unified description of ph- and pp-correlations on a mean-field level by using two average potentials: the self-consistent Hartree field which encloses all the long range ph correlations, and the pairing field $\hat{\Delta}$ which sums up the pp-correlations. In this theory the energy is a functional of two densites, the normal density $\hat{\rho}$ and the pairing density $\hat{\kappa}$

$$E[\hat{\rho}, \hat{\kappa}, \phi_m] = E_{RMF}[\hat{\rho}, \phi_m] + E_{pair}[\hat{\kappa}], \tag{5}$$

where $E_{RMF}[\hat{\rho}, \phi]$ is the RMF-functional defined in Eq. (3), and the pairing energy $E_{pair}[\hat{\kappa}]$ is given by

$$E_{pair}[\hat{\kappa}] = \frac{1}{4} \mathrm{Tr}[\hat{\kappa}^* V^{pp} \hat{\kappa}]. \tag{6}$$

V^{pp} denotes a general two-body pairing interaction. $\hat{\rho}$ and $\hat{\kappa}$ can be combined to the generalized density matrix of Valatin [12]

$$\mathcal{R} = \begin{pmatrix} \rho & \kappa \\ -\kappa^* & 1 - \rho^* \end{pmatrix}. \tag{7}$$

and the correpsonding mean field hamiltonian, the Dirac-Hartree-Bogoliubov hamiltonian is obtained as a functional derivativ of $E[\hat{\rho}, \hat{\kappa}, \phi_m]$ with respect to $\mathscr{R}$

$$\mathscr{H} = \frac{\delta E}{\delta \mathscr{R}} = \begin{pmatrix} \hat{h}_D - \lambda & \hat{\Delta} \\ -\hat{\Delta}^* & -\hat{h}_D + \lambda \end{pmatrix}. \tag{8}$$

with

$$\hat{h}_D = \frac{\delta E}{\delta \hat{\rho}} \quad \text{and} \quad \hat{\Delta} = \frac{\delta E}{\delta \hat{\kappa}}. \tag{9}$$

The Dirac-Hartree-Bogoliubov spinors are found as eigenstates of the Hamiltonian (8). They yield the density $\hat{\rho}$ and $\hat{\kappa}$. $\hat{\rho}$ is used as source in the Klein-Gordon equations for the various meson fields providing the potentials in the Dirac-hamiltonian $\hat{h}_D$ of the next iteration. $\hat{\kappa}$ determines the new pairing field $\hat{\Delta}$ (for details see [13]).

SYMMETRY CONSERVING DIRAC-HARTREE-BOGOLIUBOV THEORY

An essential advantage of the mean-field approach to nuclear structure in general, and of the relativistic Hartree-Bogoliubov framework in particular, is the possibility to break symmetries. For example, by using deformed, symmetry violating single particle states in Slater determinants, considerable mixing of configurations can be taken into account, as compared to products of states that conserve the symmetry. Of course, the exact many-body wave function $|\Psi\rangle$ has to be an eigenstate of the symmetry operators of the Lagrangian, i.e. in general it has a very small overlap with the corresponding mean field function $|\Phi\rangle$. However, since the early studies of Peierls [14], Elliot [15] and Kerman [16], it has been well known that restoring the symmetries in a deformed mean-field function by projecting onto the eigenspace of the symmetry operators, in many cases correlated many-body states are obtained that are very close to the exact eigenstates of the Hamiltonian. Over the years different techniques have been developed to restore the rotational symmetry in deformed nuclei by angular momentum projection, or the gauge symmetry related to particle number conservation for BCS or HFB wave functions [1].

In a variational calculation of nuclear ground or excited states the variation should, in principle, be performed after the projection onto the eigenspace of the symmetry operator[17]. In general the resulting equations are very complicated and difficult to solve. In cases of large symmetry violations, i.e. in the limit of strong deformation, it is often sufficient to carry out the projection only on the average. A good approximation is to solve the variational equations by using a Lagrange parameter, which guarantees that the wave function on the average carries the right quantum numbers. This is a well known procedure in the case of particle-number violation, or in the cranking approach to nuclear rotation. The resulting solutions represent intrinsic wave functions and the projection has still to be performed. For many observables, for instance the total energy, the expectation values calculated with projected states differ only slightly from those obtained with unprojected wave functions. Since the projected wave functions include additional correlations, the corresponding energies are usually somewhat lower than

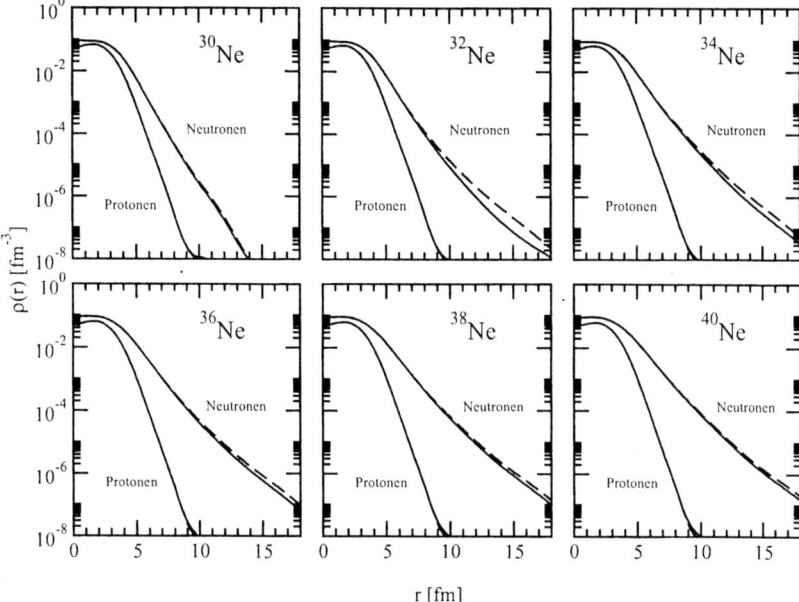

FIGURE 1. Single-nucleon density distributions of Ne isotopes. For neutrons, dashed curves correspond to results obtained without particle number projection; solid lines correspond to projected results.

those calculated with unprojected states. In many cases the additional correlation energy can be absorbed in the parameters of the density functional, adjusted to experimental data. This explains the sucess of simple mean-field calculations that do not include explicit symmetry restoration. The necessary condition, however, is that the symmetry must be relatively strongly broken. For transitional nuclei in which phase transitions occur and where the corresponding order parameters, such as the deformation parameter or pairing gap, become small, fluctuations have to be taken into account explicitly. In order to obtain accurate solutions, variation has to be performed after projection.

As a simple example, here we consider particle number projection. In the framework of non-relativistic mean-field models, a well known procedure is to evaluate the energy

$$E^N = \frac{\langle \Phi | \hat{H} P^N | \Phi \rangle}{\langle \Phi | P^N | \Phi \rangle} \tag{10}$$

and to minimize the projected energy surface within the manyfold of BCS or HFB wave functions $|\Phi\rangle$ by using the gradient method [5, 6, 18]. As we have seen in the introduction for relativistic models the gradient method cannot be applied directly, because the solutions of the stationary RMF-equations do not correspond to minima on the energy surface. Rather, in the *no-sea* approximation they correspond to saddle-point configurations. Thus it is always possible to find a gradient that leads to configurations with lower energy.

In the non-relativistic framework, symmetry projected HFB equations have been recently derived in Ref. [8]. They allow the determination of the stationary solution by successive diagonalization of the corresponding projected HFB matrix. The basic idea behind this derivation is that there is a one to one correspondence between the unprojected wave function $|\Phi\rangle$ in Eq. (10), and the corresponding unprojected densities $\hat{\rho}$ and $\hat{\kappa}$. This implies that one can derive an analytic expression for the projected energy (10), as a function of $\hat{\rho}$ and $\hat{\kappa}$: $E^N(\hat{\rho}, \hat{\kappa})$. From the projected energy functional the projected HFB equations are derived

$$
\begin{pmatrix} \hat{h}^N & \hat{\Delta}^N \\ -\hat{\Delta}^{N*} & -\hat{h}^{N*} \end{pmatrix} \begin{pmatrix} U(\mathbf{r}) \\ V(\mathbf{r}) \end{pmatrix}_k = \begin{pmatrix} U(\mathbf{r}) \\ V(\mathbf{r}) \end{pmatrix}_k E_k^N ,
\tag{11}
$$

with the projected fields

$$
\hat{h}^N = \frac{\delta E^N}{\delta \hat{\rho}} \qquad \text{and} \qquad \hat{\Delta}^N = \frac{\delta E^N}{\delta \hat{\kappa}} .
\tag{12}
$$

The wave functions $U(\mathbf{r})$ and $V(\mathbf{r})$ correspond to the unprojected (intrinsic) state $|\Phi\rangle$. They determine the unprojected densities $\hat{\rho}$ and $\hat{\kappa}$.

This method can also bee applied to the relativistic Hartree-Bogoliubov framework. In this case one starts with the energy functional of Eq. (5). It contains as variational parameters the densities $\hat{\rho}$ and $\hat{\kappa}$, and the meson fields ϕ_m. The single-particle term reads: $Tr[(\hat{h}_0 + \Gamma_m \phi_m) \hat{\rho}]$, $E_{mes}(\phi_m)$ depends only on the ϕ_m's, and a pairing term is determined by a two-body Gogny-type interaction $\langle \Phi | \hat{V}^{pp} | \Phi \rangle$. The starting point is the projected energy functional

$$
E^N(\hat{\rho}, \hat{\kappa}, \phi_m) = Tr\left[(\hat{h}_0 + \Gamma_m \phi_m) \hat{\rho}^N \right] + E_{mes}(\phi_m) + E_{pair}^N,
\tag{13}
$$

with the projected single-particle density and the projected pairing energy read

$$
\hat{\rho}_{nn'}^N = \frac{\langle \Phi | \psi_{n'}^+ \psi_n P^N | \Phi \rangle}{\langle \Phi | P^N | \Phi \rangle} \qquad \text{and} \qquad E_{pair}^N = \frac{\langle \Phi | \hat{V}^{pp} P^N | \Phi \rangle}{\langle \Phi | P^N | \Phi \rangle} ,
\tag{14}
$$

respectively. By using the techniques of Ref: [8, 9, 19], both quantities can be expressed analytically in terms of the unprojected densities $\hat{\rho}$ and $\hat{\kappa}$. Carrying out the variation with respect to $\hat{\rho}$, $\hat{\kappa}$, and with respect to the meson fields ϕ_m, one obtains the number-projected HB-equations (11) and the corresponding projected Klein-Gordon equations:

$$
-\Delta\phi_m + U'(\phi_m) = \pm Tr\left[\Gamma_m \hat{\rho}^N \right] .
\tag{15}
$$

APPLICATIONS TO HALO PHENOMENA NEAR THE NEUTRON DRIP LINE

There are, of course, many cases of weak pairing correlations where number projection becomes important [7]. They include, in particular, systems with only few valence

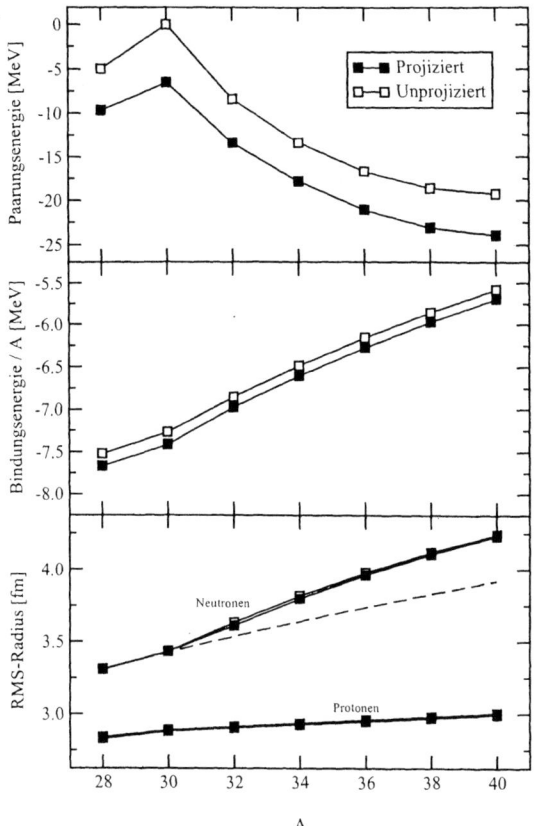

FIGURE 2. Pairing energies, binding energies per nucleon, and *rms* radii of Ne isotopes. White and black symbols correspond to calculations without and with number-projection before the variation, respectively. Virtually identical results for the radii are obtained in the two cases. The dashed line in the lower panel corresponds to the $N^{1/3}$ empirical curve.

particles. In order to illustrate the effect, we apply the number-projected RHB model for the description of the halo-phenomenon in light nuclei [20]. In the case when the halo is formed by just one or two nucleons, it is not obvious that a description in terms of unprojected RHB presents a reasonable approximation. In Fig. 1 we display the proton and neutron density distributions of Ne isotopes, obtained with and without particle number projection before variation. As it has been shown [21] the even-even Ne nuclei with $A = 32$ are predicted to display the formation of a neutron halo structure. We notice in Fig. 1 that the neutron density profiles calculated with and without number projection are almost identical, with the exception of ^{32}Ne. This nucleus is exactly at the point of phase transition, with only two valence neutrons forming the halo. In Fig. 2 the corresponding pairing energies, binding energies, and *rms* radii are shown. The calculated radii are practically identical in both cases. The energies differ by ≈ 5 MeV for the pairing energy, but only by ≈ 0.2 MeV for the binding energies. This is a well known effect. The increase in the pairing energy is compensated by the reduced mean-field potential. We notice that particle number projection produces a constant energy shift. The correlation energy does not depend on the number of neutrons and, therefore, it can be absorbed in the parameters of the effective interaction adjusted to empirical binding energies.

ACKNOWLEDGMENTS

This work has been supported by the Bundesministerium für Bildung und Forschung under the project 06 MT 193 and by the Gesellschaft für Schwerionenforschung, Darmstadt. In particular I would like to thank Emilio Lopes, Javid Sheikh and Mario Stoitsov, for essential contributions to this work.

REFERENCES

1. Ring, P., and Schuck, P., *The nuclear many-body problem*, Springer, Heidelberg, 1980.
2. Dietrich, K., Mang, H. J., and Pradal, J. H., *Z. Phys.*, **190**, 357 (1966).
3. Beck, R., Mang, H. J., and Ring, P., *Z. Phys.*, **231**, 26 (1970).
4. Valor, A., Egido, J. L., and Robledo, L. M., *Nucl. Phys.*, **A665**, 46 (2000).
5. Egido, J. L., and Ring, P., *Nucl. Phys.*, **A383**, 189 (1982).
6. Egido, J. L., and Ring, P., *Nucl. Phys.*, **A388**, 19 (1982).
7. Anguiano, M., Egido, J. L., and Robledo, L. M., *Phys. Lett.*, **B545**, 62 (2002).
8. Sheikh, J. A., and Ring, P., *Nucl. Phys.*, **A665**, 71 (2000).
9. Sheikh, J. A., Lopes, E. C., and Ring, P., *Yadernaya Fisika*, **64**, 1 (2001).
10. Gonzales-Llarena, T., Egido, J. L., Lalazissis, G. A., and Ring, P., *Phys. Lett.*, **B379**, 13 (1996).
11. Boguta, J., and Bodmer, A. R., *Nucl. Phys.*, **A292**, 413 (1977).
12. Valatin, J. G., *Phys. Rev.*, **122**, 1012 (1961).
13. Ring, P., *Progr. Part. Nucl. Phys.*, **37**, 193 (1996).
14. Peierls, R. E., and Yoccoz, J., *Proc. Phys. Soc.*, **A70**, 381 (1957).
15. Elliot, J. P., and Flowers, B. H., *Proc. Roy. Soc.*, **A242**, 57 (1957).
16. Kerman, A. K., *Ann. Phys. (N.Y.)*, **12**, 300 (1961).
17. Zeh, H. D., *Z. Phys.*, **188**, 361 (1965).
18. Anguiano, M., Egido, J. L., and Robledo, L. M., *Nucl. Phys.*, **A696**, 467 (2001).
19. Sheikh, J. A., Ring, P., and Rossignoli, R., *Phys. Rev.*, **C66**, 044318 (2002).
20. Lopes, E. C., *Symmetry projected covariant density functional theory with applications for exotic nuclei*, Phd thesis, Technical University of Munich (unpublished) (2002).
21. Pöschl, W., Vretenar, D., Lalazissis, G. A., and Ring, P., *Phys. Rev. Lett.*, **79**, 3841 (1997).

Neutrons and Protons - Vive la Différence

D. D. Warner

CCLRC Daresbury Laboratory, Daresbury, Warrington, WA44AD, UK

Abstract. Particular features and symmetries associated with the neutron-proton interaction are addressed. Coulomb energy differences provide a sensitive probe of the changes in single particle structure with increasing angular momentum. Recent results on the nuclei ^{55,56}Cr provide an example of the particular importance of spin-orbit partner orbits in explaining the changing shell structure encountered in nuclei with an increasing neutron excess. The phenomenon of neutron-proton pairing is explored within the framework of the isospin-invariant IBM, which provides a first estimate of spectroscopic factors for deuteron transfer.

MIRROR NUCLEI AT HIGH SPIN

The charge independence of the nuclear force is one of the fundamental tenets of nuclear structure. Until the last decade studies of the Coulomb energy differences (CED) in mirror partners were focused almost exclusively on the ground states of nuclei. However, in recent years the massive increases in sensitivity and resolving power which have resulted from the advent of large arrays of gamma-ray detectors have provided a possibility to study nuclei with N<Z to ever increasing excitation energy. In particular, in the $f_{7/2}$ shell it has now become possible to extract information on CED between excited states to a relatively high spin and to investigate how the Coulomb energy changes as the spin increases. Such changes are simply caused by changes in the spatial rearrangement of the protons.

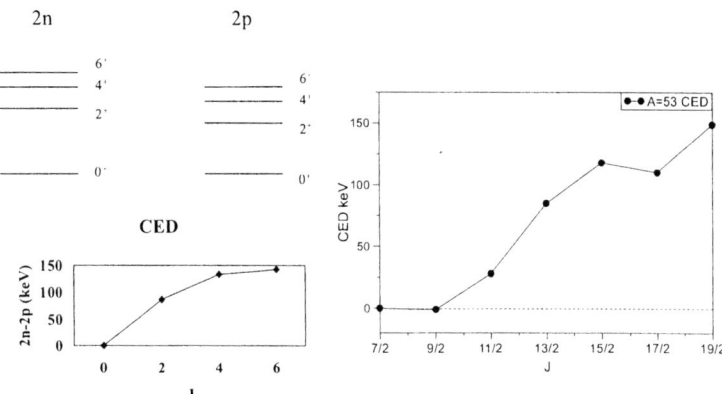

FIGURE 1. Left: Schematic illustration of CED for 2 nucleons in the $f_{7/2}$ orbital. Right: Measured CED for the A=53 mirror pair. [1]

CP726, *Nuclear Physics, Large and Small: International Conference on Microscopic Studies of Collective Phenomena*, edited by R. Bijker, R. F. Casten, and A. Frank
© 2004 American Institute of Physics 0-7354-0207-8/04/$22.00

A simple example is given in Fig. 1 for two nucleons in the $f_{7/2}$ shell. The CED involve a difference in *excitation* energies so that all features associated with the bulk Coulomb energy and with the differences in Coulomb energy of the ground states are eliminated. The breaking of the nucleon pair to form states with J>0 in each case causes a decrease in the spatial overlap of the two nucleons; in the case of the proton pair, this decreases the Coulomb energy and hence gives rise to a compressed spectrum for the 2p system. A quantitative estimate of the 2-particle CED from the harmonic oscillator wave functions for the $f_{7/2}$ orbital is included in the figure.

The most recent results have just been obtained for the A=53 mirror system [1]. The reaction used involved a ^{32}S beam impinging on a ^{24}Mg target to produce a ^{56}Ni compound nucleus. The reaction channels of interest were the 2pn and p2n channels leading to the A=53 pair. Since ^{53}Fe was well studied prior to the measurement, the real goal of the experiment was to extract the structure of the T_Z= -1/2 nucleus ^{53}Co . The experiment was performed using the Gammasphere array at the ATLAS facility at Argonne National Laboratory and the measured CED are plotted on the right in Fig. 1. The effect anticipated on the left of the figure is evident, both nuclei having been studied up to the band terminating spin of 19/2$^-$ which corresponds to full alignment of the three $f_{7/2}$ holes. There is thus only one pair of nucleons to break in each mirror partner.

Calculating the Coulomb energy differences

In using shell model wave functions to estimate the Coulomb energy contribution for each level, it is necessary to define a set of two proton Coulomb matrix elements. This has been done for A=53 using the pure $f_{7/2}$ wave functions of Kutschera, Brown and Ogawa (KBO) [2] and the result is shown in the top panel of Fig. 2 Values can also be obtained from the A=42 mirror pair [3](Fig. 2, middle). The CED between two states can be represented as $E_C(J) = BE_J(^{42}Ti) - BE_J(^{42}Ca) + constant$ where the constant accounts for the neutron-proton mass difference and the Coulomb interaction with the core. Finally, the bottom panel of Fig. 2 shows the result of a fit to the CED of the A=47 and 49 mirror pairs, using full fp- shell wave functions [1, 4]. It is immediately evident that the three methods produce almost identical values which contain an anomaly; the values *increase* in going from the J=0 to the J=2 state in each case. This is clearly unphysical on a simple intuitive basis, since breaking a proton pair must decrease the Coulomb energy. In contrast, matrix elements extracted from the A=42 isobaric triplet behave as expected, as do the pure harmonic oscillator values.

The source of the anomaly remains unclear. It is probably not surprising that the matrix elements need to be renormalised; it is perhaps more surprising that a single set of values can give such good agreement across the entire $f_{7/2}$ shell. In a recent study [4], the origin of this anomaly is attributed to an isospin non-conserving contribution to the nuclear interaction for J=2 for mirror pairs. However, the behaviour of the A=42 nuclei suggests core excitation as perhaps a more likely source.

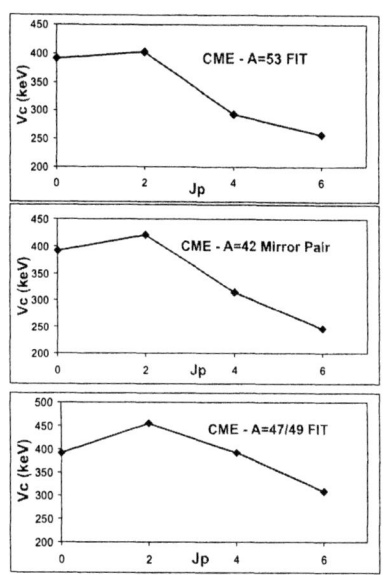
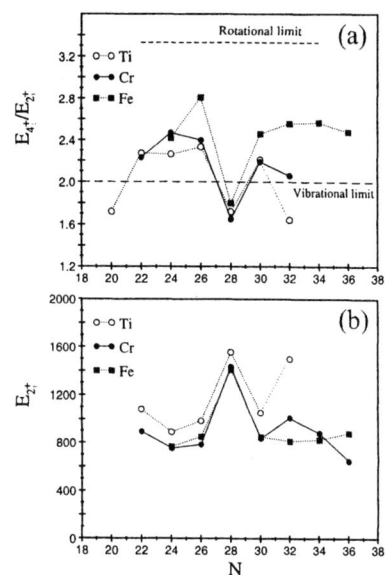

FIGURE 2. (Left) 2-proton Coulomb matrix elements extracted using three methods (see text).
FIGURE 3. (Right) $E(4_1^+)/E(2_1^+)$ (a) and $E(2_1^+)$ (b) for Ti, Cr and Fe isotopes vs. neutron number N.

NEW SUBSHELL FOR N=32

Recent studies have shown the shell model magic numbers to be remarkably fragile, appearing and disappearing over relatively small regions of the nuclear chart. Such structural changes stem principally from the dominant attractive strength of the spin-isospin part of the effective nucleon-nucleon interaction which manifests itself by shifts in the effective single particle energies through the monopole part of the shell model Hamiltonian [5]. The importance of the spin-orbit partner contribution to the neutron-proton interaction and its role in inducing rapid shape changes in nuclei nearer to stability was first recognised by Federman and Pittel [6].

Empirically, the onset of a breakdown in the anticipated shell structure is signalled by a change in the energies and electromagnetic properties of the low-lying collective states across a chain of isotopes or isotones. The different isotopes of Cr (Z=24) are a case in point. The reaction ^{48}Ca(^{11}B,pxn) has been used to populate states in the nuclei ^{55,56}Cr at the Yale ESTU Tandem Van de Graaff accelerator using the YRASTBall array [7].

The behaviour of $E(2_1^+)$ and the energy ratio $E(4_1^+)/E(2_1^+)$ are plotted Fig. 3. It is clear that none of the nuclei plotted reach the rotational limit but rather reach maximum values of $E(4_1^+)/E(2_1^+)$ which are characteristic of gamma softness. Above N=28, the iron isotopes return to a constant value, still reflecting the nearness of the Z=28 shell closure. Yet in the chromium isotopes, where the number of valence protons is at its maximum, after a slight increase the ratio $E(4_1^+)/E(2_1^+)$ returns towards the vibrational limit in ^{56}Cr. Such behaviour signals a significant and unexpected drop in collectivity in this nucleus, a conclusion reinforced by the concomitant rise in $E(2^+_1)$ shown in Fig. 3(b). This latter feature led to the initial suggestion [8] of a sub-shell gap at N=32 for the Cr isotopes. The N=32 sub-shell effect is not present in Fe and therefore appears to weaken with increasing Z. Very recent data on ^{52}Ti [9], included in Fig. 3, seems to confirm this behaviour.

The presence of subshell gaps at both N=32 and 34 have been predicted in this region [8, 10]. In ^{49}Ca, with Z=20 and 1 neutron in the fp orbits, relatively large energy gaps exist between both the $p_{3/2}$ - $p_{1/2}$ and $p_{1/2}$-$f_{5/2}$, suggesting the possibility of subshell effects occurring at both N=32 and 34. In ^{57}Ni, however, the $f_{5/2}$ state has come down in energy to lie between the $p_{3/2}$ and $p_{1/2}$, eliminating both gaps. This transition can be ascribed to an increasing $f_{7/2}$-$f_{5/2}$ monopole contribution to the shell model effective interaction as protons are added to the $f_{7/2}$ orbital, and can be described in terms of an effective single particle energy [11] of the $\nu f_{5/2}$ state which decreases as $f_{7/2}$ protons are added. The behaviour of the Cr isotopes and the differences in the theoretical predictions probe the *details* of the transition in single particle energies across the shell. In one case [8], at Z=24, N=32 a significant gap still remains between the $p_{3/2}$ and $p_{1/2}$ states, sufficient to explain the observed subshell effect; in the other [10], a large gap exists between the $p_{1/2}$ and $f_{5/2}$ neutron orbits for the Ca and Ti isotopes, resulting in a N=34 subshell, which is gradually eroded with increasing Z.

DEUTERON TRANSFER ON THE N=Z LINE

The interacting boson model (IBM) [12] provides a description of nuclei in terms of correlated nucleon-pair excitations which are treated as bosons. As such, it offers a natural framework to discuss the issue of two-nucleon transfer. A description of deuteron transfer requires a version of IBM which involves bosons corresponding to neutron-proton (np) pairs and of particular relevance is the IBM-4 [13] since it contains np pairs with isospin $T = 0$ and $T = 1$. The full IBM-4 employs bosons with orbital angular momentum $L = 0$ (s boson) or $L = 2$ (d boson), with intrinsic spin $S = 0$ or $S = 1$, and with isospin $T = 0$ (if $S = 1$) or $T = 1$ (if $S = 0$).

To avoid the complexity of the full IBM-4, it is instructive to confine the analysis to $L = 0$ bosons. This simplification preserves the complete spin-isospin structure of the model—crucial for the study of deuteron-transfer properties—and can be put to use in the analysis of the competition between isoscalar and isovector pairing in self-conjugate nuclei [14].

Two different symmetry classifications occur in the $L = 0$ IBM-4:

$$U(6) \supset \left\{ \begin{array}{c} SU(4) \\ U_T(3) \otimes U_S(3) \end{array} \right\} \supset SO_T(3) \otimes SO_S(3). \tag{1}$$

The total number of bosons $[N_b]$ labels U(6) while $SO_T(3)$ and $SO_S(3)$ are associated with the total isospin T and the total spin S of the bosons. A simple Hamiltonian that describes the transition from one limit of (1) to the other is of the form

$$H = aC_2[SU(4)] + bC_1[U_S(3)], \tag{2}$$

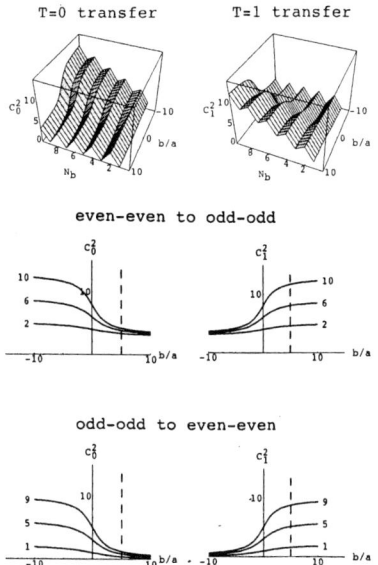

FIGURE 4. The $T = 0$ and $T = 1$ d-transfer intensities C_T^2 between $N = Z$ nuclei vs. b/a and boson number N_b for the lowest $T = 0$ and $T = 1$ eigenstates of the Hamiltonian (2). Upper part: the entire surface for $1 \leq N_b \leq 10$ and $-10 \leq b/a \leq 10$. Middle and lower parts: the even-even to odd-odd and odd-odd to even-even intensities for specific N_b-values. In odd-odd nuclei (N_b odd) the figure shows the isospin-allowed intensity to $T = 0$ for C_0^2 and to $T = 1$ for C_1^2. The dashed line indicates the value of b/a obtained from nuclear masses.

where $C_n[G]$ denotes a linear or quadratic ($n = 1, 2$) Casimir operator of the algebra G. The first term in (2) is associated with SU(4) and implies equal single-boson energies and boson-boson interactions in the two isospin-spin channels $(S, T) = (0, 1)$ and $(1, 0)$. The second term breaks this equivalence, its main physical origin [15] being the one-body spin-orbit term of the shell model. The transition from SU(4) to $U_T(3) \otimes U_S(3)$ is governed by the single parameter b/a and results can be obtained from a simple diagonalization [14].

Deuteron transfer is described in this model by the operators $b_{01}^\dagger$ ($b_{10}^\dagger$) for $T = 0$ ($T = 1$) transfer, where $b_{01}^\dagger$ ($b_{10}^\dagger$) creates a boson with $T = 0$ and $S = 1$ ($T = 1$ and $S = 0$), both with orbital angular momentum $L = 0$. The full derivation [16] illustrates that the transfer intensity between the states $|[N_b]\phi_A T_A S_A\rangle$ and $|[N_b + 1]\phi_B T_B S_B\rangle$ (where ϕ_A and ϕ_B are additional labels) is proportional to the quantity

$$C_T^2 \equiv \langle [N_b + 1]\phi_B T_B S_B \| b_{TS}^\dagger \| [N_b]\phi_A T_A S_A \rangle^2, \tag{3}$$

By varying the ratio b/a one can study the qualitative features of deuteron transfer with changing $T = 0$ versus $T = 1$ pairing correlations. The result for $N = Z$ nuclei is shown in Fig. 4. For the transfer starting from an even-even $N = Z$ nucleus (which has a ground state with $T = 0$) one finds two states excited in the low-energy region of the odd-odd nucleus corresponding to $T = 0$ and $T = 1$ transfer, respectively. Not surprisingly, the two states are equally excited in the SU(4) limit ($b/a = 0$) while otherwise the sign of b/a determines which of the two transfer intensities is strongest. The ground state of the odd-odd nucleus has $T = 0$ for $b/a < 0$ and $T = 1$ for $b/a > 0$ and this may lead to an isospin selection rule in the odd-odd to even-even case if one only considers transfer to the low-energy states of the even-even nucleus which have $T = 0$. It is seen from Fig. 4 that the transfer intensities rapidly change around the SU(4) limit but saturate quickly at large values of $|b/a|$.

What are appropriate values of b/a in actual nuclei? Qualitative arguments indicate that it should be positive, while a more quantitative estimate can be given based on nuclear masses. The $L = 0$ IBM-4 can be used to calculate binding energies of $N = Z$ nuclei [17]. From the fit to the nuclear masses the value $b/a \approx 5$ is obtained. Even if there is considerable uncertainty in the value of this ratio, the fact that the deuteron-transfer intensities quickly saturate for large b/a leads to a clear prediction of this analysis: The favoured deuteron-transfer mode in this mass region has $T = 1$ rather than $T = 0$ character. Some appreciable strength of the latter can only be expected in the transfer from an even-even to an (excited) $T = 0$ state of an odd-odd nucleus.

ACKNOWLEDGMENTS

I wish to thank many colleagues who have contributed to these studies, in particular Mike Bentley, Scott Williams, Katie Chandler, Duncan Appelbe, Alejandro Frank and Piet Van Isacker. I would also like to thank Stu Pittel for some 20 years of accumulated interaction and discussion, most of it stimulating and all of it enjoyable.

REFERENCES

1. Williams, S. J., et al., *Phys. Rev.*, **C68**, 011301 (2003).
2. Kutschera, W., Brown, B. A., and Ogawa, K., *Riv. Nuovo. Cimento Soc. Ital. Fis.*, **31**, 1 (1978).
3. Poves, A., et al., *Nucl. Phys.*, **A694**, 157 (2001).
4. Zuker, A. P., et al., *Phys. Rev. Lett.*, **89**, 142502 (2002).
5. Otsuka, T., et al., *Phys. Rev. Lett.*, **87**, 082502 (2001).
6. Federman, P., and Pittel, S., *Phys. Lett.*, **B69**, 385 (1977).
7. Appelbe, D. E., et al., *Phys. Rev.*, **C67**, 034309 (2003).
8. Prisciandaro, J. I., et al., *Phys. Lett.*, **B510**, 17 (2001).
9. Janssens, R. V. F., et al., *Phys. Lett.*, **B546**, 55 (2002).
10. Honma, M., et al., *Phys. Rev.*, **C65**, 061301 (2002).
11. Utsuno, Y., et al., *Phys. Rev.*, **C60**, 054315 (1999).
12. Iachello, F., and Arima, A., *The Interacting Boson Model*, Cambridge University Press, Cambridge, 1987.
13. Elliott, J., and Evans, J., *Phys Lett.*, **B101**, 216 (1981).
14. Isacker, P. V., and Warner, D. D., *Phys. Rev. Lett.*, **78**, 3266 (1997).
15. Juillet, O., and Josse, S., *Eur. Phys. J.*, **A8**, 291 (2000).
16. Isacker, P. V., Frank, A., and Warner, D. D., *to be published* (2004).
17. Baldini-Neto, E., et al., *Phys. Rev.*, **C65**, 064303 (2002).

The transition between axial and triaxial structure in the IBM-2

M. A. Caprio

Center for Theoretical Physics, Sloane Physics Laboratory, Yale University, New Haven, Connecticut 06520-8120, USA

Abstract. The phase diagram of the two-fluid proton-neutron interacting boson model (IBM-2) is investigated.

The phase diagrams of one-fluid algebraic models, and in particular the interacting boson model (IBM) [1], have been well studied [2, 3]. In the present work, the phase diagram of a two-fluid algebraic model, the proton-neutron interacting boson model (IBM-2) [4, 5], is investigated.

The phase structure of the Hamiltonian

$$H = \varepsilon(n_d^\pi + n_d^\nu) + \kappa(Q^{\pi,\chi_\pi} + Q^{\nu,\chi_\nu}) \cdot (Q^{\pi,\chi_\pi} + Q^{\nu,\chi_\nu}), \qquad (1)$$

is determined in the classical limit using the coherent state formalism. It is convenient to define $\chi_s \equiv \frac{1}{2}(\chi_\pi + \chi_\nu)$ and $\chi_v \equiv \frac{1}{2}(\chi_\pi - \chi_\nu)$ and to let $\varepsilon = (1 - \xi')/N$ and $\kappa = -\xi'/N^2$, with $0 \leq \xi' \leq 1$. The IBM-2 exhibits four dynamical symmetries, and thus the phase diagram showing transitions between these is inherently three-dimensional (Fig. 1). Three of the IBM-2 symmetries — $U_{\pi\nu}(6)$, $SU_{\pi\nu}(3)$, and $SO_{\pi\nu}(6)$ — lie in the plane $\chi_v = 0$ and have direct analogues in the one-fluid IBM [6]. The fourth, $SU^*_{\pi\nu}(3)$ [7], is obtained for $\xi' = 1$, $\chi_s = 0$, and $\chi_v = -\sqrt{7}/2$. The geometric interpretation is that a proton fluid with axially symmetric prolate deformation and a neutron fluid with axially symmetric oblate deformation are coupled with their symmetry axes orthogonal to each other, resulting in an overall rigid triaxial nuclear deformation [7, 8, 9]. (For $\chi_v = +\sqrt{7}/2$, denoted here by $\overline{SU^*_{\pi\nu}(3)}$, the proton and neutron fluid deformations are interchanged.)

The coherent state energy surface for H is a complicated function of four deformation parameters (β_π, γ_π, β_ν, and γ_ν) and the three relative Euler angles between the proton and neutron intrinsic frames. In the present analysis, to render the minimization tractable, γ_π and γ_ν have been restricted to 0 or $\pi/3$. An alternative analysis by Arias, Dukelsky, and García-Ramos [10] instead takes all relative Euler angles to be zero.

The transition between axial and triaxial configurations can be studied analytically for $\xi' = 1$, yielding expressions for the boundary curve, equilibrium deformations, and ground state energy. The transition is everywhere first-order. Limited analytic results can also be obtained for the boundary between undeformed and deformed configurations. A line of second-order phase transition points, with $\xi' = 1/5$ and $\chi_\pi/\chi_\nu = -N_\nu/N_\pi$, is embedded in a surface of first-order phase transition. The remainder of the phase diagram (Fig. 1) is obtained by numerical minimization of the energy surface.

CP726, *Nuclear Physics, Large and Small: International Conference on Microscopic Studies of Collective Phenomena*, edited by R. Bijker, R. F. Casten, and A. Frank
© 2004 American Institute of Physics 0-7354-0207-8/04/$22.00

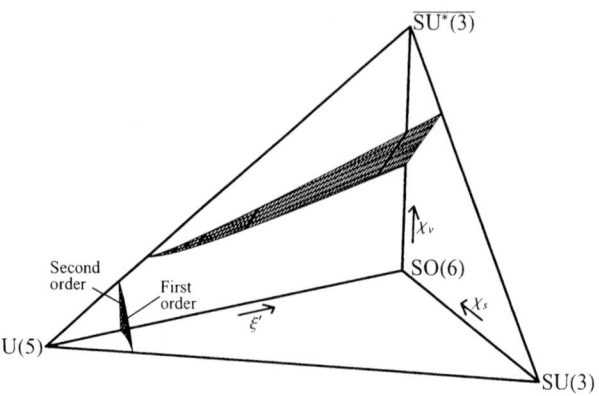

FIGURE 1. Phase diagram of the IBM-2 parameter space for the Hamiltonian of (1), with $N_\pi/N_\nu=1$, obtained as described in the text. The boundary surfaces between undeformed, axially symmetric deformed, and triaxial equilibrium configurations are shown.

Numerical diagonalization of the IBM-2 Hamiltonian at the $\mathrm{SU}_{\pi\nu}(3)$-$\mathrm{SU}_{\pi\nu}^*(3)$ transition point provides some insight into observable signatures. The ground state and 0_2^+ state undergo an avoided crossing. Two distinct low-lying families of bands are present, based upon these states, with suppressed $E2$ transitions between them. Two close-lying 1^+ excitations arise, apparently from the avoided crossing of the regular [1] and triaxial [8] scissors modes.

Analysis of the IBM-2 phase diagram provides a framework for studying the transition between axial and triaxial structure in the IBM-2. An example of structure near the axial-triaxial boundary was proposed in Ref. [7], and data on nuclei away from stability may yield further examples. Similar analyses may be applied to other two-fluid or multi-fluid algebraic models, such as the $\mathrm{U}_{core}(6) \otimes \mathrm{U}_{skin}(6)$ description of core-skin interaction in neutron rich nuclei [11] or the $\mathrm{U}(4) \otimes \mathrm{U}(4)$ vibron model for molecules.

Discussions with F. Iachello, J. M. Arias, and R. Bijker are gratefully acknowledged. This work was supported by the US DOE under grant DE-FG02-91ER-40608.

REFERENCES

1.   Iachello, F., and Arima, A., *The Interacting Boson Model*, Cambridge Univ. Press, Cambridge, 1987.
2.   Dieperink, A. E. L., Scholten, O., and Iachello, F., *Phys. Rev. Lett.*, **44**, 1747 (1980).
3.   Feng, D. H., Gilmore, R., and Deans, S. R., *Phys. Rev. C*, **23**, 1254 (1981).
4.   Arima, A., Otsuka, T., Iachello, F., and Talmi, I., *Phys. Lett. B*, **66**, 205 (1977).
5.   Otsuka, T., Arima, A., Iachello, F., and Talmi, I., *Phys. Lett. B*, **76**, 139 (1978).
6.   Van Isacker, P., Heyde, K., Jolie, J., and Sevrin, A., *Ann. Phys. (N.Y.)*, **171**, 253 (1986).
7.   Dieperink, A. E. L., and Bijker, R., *Phys. Lett. B*, **116**, 77 (1982).
8.   Leviatan, A., and Kirson, M. W., *Ann. Phys. (N.Y.)*, **201**, 13 (1990).
9.   Ginocchio, J. N., and Leviatan, A., *Ann. Phys. (N.Y.)*, **216**, 152 (1992).
10. Arias, J. M., Dukelsky, J., and García-Ramos, J. E. (in these proceedings).
11. Warner, D. D., and Van Isacker, P., *Phys. Lett. B*, **395**, 145 (1997).

How to Measure the Spreading Width of Superdeformed Nuclei

D. M. Cardamone*, C. A. Stafford* and B. R. Barrett*

*Department of Physics, University of Arizona, Tucson, AZ 85721

Abstract. A new expression for the branching ratio for the decay via the $E1$ process in the normal-deformed band of superdeformed nuclei is given within a simple two-level model. Using this expression, the spreading or tunneling width $\Gamma^{\downarrow}$ for superdeformed decay can be expressed entirely in terms of experimentally known quantities. We show how to determine the tunneling matrix element V from the measured value of $\Gamma^{\downarrow}$ and a statistical model of the energy levels.

Since their first experimental observation [1], the mechanism by which superdeformed (SD) nuclei decay into the normal deformed (ND) band has been a subject of considerable interest. The most common theoretical approach [2] has been to model the nucleus as a particle existing in a spin-dependent double-well potential in deformation space. Despite the popularity of this model, it has not been clear how to connect the experimentally observable parameters (i.e. rates and branching ratios) with the nuclear-stucture physics reflected in the shape of the potential well.

In Ref. [3], two of us first presented the two-level model for SD decay. The assumption of this model is that only the nearest ND neighbor to the decaying SD state plays an important role in the decay process, and thus other ND states may be neglected. We give a schematic diagram of the two-level system in Fig. 1.

Such a two-level model is exactly solvable using Green function techniques. The energy-domain retarded Green function $G(E)$ is given by its inverse in the SD, ND basis:

$$G^{-1}(E) = \begin{pmatrix} E - \varepsilon_S + i\Gamma_S/2 & -V \\ -V & E - \varepsilon_N + i\Gamma_N/2 \end{pmatrix}, \tag{1}$$

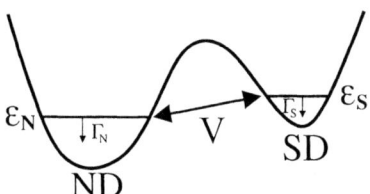

FIGURE 1. Schematic diagram of the two-level problem. V is the tunneling matrix element connecting the two states. Electromagnetic decay within each band gives the states their finite widths Γ_N and Γ_S. ε_S and ε_N are the energies of the two levels in the absence of V.

CP726, *Nuclear Physics, Large and Small: International Conference on Microscopic Studies of Collective Phenomena,* edited by R. Bijker, R. F. Casten, and A. Frank
© 2004 American Institute of Physics 0-7354-0207-8/04/$22.00

where H is the two-well Hamiltonian, with energies broadened by the decay widths Γ_S and Γ_N. The ND branching ratio is then given by Parseval's theorem as [4]

$$F_N = \Gamma_N \int_{-\infty}^{\infty} \frac{dE}{2\pi} |G_{NS}(E)|^2 = \frac{\Gamma_N \Gamma^\downarrow / (\Gamma_N + \Gamma^\downarrow)}{\Gamma_S + \Gamma_N \Gamma^\downarrow / (\Gamma_N + \Gamma^\downarrow)}, \tag{2}$$

where the spreading width for tunneling through the barrier is

$$\Gamma^\downarrow = \frac{(\Gamma_N + \Gamma_S) V^2}{(\varepsilon_N - \varepsilon_S)^2 + (\Gamma_N + \Gamma_S)^2 / 4}. \tag{3}$$

This exact two-level result for $\Gamma^\downarrow$ is consistent with Fermi's Golden Rule [3]. We note that Eq. 2 is exactly what one would expect for a two-stage quantum tunneling problem. Furthermore, inversion of this result yields $\Gamma^\downarrow$ purely as a function of experimentally known values:

$$\Gamma^\downarrow = \frac{F_N \Gamma_N \Gamma_S}{\Gamma_N - F_N (\Gamma_N + \Gamma_S)}. \tag{4}$$

Without knowing the theoretical quantity $\varepsilon_N - \varepsilon_S$, it is impossible to exactly determine the tunneling matrix element V from experimental data. We can nonetheless treat the detuning probabilistically by the assumption that the states of the ND band are distributed according to the Gaussian Orthogonal Ensemble. This approach yields a probability distribution for V, with an ensemble average

$$\langle V \rangle = \sqrt{\frac{\Gamma^\downarrow}{\Gamma_S + \Gamma_N}} \left[\frac{D_N}{4} + \mathcal{O}\left(\frac{(\Gamma_S + \Gamma_N)^2}{D_N} \right) \right], \tag{5}$$

where D_N is the average level spacing in the ND well.

The two-level approximation thus provides an exactly solvable model for SD decay. It has been shown [4, 5] that the effect of the other ND levels is negligable in the $A \approx 190$ mass region, and only moderate in the $A \approx 150$ region. The merit of this simple, exactly solvable model is that it allows relatively easy extraction of parameters related to nuclear structure.

We thank T. L. Khoo and S. Åberg for helpful discussions and acknowledge support from NSF grant PHY-0210750. B. R. B. acknowledges partial support from NSF grants PHY-0070858 and PHY-0244389. We also thank the Institute of Nuclear Theory at the University of Washington for its hospitality and the Department of Energy for partial support during the formulation and development of this work.

REFERENCES

1. P. J. Twin et al., Phys. Rev. Lett. 57(1986), 811.
2. E. Vigezzi, R. A. Broglia, and T. Dossing, Phys. Lett. B 249(1990), 163; Nucl. Phys. A520(1990), 179c.
3. C. A. Stafford and B. R. Barrett, Phys. Rev. C 60(1999), 051305.
4. D. M. Cardamone, C. A. Stafford, and B. R. Barrett, Phys. Rev. Lett. 91(2003), 102502.
5. A. Ya. Dzyublik and V. V. Utyuzh, Phys. Rev. C 68(2003), 024311.

Role of the intruder level in upper-pf shell nuclei [1]

K. P. Drumev, C. Bahri, V. G. Gueorguiev and J. P. Draayer

Department of Physics and Astronomy, Louisiana State University, Baton Rouge, LA 70803 USA

Abstract. Shell-model calculations for ^{58}Cu and ^{64}Ge in the $pf_{5/2}g_{9/2}$ model space using a realistic interaction are reported and compared to those generated using an appropriately renormalized counterpart of the interaction in the truncated $pf_{5/2}$ subspace. The results suggest that reliable computations can be performed in a space that does not explicitly include the intruder level so long as the interaction as well as the transition operators are renormalized appropriately.

The role of intruder levels that penetrate down into lower-lying shells in atomic nuclei has been the focus of many studies and debates. These levels are found in heavy deformed nuclei where the strong spin-orbit interaction destroys an underlying harmonic oscillator symmetry of the nuclear mean-field potential. In this contribution we report on calculations that consider the occupancy of these levels, their contribution to the nuclear deformation, and the role they play in the overall dynamics.

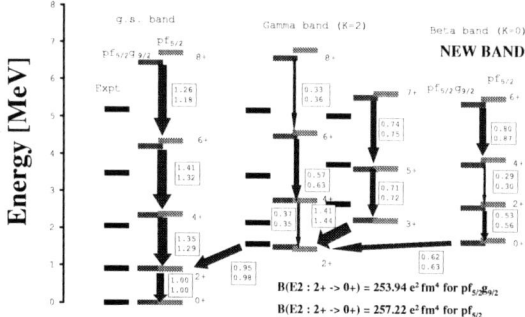

FIGURE 1. Energy spectrum and B(E2) transition strengths for ^{64}Ge. The width of the arrows in the figure represent the relative B(E2) strengths, normalized to unity for the ground band $2^+ \rightarrow 0^+$ transition. Numbers in each box are for the $pf_{5/2}g_{9/2}$ and the more restricted $pf_{5/2}$ model spaces, respectively.

We carried out m-scheme shell-model calculations for the ^{58}Cu and ^{64}Ge nuclei in the $pf_{5/2}g_{9/2}$ model space assuming the occupancy of the $f_{7/2}$ orbital to be 'frozen'. This choice was motivated by the $f_{7/2}$ orbit's high occupation as reported elsewhere [1]. The Hamiltonian we used is a G-matrix with a phenomenologically adjusted monopole part [2]. A renormalized version of this interaction in the $pf_{5/2}$ space has been introduced for describing beta decays [3].

---

[1] Support provided by the U.S. National Science Foundation under grant No: 0140300.

CP726, *Nuclear Physics, Large and Small: International Conference on Microscopic Studies of Collective Phenomena*, edited by R. Bijker, R. F. Casten, and A. Frank
© 2004 American Institute of Physics 0-7354-0207-8/04/$22.00

Results for the energy spectrum and B(E2) strengths of ^{64}Ge are shown in Figure 1 for both model spaces. The renormalized version of the theory in the $pf_{5/2}$ subspace not only reproduces the excitation energies obtained in the larger $pf_{5/2}g_{9/2}$ space, but also gives very similar values for the B(E2) transition strengths. These results also confirm those from a study using a schematic interaction [4]. Similar behavior was observed for ^{58}Cu. Besides the ground state and gamma bands for ^{64}Ge, which were previously observed and discussed, a new (possibly beta) band is identified.

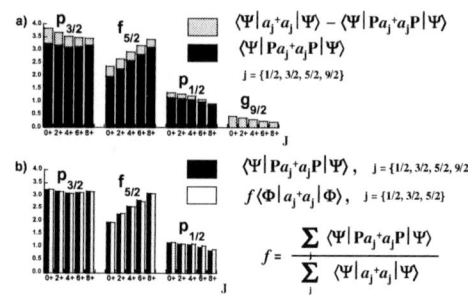

FIGURE 2. (a) Occupancies of single-particle orbitals for yrast states of ^{64}Ge. (b) Comparisons of $pf_{5/2}g_{9/2}$ occupancies projected into the $pf_{5/2}$ space with the corresponding rescaled $pf_{5/2}$ space results.

Figure 2 shows calculated occupation numbers of single-particle orbitals for states belonging to the ground-state band of ^{64}Ge. The lower (solid) bars in Figure 2a represent the occupancy in the $pf_{5/2}g_{9/2}$ space calculation from the many-particle basis states with no particles in the intruder level while the upper (gray) bars are related to those that count the occupancy when the intruder level is occupied. After rescaling occupancies of the states obtained in the $pf_{5/2}$ subspace (white), a pattern that is very similar to that of the occupancies in the larger space is obtained, as shown in Figure 2b.

In statistical spectroscopy, a quick test for the goodness of a symmetry is given by the correlation coefficients between a linear combination of the symmetry's invariant operators and the Hamiltonian of the system. This procedure has been applied to sd and pf shell nuclei and we also did it for the upper-pf shell case with similar findings [5]. Another way to probe the goodness of a symmetry is by calculating the strength distribution of the invariant operators. Results for the distribution of the second order Casimir operator C_2 of SU(3) for the first five states from the yrast band in ^{64}Ge indicate contribution of 50-60% from the leading SU(3) representation, a clear signal that the symmetry is good. In summary, these observations suggest that very good results can be obtained using a symmetry-adapted, truncated set of basis states.

REFERENCES

1. M. Honma, T. Mizusaki, and T. Otsuka, Phys. Rev. Lett. **77**, 3315 (1996).
2. E. Caurier, F. Novacki, A. Poves, and J. Retamosa, Phys. Rev. Lett. **77**, 1954 (1996).
3. P. Van Isacker, O. Juillet, and F. Nowacki, Phys. Rev. Lett. **82**, 2060 (1999).
4. K. Kaneko, M. Hasegawa, and T. Mizusaki, Phys. Rev. C **66**, 051306(R) (2002).
5. K. P. Drumev, V. G. Gueorguiev, C. Bahri, and J. P. Draayer, in preparation.

Global and Local Behaviour of Nuclear Ground-State Properties as fingerprints to Shape Coexistence in the Lead Isotopes

R. Fossion, V. Hellemans, S. De Baerdemacker and K. Heyde

Laboratory for Subatomic and Radiation Physics (INW),
Gent University, Proeftuinstraat 86, B-9000 Gent, Belgium

Abstract. A three-configuration mixing calculation is presented in the context of the Interacting Boson Model (IBM), with the aim to describe recently observed collective bands built on low-lying 0^+ states in the neutron-deficient lead isotopes. Possible effects on the nuclear binding energy are addressed, caused by mixing of these low-lying 0^+ intruder states into the ground state, and a new method is described in order to provide a consistent description of both ground-state and excited-state properties.

Ample evidence has been accumulated for the presence of nuclear shape coexistence phenomena throughout the whole table of isotopes, especially at and near closed shells [1, 2]. The neutron-deficient lead isotopes in particular, with a closed proton shell at $Z = 82$, show very rich excitation spectra. Three "families" of excited states are observed, with different spectroscopic properties, and with a behaviour that strongly depends on the neutron number [3]. The low-lying excited 0^+ states have been interpreted within two different frameworks: the mean field and the shell model. In a mean-field approach [4], the spectrum is understood as reflecting several competing minima in the potential energy surface (PES), corresponding to spherical, oblate and prolate deformations. In a shell-model picture, the excited 0^+ states are generated by multi-particle multi-hole (mp-mh) proton excitations across the $Z = 82$ shell gap. The excitation energies of these intruder states are lowered by the residual proton-neutron interaction. mp-mh excitations cannot be easily handled in full-scale shell model calculations, in particular for the large model space required for the description of heavy open-shell nuclei. They are, therefore, treated with the help of algebraic models, such as the Interacting Boson Models (IBM). In the first part of this poster, an IBM1-mixing calculation is proposed, that describes the three different intrinsic "shape" configurations. In order to reduce the number of parameters that appear in such a configuration-mixing calculation, use is made of the concept of intruder-spin symmetry, relating configurations with different numbers of particle (N_p) and hole (N_h) bosons (i.e., fermion pairs), but with a constant total number of bosons ($N = N_p + N_h$). In this way, experimental excitation energies in adjacent Pt and W nuclei are used to fix the essential IBM parameters [5]. Apart from mean-field and shell model, a third, purely phenomenological approach has also been used in order to interpret the experimental findings: the shape-mixing picture [6]. In this model, the physical observed states are the result of interactions between the several configurations. They result as a superposition of spherical, oblate and prolate configurations, the relative

CP726, *Nuclear Physics, Large and Small: International Conference on*
Microscopic Studies of Collective Phenomena, edited by R. Bijker, R. F. Casten, and A. Frank
© 2004 American Institute of Physics 0-7354-0207-8/04/$22.00

weights in the mixing being determined by a fit to the experimental data.

The possibility to study nuclear masses with the highest possible precision has become available over the last years, in particular at the ISOLTRAP and MISTRAL set-ups at ISOLDE/CERN. Here, precisions of the order of 30 keV on a total mass of a heavy Pb nucleus (≈ 1600 MeV) are reached. Deviations from the global trend (liquid-drop behaviour) are showing up in the nuclear masses in various localised regions over the chart of isotopes. In the Pb region, it is most probably the effect of mixing of low-lying intruder configurations (oblate and/or prolate shape configurations) in the ground state that turns out to be responsible for increased binding energies in the neutron-deficient region [7]. Using configuration mixing in the IBM1, detailed studies can be carried out, that give a consistent description of both the excited-state properties and the nuclear ground state (nuclear mass and nuclear binding energy) [8]. Moreover, it has been observed that in a consistent study of long chains of isotopes, one has to treat the ground state (through its binding energy) on equal footing with the excited states (relative energy spectrum). It turns out that parameters producing very similar energy spectra can still result in important differences in the ground-state binding energy (order of 1 MeV) [9]. This is also the topic of the second part of the poster.

ACKNOWLEDGMENTS

Financial support of the IWT-Flanders is acknowledged. RF also receives financial support from a Marie Curie Fellowship of the European Community (contract number 2000-00084). The authors are grateful to P.H. Heenen, M. Bender, G. Dracoulis, A. Dewald, J.L. Wood, P. Van Duppen, M. Huyse, P. Van Isacker and J. Jolie.

REFERENCES

1. K. Heyde, P. Van Isacker, M. Warpquier, J.L. Wood and R.A. Meyer, Phys. Rep. **102** (1983) 291.
2. J.L. Wood, K. Heyde,.W. Nazarewicz, M. Huyse and P. Van Duppen, Phys. Rep. **215** (1992) 101.
3. A. N. Andreyev, M. Huyse, P. Van Duppen, L. Weissman, D. Ackermann, J. Gert, F.P. Heßberger, S. Hofmann, A. Kleinböhl, G. Münzenberg, S. Reshitko, C. Schlegel, H. Schaffner, P. Cagarda, M. Matos, S. Saro, A. Keenan, C. Moore, C.D. O'Lears, R.D. Page, M. Taylor, H. Kettunen, M. Leino, A. Lavrentiev, R. Wyss and K. Heyde, Nature **405** (2000) 430.
4. M. Bender, P. Bonche, T. Duguet and P.-H. Heenen, accepted in March 2004, to be published in Phys. Rev. C., and references therein.
5. R. Fossion, K. Heyde, G. Thiamova and P. Van Isacker, Phys. Rev. **C67** (2003) 024306.
6. G.D. Dracoulis, G.J. Lane, A.P. Byrne, T. Kibédi, A.M. Baxter, A.O. Macchiavelli, P. Fallon and R.M. Clark, accepted in March 2004, to be published in Phys. Rev. C., and references therein.
7. S. Schwarz, F. Ames, G. Audi, D. Beck, G. Bollen, C. De Coster, J. Dilling, O. Engels, R. Fossion, J. E. García Ramos, S. Henry, F. Herfurth, K. Heyde, A. Kellerbauer, H.-J. Kluge, A. Kohl, E. Lamour, D. Lunney, I. Martel, R. B. Moore, M. Oinonen, H. Raimbault-Hartmann, C. Scheidenberger, G. Sikler, J. Szerypo, C. Weber, Nucl. Phys. **A693** (2001) 533.
8. R. Fossion, C. De Coster, J.E. García-Ramos, T. Werner and K. Heyde, Nucl. Phys. **A697** (2002) 703.
9. J.E. García-Ramos, C. De Coster, R. Fossion and K. Heyde, Nucl. Phys. **A688** (2001) 735.

Open Questions in Stellar Nuclear Physics: II [1]

Moshe Gai

Laboratory for Nuclear Science at Avery Point,
University of Connecticut, 1084 Shennecossett Road, Groton, CT 06340.
moshe.gai@yale.edu, http://www.phys.uconn.edu

Abstract. No doubt, among the most exciting discoveries of the third millennium thus far are oscillations of massive neutrinos and dark energy that leads to an accelerated expansion of the Universe. Accordingly, Nuclear Physics is presented with two extraordinary challenges: the need for precise (5% or better) prediction of solar neutrino fluxes within the Standard Solar Model, and the need for an accurate (5% or better) understanding of stellar evolution and in particular of Type Ia super nova that are used as cosmological standard candle. In contrast, much confusion is found in the field with contradicting data and strong statements of accuracy that can not be supported by current data. We discuss an experimental program to address these challenges and disagreements.

HELIUM BURNING AND THE C/O RATIO

The C/O ratio at the end of helium burning is still poorly known, twenty years after it was declared by Willie Fowler the "holy grail" of Nuclear Astro-Physics [1]. This parameter is essential for almost all aspects of stellar evolution of massive stars, and most recently it was also suggested to be essential for understanding the light curve of SNeIa [2]. The finding of Hoeflich were recently challenged [4], but the C/O ratio is most certain to play a major role in our understanding of the Phillips empirical relationship of peak luminosity and the shape of the light curve of Type Ia supernova [3]. Since the Phillips relationship is at the very foundation of using SNeIa as standard cosmological candle it is essential to understand it. The new generation of dedicated space telescopes that will solely measure Type Ia supernova makes it very important to understand SNeIa.

In order to measure the C/O ratio at the end of helium burning the cross section of the $^{12}C(\alpha,\gamma)^{16}O$ needs to be known at approximately 300 keV, but thus far it was measured only down to approximately 1.2 MeV. The extrapolation of this cross section to stellar energies (300 keV) is particularly difficult due to the substantial contribution from bound states. The properties of the bound states and their interference with quasi-bound states were thus far determined with the use of R-matrix theory. However, it now appears that the claimed accuracy of the R-matrix fits can not be substantiated. While the TRIUMF group quote an E1 astrophysical cross section factor with 25% uncertainty [5], Hale extracts a value that is eight times smaller [6]. Similarly elastic scattering data was used by the Notre Dame group to extract the E2 S-factor with the claimed 20% accuracy [7]. But this analysis in of itself was criticized for lack of theoretical foundation [8], and the

---

[1] Work Supported by USDOE Grant No. DE-FG02-94ER40870.

CP726, Nuclear Physics, Large and Small: International Conference on
Microscopic Studies of Collective Phenomena, edited by R. Bijker, R. F. Casten, and A. Frank
© 2004 American Institute of Physics 0-7354-0207-8/04/$22.00

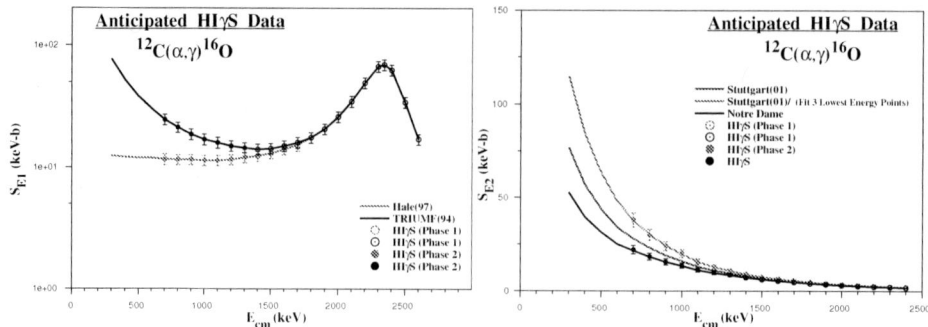

FIGURE 1. Anticipated HIγS data on the E1 S-factor as compared to the values quoted by the TRIUMF collaboration [5] and Hale [6], and on the E2 S-factor as compared to the results of the Notre Dame [7] and the Stuttgart groups [9].

result turned out to be a factor of 2.5 smaller than extracted by the Stuttgart group [9] that used R-matrix theory to extrapolate angular distribution data of the $^{12}C(\alpha, \gamma)^{16}O$ capture reaction itself.

A most promising new approach to measure both the E1 and E2 astrophysical cross section factors of the $^{12}C(\alpha, \gamma)^{16}O$ reaction at energies as low as 700 keV emerged with the use of the High Intensity γamma Source (HIγS) at the TUNL lab at Duke [10]. In this experiment one will study the photodisintegration of ^{16}O with an Optical Readout Time Projection Chamber (TPC). The anticipated results of the HIγS facility are shown in Fig. 1, as compared to the disagreeing results discussed above.

REFERENCES

1. W.A. Fowler, Rev. Mod. Phys. **56**(1984)149.
2. P. Hoeflich, J. C. Wheeler, and F. K. Thielemann; ApJ **495**(1998)617.
3. M. Phillips, Astrophys. J. Lett. **413**(1993)105. M. Hamuy *et al.*, Astron. J. **106**(1993)2392; ibid **109**(1995)1. A. Riess, W. Press, R. Kirshner, Astrophys. J. **473**(1996)88.
4. F. K. Roepke, and W. Hillebrandt; astro-ph/0403509.
5. R.E. Azuma *et al.*; Phys. Rev. **C50**(1994)1194.
6. G.M. Hale; Nucl. Phys. **A621**(1997)177č.
7. P. Tischhauser *et al.*; Phys. Rev. Lett. bf 88(2002)072501.
8. J.M. Sparenberg; Phys. Rev. **C69**(2004)034601.
9. R. Kunz *et al.*; Phys. Rev. Lett. **86**(2001)3244.
10. M. Gai, H. Weller, A. Breskin and V. Dangendorf for the UConn-Duke-Weizmann-PTB-UHartford-GCSU-LLN Collaboration.

Mixed-Symmetry States in ^{96}Mo

S. R. Lesher*, D. Bandyopadhyay*, N. Boukharouba*, C. Fransen*[†], M. Mynk*, C.J. McKay*, M.T. McEllistrem* and S.W. Yates*

*University of Kentucky, Lexington, KY 40506-0055, USA
[†]University of Cologne, Cologne, Germany

Abstract.
Mixed-symmetry states in ^{96}Mo have been identified using the (n,n′γ) reaction. From the lifetime measured using the Doppler-shift attenuation method and multipole mixing ratios, B(E2) and B(M1) values were determined. The 2^+_{ms} state was found at 2095.77 keV with B(M1: $2^+_{ms} \rightarrow 2^+_1$) = $0.18\mu^2_N$, while the 1^+_{ms} strength lies primarily in states at 2794.45 and 3300.23 keV.

INTRODUCTION & EXPERIMENT

Since the experimental discovery of the "scissors-mode" state, 1^+_{sc}, in ^{156}Gd in 1984 [1], these excitations have been identified in many nuclei and their systematics are well known [2, 3]. The IBM-2 has been used to explain scissor-mode states as members of a larger classification of "mixed-symmetry" states. Unlike the IBM-1, which only predicts symmetric states, this model allows for the exchange of proton and neutron labels in the boson wave function, giving rise to a new quantum number, F-spin, the equivalent of boson isospin. Fully symmetric states have a maximum F-spin value, $F_{max} = \frac{N_\pi + N_\nu}{2}$, where N_π and N_ν are the number of protons and neutrons, respectively, and mixed-symmetry states have $F < F_{max}$.

In deformed and γ-unstable nuclei, the $J^\pi = 1^+_{sc}$ state is the lowest-lying mixed-symmetry state and is characterized by a strong M1 transition to the ground state. In nearly spherical or weakly deformed nuclei, the lowest-lying mixed-symmetry state has $J^\pi = 2^+$. These states are characterized by a strong M1 transition to the 2^+_1 state and a weak E2 branch to the ground state [4]. The higher-lying mixed-symmetry states can be identified by strong M1 decays to symmetric states, weak E2 transitions to non-symmetric states, and strong E2 decays to the 2^+_{ms} state. One would expect the matrix element $|\langle J_s |M1| J_{ms}\rangle|$ to be $\approx 1\mu_N$ and B(E2) to symmetric states on the order of a few single particle units at most. Most of these levels have been observed in ^{94}Mo [5].

The ^{96}Mo(n,n′γ) experiments were performed at the University of Kentucky accelerator facility, with neutron production by the ^{3}H(p,n) reaction. The scattering sample was 40.05 g of 96.69% enriched ^{96}Mo in a polyethylene container 5.2 cm high and 2.6 cm in diameter. The emitted γ rays were detected with a 50% efficient HPGe detector, using time-of-flight gating and a BGO shield for Compton suppression and reduction of background effects. An excitation function measurement was performed in 0.1-MeV steps from 2.0 to 4.0 MeV, with the detector at an angle of 90°. Angular distribution measurements were performed at neutron energies of 2.5, 3.0, 3.5, and 4.0 MeV over

CP726, *Nuclear Physics, Large and Small: International Conference on Microscopic Studies of Collective Phenomena*, edited by R. Bijker, R. F. Casten, and A. Frank
© 2004 American Institute of Physics 0-7354-0207-8/04/$22.00

the angular range from 40° to 150°. The four neutron energies were chosen to reduce $\gamma-$ray feeding of the desired levels for obtaining accurate level lifetimes.

DISCUSSION & CONCLUSION

The search for the 2_{ms}^+ state began with an examination of the B(M1) values of decays from the 2^+ states below 3 MeV to the first excited state [4]. The fourth 2^+ state (2_4^+) has the largest B(M1) value. From the lifetime of $\tau = 138_{-7}^{+8}$ fs and multipole mixing ratio of δ=-0.009, a B(M1;$2_4^+ \to 2_1^+$) of 0.178(10) μ_N^2 was determined. The reduced matrix element, $|\langle 2_1^+ | M1 | 2_4^+ \rangle| = 0.94(22)\mu_N$, is in agreement with that expected for a mixed-symmetry state.

After the identification of the lowest MS state, the higher states can be sought. The $2_1^+ \otimes 2_{ms}^+$ coupling should lead to a quintuplet of states near 2.9 MeV. In this contribution, we will only concentrate on the 1_{ms}^+ state.

Levels at 2794.45 and 3300.23 keV were identified in previous measurements with the $J^\pi = 1^+$ firmly assigned [6, 7]. We observed decays to both the ground state and 2_2^+ state from each level. The 2794.45-keV ground-state transition is, of course, pure M1 and has a B(M1;$1_2^+ \to 0_1^+$) of 0.047(4) μ_N^2; however, we were unable to obtain a multiple mixing ratio for the weak $1_1^+ \to 2_2^+$ transition. Mixed-symmetry characteristics are found when the transition is taken in the δ =0 limit to be a pure M1, B(M1:$1_1^+ \to 2_2^+$) = $0.131\mu_N^2$; though a pure E2 would obtain B(E2; $1_1^+ \to 2_2^+$) = 43 W.u. The other 1^+ level at 3300.23 keV has a stronger ground-state transition, B(M1;$1_3^+ \to 0_1^+$)=0.131 μ_N^2. Again, the δ value of the $1^+ \to 2_2^+$ transition was unobtainable, but the values in the limits of the transition were calculated. If a pure M1 ($\delta = 0$) to the 2_2^+ state, B(M1; $1_3^+ \to 2_2^+$) = $0.076\mu_N^2$; if it is assumed to be a pure E2, B(E2; $1_3^+ \to 2_1^+$) = 12.9 W.u. Therefore, it appears the primary component of the 1_{ms}^+ lies in the 1_1^+ state at 2794.45 keV.

By using the (n,n′γ) reaction we were able to study ^{96}Mo and search for mixed-symmetry states. We were able to identify the major components of 2_{ms}^+ state at 2095.77 keV. Although the 2_{ms}^+ assignment is clear, we begin to see significant fragmentation of the higher-lying 1^+ mixed symmetry state. The MS strength is primarily contained in levels at 2794.45 and 3300.32 keV. We must now search for the other members of the $2_{ms}^+ \otimes 2^+$ quintet and compare these values to the well-known ^{94}Mo.

REFERENCES

1. Bohle, D., Richter, A., Steffen, W., Dieperink, A., LoIudice, N., Palumbo, F., and O.Scholten, *Phys. Lett.*, **137B**, 27 (1984).
2. Richter, A., *Prog. Part. Nucl. Phys.*, **34**, 261 (1995).
3. Kneissl, U., Pitz, H., and Zilges, A., *Prog. Part. Nucl. Phys.*, **37**, 349 (1996).
4. Iachello, F., *Phys. Rev. Lett.*, **53**, 1427 (1984).
5. Fransen, C., *et al.*, *Phys. Rev. C*, **67**, 024307 (2003).
6. Werner, V. *Thesis, University of Cologne* (2000).
7. Fransen, C. *et al.*, to be published (2004).

On solvable Bohr Hamiltonians

G. Lévai

Institute of Nuclear Research of the Hungarian Academy of Sciences (ATOMKI),
P. O. Box 51, H-4001 Debrecen, Hungary

Abstract. The sextic oscillator is discussed as a two-parameter potential for which the Bohr Hamiltonian can be solved analytically in a number of situations. In the simplest case it can be considered as a γ-independent potential, which can have a minimum at $\beta = 0$ and/or at $\beta > 0$. Preliminary results are presented from a study in which the potential parameters were determined by fitting the low-lying energy spectrum of even Ru isotopes, and the evolution of the corresponding potential shape along the Ru chain was analyzed. The results are consistent with those of other approaches.

In certain situations the Bohr Hamiltonian describing the collective motion in nuclei in terms of the shape variables (β, γ) [1] can be reduced to a Schrödinger-like equation in the β variable. This problem has received considerable attention recently after the introduction of the so-called critical point symmetries, which are related to transitions from one shape phase to another one. The first symmetry of this kind, E(5) [2] is expected to occur when a transition from the spherical to deformed γ-unstable shape takes place, and the corresponding potential is expected to be independent of γ and to have a relatively flat shape in the β variable. The purpose of the present contribution is to extend the range of solvable Bohr Hamiltonians (see Ref. [3] and references) with the sextic oscillator [4] and to promote its application in realistic calculations.

Assuming that the $V(\beta, \gamma)$ potential is γ-independent, the substitution $\Psi(\beta, \gamma, \theta_i) = \beta^{-2}\phi(\beta)\Phi(\gamma, \theta_i)$ leads to the Schrödinger-like equation

$$-\frac{\mathrm{d}^2\phi}{\mathrm{d}\beta^2} + \left(\frac{(\tau+1)(\tau+2)}{\beta^2} + u(\beta)\right)\phi = \varepsilon\phi, \tag{1}$$

where the $\varepsilon = \frac{2B}{\hbar^2}E$ energy and the $u(\beta) = \frac{2B}{\hbar^2}U(\beta)$ potential are expressed in reduced units. As a new example to be applied in (1) we proposed [5] the sextic oscillator

$$u(\beta) = (b^2 - 4ac)\beta^2 + 2ab\beta^4 + a^2\beta^6 + u_0. \tag{2}$$

This potential is quasi-exactly solvable [4], which means that exact solutions can generally be found for the lowest few levels. The wavefunctions are obtained in the form

$$\phi_n(x) = N_n P_n(x^2)(x^2)^{s-\frac{1}{4}}\exp\left(-\frac{a}{4}x^4 - \frac{b}{2}x^2\right) \qquad n = 0, 1, 2, \ldots \tag{3}$$

where P_n is a polynomial of order n. Normalizability requires $a \geq 0$, while $a = 0$ reduces (2) to the harmonic oscillator. The integrals necessary to evaluate the normalization

CP726, Nuclear Physics, Large and Small: International Conference on
Microscopic Studies of Collective Phenomena, edited by R. Bijker, R. F. Casten, and A. Frank
© 2004 American Institute of Physics 0-7354-0207-8/04/$22.00

TABLE 1. Parameters a and b fitted to the ^{A}Ru spectra.

A	98	100	102	104	106	108	110	112
a	[0]	[0]	[0]	1496	4190	5154	14684	11563
b	347	318	283	216	143	114	-63	-36

constants N_n and the the E2 transition matrix elements can be obtained in closed form [5]. The parameters appearing in (3) have to satisfy $s + M + 1/2 = (\tau + 2M + 7/2)/2 \equiv c$, where M=0, 1, 2, ... and its maximum number specifies the number of exact solutions. Since c depends on the $\tau + 2M$ combination, $u(\beta)$ in (2) is slightly different for even and odd values of τ. (Taking solutions with $M \leq 1$ c is 11/4 and 13/4 in the respective two cases.) The shape of the potential is rather flexible: it can have a minimum at $\beta = 0$ or at $\beta = \beta_{min} > 0$, and in addition, it can also have a local maximum at $\beta_{max} < \beta_{min}$.

In Ref. [5] analysis of the energy eigenvalues and wavefunctions belonging to the lowest-lying states, i.e. those with $\xi, \tau = 1,0; 1,1; 1,2; 1,3; 2,0; 2,1$ (corresponding to $M = 0$ and 1) has been given. As an illustration the ^{134}Ba nucleus was analyzed, and it was found that the sextic oscillator with $a = 40000$ and $b = 200$ reproduces the energy and $B(E2)$ ratios more successfully than other potentials [5]. With these parameters the potential (2) is rather flat near $\beta = 0$, but it has a minimum at $\beta > 0$.

In order to test the performance of the sextic oscillator as a realistic model we fitted the low-lying energy spectrum of the even Ru isotopes with A =98 to 112 and plotted the corresponding potentials. The resulting parameters are displayed in Table 1. When (small) negative, i.e. unphysical value was obtained for a, we took the harmonic limit $a = 0$. In these cases (A=98, 100 and 102) the energy spectrum was indeed rather close to a harmonic situation. The potential was rather flat for ^{104}Ru, which is in accordance with an expected phase transition and E(5) symmetry here [6]. The potential was still very flat for ^{106}Ru, but it had a finite minimum that became increasingly deeper and got further away from $\beta = 0$ as A increased to 112. These findings are in reasonable agreement with the expectations and confirm the applicability of the sextic potential in the Bohr Hamiltonian. Work is in progress to calculate the intensity of E2 transitions.

Given the flexible shape of the sextic oscillator this potential might be useful in other regions too. With some modification of the formalism dependence on the γ variable can be built in and nuclei close to the X(5) symmetry [7] might be discussed.

This work was supported by the OTKA grant No. T37502 (Hungary).

REFERENCES

1. A. Bohr and B. Mottelson, *Nuclear structure* (Benjamin, Reading, MA, 1975) Vol. II.
2. F. Iachello, *Phys. Rev. Lett.* **85**, 3580 (2000).
3. L. Fortunato and A. Vitturi, *J. Phys. G* **29**, 1341 (2003).
4. A. G. Ushveridze, *Quasi-exactly solvable models in quantum mechanics* (IOPP, Bristol, 1994).
5. G. Lévai and J. M. Arias, *Phys. Rev. C* **69**, 04304 (2004).
6. A. Frank, C. E. Alonso and J. M. Arias, *Phys. Rev. C* **65**, 014301 (2001).
7. F. Iachello, *Phys. Rev. Lett.* **87**, 052502 (2001).

Evolution of the N = 32, 34 Shell Closures

S. N. Liddick[1][†]

National Superconducting Cyclotron Laboratory, Michigan State University, East Lansing,
Michigan 48824
Department of Chemistry, Michigan State University, East Lansing, Michigan 48824

Abstract. β-decay studies for neutron-rich $\pi f_{7/2} - \nu f p$ shell nuclei have been systematically performed at the National Superconducting Cyclotron Laboratory. The $E(2_1^+)$ values display a peak at N = 32 compared to neighboring isotopes revealing a subshell closure for this neutron number as predicted by pf-shell model calculations. However, a value of 1127 keV for $E(2_1^+)$ in $^{56}Ti_{34}$, which is lower than that of neighboring $^{54}Ti_{32}$ suggests the absence of a shell closure at N = 34 in the Ti isotopes, contrary to predictions.

INTRODUCTION

The development of an N = 32 subshell closure and the possible existence of an N = 34 shell closure in the Ca to Ni region has undergone systematic study in recent years. The presence of an N = 32 subshell closure located in the neutron-rich isotopes of Ca, Ti, and Cr [2, 3, 4, 5, 6, 7, 8] has been attributed to a strong proton-neutron monopole interaction between protons in the $\pi 1 f_{7/2}$ orbital and neutrons in the $\nu 1 f_{5/2}$ state [2]. The monopole migration of the $\nu 1 f_{5/2}$ orbital with the removal of protons from the $\pi 1 f_{7/2}$ state and a large spin-orbit splitting between the $\nu 2 p_{3/2}$ and $\nu 2 p_{1/2}$ levels results in the formation of gap at N = 32. Shell model calculations using the new effective pf-shell interaction, labelled GXPF1 [9], reproduce the observed subshell closure at N = 32, and predict that, in the Ti and Ca isotopes, the monopole migration of the $\nu 1 f_{5/2}$ state results in a large energy gap between the $\nu 1 f_{5/2}$ and $\nu 2 p_{1/2}$ orbitals and the formation of a new N = 34 shell closure. However, a recent experiment on ^{56}Ti [6] determined the energy of the first excited 2^+ state $[E(2_1^+)]$ to be 1127 keV, 400 keV below shell model predictions based on the GXPF1 interaction, and lower than the $E(2_1^+)$ found for ^{54}Ti [4]. The systematics of the $E(2_1^+)$ for the pf-shell nuclei show the evolution of the subshell closure located at N = 32 and the apparent absence of a shell closure at N = 34 in the $_{22}Ti$ isotopes.

EXPERIMENTS

The β-decay properties of $^{56,57,58}V$, $^{55,56}Ti$, and $^{54,55,56}Sc$ have been studied at the NSCL as part of a systematic investigation of nuclei around the N = 32 subshell and the expected N = 34 shell closure, experimental details and results, especially for the $E(2_1^+)$ of even-even daughter nuclei, have been presented in Refs. [2, 3, 4, 5, 6].

CP726, *Nuclear Physics, Large and Small: International Conference on
Microscopic Studies of Collective Phenomena*, edited by R. Bijker, R. F. Casten, and A. Frank
© 2004 American Institute of Physics 0-7354-0207-8/04/$22.00

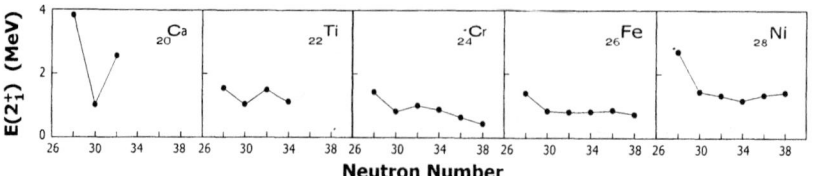

FIGURE 1. Experimental energies of $E(2_1^+)$ for pf-shell nuclei.

RESULTS

Fig. 1 shows the experimental first excited 2_1^+ state for the isotopes of Ca, Ti, Cr, Fe and Ni. The rise in $E(2_1^+)$ at N = 32 with the removal of protons from the $\pi 1 f_{7/2}$ follows an expected increase in the single-particle energy gap between the $\nu 1 f_{5/2}$ and $\nu 2 p_{3/2}$ states. In the Cr isotopes, the increase in energy of $\nu 1 f_{5/2}$ level combined with the large spin-orbit splitting between the $\nu 1 p_{3/2}$ and $\nu 1 p_{1/2}$ state has produced a significant energy separation between the $\nu 1 f_{5/2}$ and $\nu 1 p_{3/2}$ levels resulting in a subshell closure at N = 32 evidenced by the systematically higher $E(2_1^+)$ at N =32. Shell-model calculations using the GXPF1 interaction predict that with the further removal of protons from the $\pi 1 f_{7/2}$ orbit between Ti and Ca the single-particle energy of the $\nu 1 f_{5/2}$ state would continue to rise eventually resulting in a large energy separation between the the $\nu 1 f_{5/2}$ and $\nu 1 p_{1/2}$ levels and the development of a shell closure located at N = 34. Such a shell closure results in an expected $E(2_1^+)$ in ^{56}Ti comparable to the value in ^{54}Ti, around 1500 keV. However, the energy of the first 2_1^+ state in ^{56}Ti was observed at 1127 keV, 400 keV lower than predicted by GXPF1 calculations.

With the removal of the last two protons from the $\pi 1 f_{7/2}$ level, the monopole migration of the $\nu 1 f_{5/2}$ state would continue, and it is possible that an N = 34 shell gap still develops in the Ca isotopes. More experimental work is needed on the low-energy structure of neutron-rich Ca nuclei to confirm such predictions.

This work was supported in part by the National Science Foundation Grant Nos. PHY-01-10253, PHY-97-24299, PHY-01-39950, and PHY-02-44453, the U.S. Department of Energy, Nuclear Physics Division, under Contract No. W31-109-ENG-38, and the Polish Scientific Committee Grant No. 2P03B-074-1.

REFERENCES

1. In collaboration with the NSCL β group, Florida State University, Argonne National Laboratory and the Niewodniczanski Institute of Nuclear Physics.
2. J. I. Prisciandaro et al., Phys. Lett. B **510**, 17 (2001).
3. P. F. Mantica et al., Phys. Rev. C **67**, 014311 (2003).
4. R. V. F. Janssens et al., Phys. Lett. B **546**, 55 (2002).
5. P. F. Mantica et al., Phys. Rev. C **68**, 044311 (2003).
6. S. N. Liddick et al., Phys. Rev. Lett. **69**, 072502 (2004)
7. D. E. Appelbe et al., Phys. Rev. C **67**, 034309 (2003).
8. A. Huck et al., Phys. Rev. C **31**, 2226 (1985)
9. M. Honma, T. Otsuka, B. A. Brown, and T. Mizusaki, Phys. Rev. C **69**, 034335 (2004).

Properties of $^{68-76}$Ni isotopes with new effective interaction

A. F. Lisetskiy*, B. A. Brown*, M. Horoi† and H. Grawe**

*National Superconducting Cyclotron Laboratory, Michigan State University, East Lansing, Michigan 48824-1321
† Physics Department, Central Michigan University, Mount Pleasant, Michigan 48859
** Gesellschaft für Schwerionenforschung mbH, D-64291 Darmstadt, Germany

Abstract. The results of the shell model calculations for even $^{68-76}$Ni isotopes with newly derived effective interaction for the $f_{5/2}p_{3/2}p_{1/2}g_{9/2}$ model space are presented. The calculated and experimental properties of neutron-rich $^{68-78}$Ni isotopes and corresponding valence mirror symmetry partners, A= 90 − 98 N= 50, isotones are compared. The similarities and differences of the proton and neutron effective interactions are emphasized and its consequences for the nuclear structure are illustrated by the example of ^{72}Ni and ^{94}Ru nuclei.

Recently we have derived new effective interaction for the $f_{5/2}p_{3/2}p_{1/2}g_{9/2}$ model space from a fit to updated experimental data for Ni isotopes from A= 57 to A=78 and N= 50 isotones from ^{79}Cu to ^{100}Sn [1]. The new interaction represents $T = 1$ part of the interaction for the full proton-neutron $f_{5/2}p_{3/2}p_{1/2}g_{9/2}$ model space covering more than 200 nuclei of the nuclear chart triangle formed by doubly-magic ^{56}Ni, ^{78}Ni and ^{100}Sn nuclei.

The properties of nickel isotopes between ^{68}Ni and ^{78}Ni nuclei are of primary interest. It is useful to look at the structure of these nuclei referring to the Valence Mirror Symmetry (VMS) concept [2]. The fact that the model space for Ni isotopes and $N = 50$ isotones is the same supposes that structure of $^{68-76}$Ni isotopes and $A = 90 − 98$ $N = 50$ isotones may be similar. Indeed, the calculations show that the contribution of the $(f_{5/2}^6 p_{3/2}^4 p_{1/2}^2)_{0^+}(g_{9/2}^n)_{0^+}$ component ($n = 0 − 8$) to the ground 0^+ states of even $^{68-76}$Ni isotopes and A= 90 − 98, $N = 50$ isotons differ only by a few percent. However, the 2^+ states in Ni isotopes lay lower than in $N = 50$ isotones, while an opposite is expected for the same structures with decrease of mass number. The origin of this difference is mainly in $(g_{9/2}^2)_J$ TBMEs: neutron effective interaction in $J = 2$ and $J = 4$ states is by 300 keV more attractive than proton one. But in both cases the largest contribution to the energy gap between the 2^+ and the 0^+ states (0.8-0.7 MeV) is due to configuration mixing. Further analysis of the wave functions allows to conclude that the Z=28 shell gap near ^{68}Ni is relatively weak (closed shell $(f_{5/2}^6 p_{3/2}^4)_{0^+}$ component is only 21 % for the ground state of ^{66}Ni) as compared to the N= 50 neutron shell gap near ^{88}Sr (closed shell $(f_{5/2}^6 p_{3/2}^4)_{0^+}$ component amounts to 60 % for the ground state of ^{88}Sr). Thus Ni isotopes have substantial amount of proton core excitations that are implicitly taken into account by the effective interaction.

CP726, Nuclear Physics, Large and Small: International Conference on
Microscopic Studies of Collective Phenomena, edited by R. Bijker, R. F. Casten, and A. Frank
© 2004 American Institute of Physics 0-7354-0207-8/04/$22.00

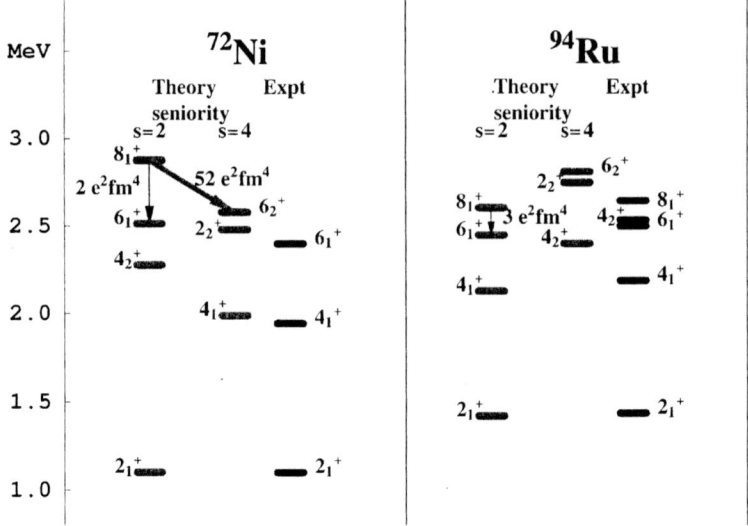

FIGURE 1. Calculated and experimental spectra of ^{72}Ni and ^{94}Ru. States with seniority $s = 2$ and $s = 4$ are shown in left and right columns, respectively.

The enhancement of the $(g_{9/2}^2)_{J=2,4}$ TBMEs has important consequences for non-yrast states in ^{72,74}Ni isotopes [3]. For example the spectra of VMS partner nuclei, ^{72}Ni and ^{94}Ru, are compared in Fig.1. Qualitative difference is that seniority $s = 4$ states are strongly pushed down in ^{72}Ni. This drastically changes decay properties of the $8^+_{s=2}$ state. According to the seniority scheme the E2 transition strength between the $(g_{9/2}^4)_{J=8,s=2}$ and the $(g_{9/2}^4)_{J=6,s=2}$ states is nine times weaker than the one between the $(g_{9/2}^4)_{J=8,s=2}$ and the $(g_{9/2}^4)_{J=6,s=4}$ states. The seniority selection rule results in isomeric $8^+_{s=2}$ state in ^{94}Ru. However, both the $6^+_{s=2}$ and the $6^+_{s=4}$ states are below the $8^+_{s=2}$ state in ^{72}Ni. This results in strong E2 transition from the $8^+_{s=2}$ state to the $6^+_{s=4}$ state and corresponding 6.1 ns lifetime that is three orders of magnitude smaller than for the $8^+_{s=2}$ state in ^{94}Ru. Another unexpected feature is that the $4^+_{s=4}$ state is below the $4^+_{s=2}$ state in ^{72}Ni. This drastically reduces the lifetime of the $6^+_{s=2}$ state and indicates a gradual transition from the seniority picture to collective one in neutron-rich Ni isotopes. Therefore this is of great scientific interest to investigate neutron-rich nickel isotopes and to learn more about the exotic phenomena indicated above.

REFERENCES

1. Lisetskiy, A., Brown, B., Horoi, M., and Grawe, H., *preprint nucl-th/0402082* (2004).
2. Wirowski, R., Yan, J., von Brentano, P., Dewald, A., and Gelberg, A., *J.Phys.G: Nucl.Phys.*, **14**, L195 (1988).
3. Grawe, H., *Nucl.Phys.A*, **704**, 211c (2002).

Influence of diffractive interactions on cosmic ray air showers

R. Luna *, C. A. García Canal†, S. J. Sciutto† and A. Zepeda*

*Depar tamento de Física, CINVESTAV-IPN, Av. IPN 2508, Col. San Pedro Zacatenco, 07360
México DF, México.
†Departamento de Física and IFLP/CONICET, Universidad Nacional de La Plata, C. C. 67 - 1900
La Plata, Argentina.

INTRODUCTION

The measurement of extensive of air showers (EAS) is presently the way to study cosmic rays with energies above several hundreds of tera electrons volts. The properties of primary cosmic rays have to be deduced from the development of the shower in the atmosphere, and from the caracteristics of the secundary detected at the observation levels.

COMPARATIVE ANALYSIS OF HADRONIC MODELS AND EFFECTS OF SHOWERS OBSERVABLES

A hadronic collision can be described as a process where an incident particle P, called the projectile, interacts with a target A –normally a nucleus of $A = Z + N$ nucleons (Z protons and N neutrons)– to produce N_{sec} secondary particles $S_1, \ldots, S_{N_{sec}}$, the secondaries verify $\sum_{i=1}^{N_{sec}} E_{S_i} < E_P$. We have performed a comparative analysis of the output coming from different hadronic packages when running them with a common input. We have run batchs of N_{coll} events ($N_{coll} = 10,000$ unless otherwise specified) for each combination of primary type, primary energy, and hadronic package. After each call to the hadronic procedures, a list of secondaries was obtained, with short-lived products (resonances) forced to decays. These secondaries were then processed to identify if the kind of collisions "diffractive" or "non-diffractive". We can see that there are evident differences among the plots corresponding to different models, especially when comparing QGSJET with the other models.

An outstanding feature is the well known fact that QGSJET [1] produces substantially more secondaries than SIBYLL [2] or DPMJET2 [?], especially at very high energies. This fact shows up clearly in the left hand side of figure 1 where the average number of secondaries is plotted versus the primary energy. In right hand side of figure 1, the fractions of diffractive events registered in our runs is plotted as a function of primary energy, in the case of proton primaries, These results indicate that in QGSJET the ratio between the diffrative and total cross sections. We have used the AIRES program [4] to simulate proton and iron induced showers with different primary energies, and using QGSJET01 and SIBYLL 2.1 to process the high energy hadronic interactions. Due to the suppression of diffractive interactions is significative at all primary energies. This is illustrated in figure 2 where X_{max} is plotted versus the primary energy.

CP726, Nuclear Physics, Large and Small: International Conference on
Microscopic Studies of Collective Phenomena, edited by R. Bijker, R. F. Casten, and A. Frank
© 2004 American Institute of Physics 0-7354-0207-8/04/$22.00

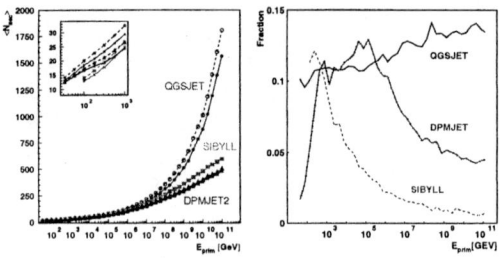

FIGURE 1. On the left hand side Average number of secondaries in proton-air collisions versus primary energy. The solid (open) symbols correspond to averages over all (non-diffractive) events.The low energy region is shown in more detail in the inset, and on the right hand side Fraction of diffractive events versus primary energies for the case of proton-air collisions.

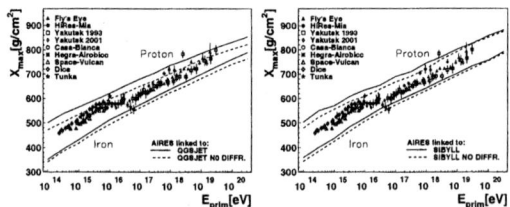

FIGURE 2. Average shower maximum versus primary energy.

The lines represent simulations of proton and iron showers enabling (solid lines) or disabling (dashed lines) the diffractive interactions. We have also plotted some available experimental data for reference.

CONCLUSIONS

We had showed for the models used that the variations in the showers observables have a very different dependence over diffracions interactions. The production of particles by the cosmic rays at ground level change between 10%-15% (10%-20%) for SIBYLL (QGSJET). An appropiate understandingg of air shower development is essential for knowing the adequate assignments of energy and composition in the experiment at ground level anf fluorecence detectors.

REFERENCES

1. N. N. Kalmykov, S. S. Ostapchenko, A. I. Pavlov,*Nucl. Phys. B (Proc. Suppl.)*, **52B**, 17 (1997).
2. R. Engel, T. K. Gaisser, T. Stanev, *Proc. 26th ICRC (Utah)*, **1**, 415 (1999).
3. S. J. Sciutto, Talk presented at the Pierre Auger Collaboration meeting, Malargüe, Argentina, October 2001.
4. S. J. Sciutto, *Proc. 27th ICRC (Hamburg)*, **1**, XXX (2001); see also **www.fisica.unlp.edu.ar/auger/aires**.

Exciting Effects of Excited 0^+ Energies

E.A. McCutchan, N.V. Zamfir, and R.F. Casten

Wright Nuclear Structure Laboratory, Yale University, New Haven, CT 06520

Abstract. The significance of considering excited 0^+ states is discussed in both an experimental and theoretical context. Experimental results on possible phase transitional nuclei are presented. New calculations on Yb and Hf nuclei in the framework of the IBA-1 model are presented and compared to previous fits in this region.

The recent introduction of a set of analytic solutions, X(5) [1] and E(5) [2], to describe nuclei at the critical point of first and second order phase transitions has led to an increased interest in the properties of 0^+ excited states. Validation of these models requires the experimental observation of nuclei exhibiting the properties of their predictions. The search for nuclei close to the X(5) model requires accurate knowledge of the ground and excited K=0^+ bands. One key signature of X(5) is the ratio of ground state energies, $R_{4/2} \equiv E(4_1^+)/E(2_1^+)$, predicted to equal 2.9. Another important observable is the location of the first excited 0^+ state at 5.7 times the energy of the ground state 2^+ energy. These two quantities are plotted in Fig. 1 as a function of neutron number for Yb and Hf nuclei along with the X(5) predictions.

As evident from Fig. 1, the $R_{4/2}$ ratio is in good agreement with the X(5) predictions for the nuclei ^{162}Yb (N = 92) and ^{166}Hf (N=94), while the energy ratio $E(0_2^+)/E(2_1^+)$ deviates from the model and also from a smooth evolution compared with the neighboring nuclei. Since the previous experimental evidence on these states [3] was somewhat ambiguous, new experiments were devised to confirm the location of the low-lying 0^+ states in these two nuclei.

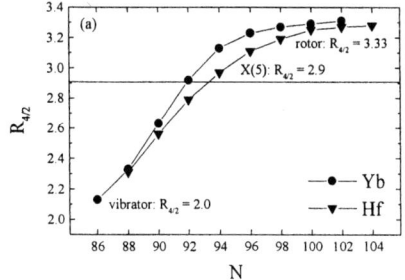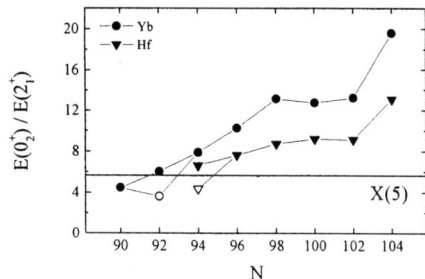

FIGURE 1. (a) $R_{4/2}$ and (b) $E(0_2^+)/E(2_1^+)$ ratios for the Yb and Hf isotopic chains as a function of neutron number. Open symbols correspond to previous values, solid symbols correspond to current values.

CP726, Nuclear Physics, Large and Small: International Conference on
Microscopic Studies of Collective Phenomena, edited by R. Bijker, R. F. Casten, and A. Frank
© 2004 American Institute of Physics 0-7354-0207-8/04/$22.00

Low-lying states in ^{162}Yb and ^{166}Hf were populated through ε decay and studied through γ-ray spectroscopy at the Yale Moving Tape Collector. New, high statistics coincidence data obtained in the present experiments allowed for the elimination of the previously reported low-lying 0^+ states (open symbols in Fig. 1) and the confirmation of higher lying states (solid symbols in Fig. 1). The confirmed excited 0^+ states agree well with the X(5) predictions and also the systematics of the region.

Previous IBA-1 calculations [4] in this region gave significant deviations from the data in their predictions for the lowest 0^+ excited sequence, not surprisingly since excited 0^+ states have long proved difficult to understand. The new emphasis on the excited 0^+ sequence prompted the reexamination of this region, seeking sets of IBA-1 parameters that reproduce the properties of *all* low-lying, positive parity excitations.

Calculations were performed in the framework of the IBA-1, using the extended consistent Q formalism [5] with the Hamiltonian

$$H(\zeta) = c[(1-\zeta)\hat{n}_d - \frac{\zeta}{4N_B}\hat{Q}^\chi \cdot \hat{Q}^\chi] \tag{1}$$

Parameters for each nucleus were determined by considering contour plots [4] of key observables relating to the ground, 0_2^+, and quasi-2_γ^+ bands. In almost all cases, a small range of parameters was able to reproduce the experimental energy ratios, $R_{4/2}$, $E(0_2^+)/E(2_1^+)$, and $E(2_\gamma^+)/E(2_1^+)$, to within 5% [6]. Using a set of polar coordinates, ρ~ζ and θ~χ, the parameters can be mapped into the IBA triangle. An equal emphasis on fitting the first excited 0^+ state results in a significantly different set of parameters for the Yb and Hf isotopic chains and thus very different trajectories as shown in Fig. 2(a). Trajectories corresponding to several isotopic chains in the rare-earth region are given in Fig. 2(b) for comparison [6].

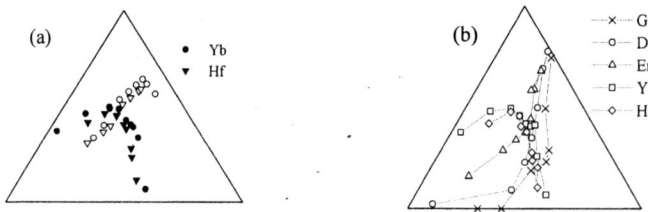

FIGURE 2. (a) Trajectories for the Yb and Hf isotopes using parameters from [4] (open symbols) and the current work [6] (filled symbols). (b) Trajectories [6] for several rare earth isotopic chains.

REFERENCES

1. F. Iachello, Phys. Rev. Lett. **87**, 052502 (2001).
2. F. Iachello, Phys. Rev. Lett. **85**, 3580 (2000).
3. http://www.nndc.bnl.gov/nudat2/index.jsp
4. W. –T. Chou, N.V. Zamfir, and R. F. Casten, Phys. Rev. C **56**, 829 (1997).
5. P.O. Lipas, P. Toivonen, and D.D. Warner, Phys. Lett. B **155**, 295 (1985).
6. E.A. McCutchan, N.V. Zamfir, and R.F. Casten, submitted Phys. Rev. C

Measurement of the Efficiency of a Neutron Detector Array

R. Monroy[†], E. Chávez[†], A. Huerta[†], R.Macias[†], M. E. Ortiz[†], L. Barrón[†], A. Ibañez[†], A. Varela[*], G. Murillo[*], R. Policroniades[*] and E. Moreno[*]

[†]Instituto de Física UNAM Ap. Po. 20-364 México D. F.
[*]Laboratorio del Acelerador ININ Ap. Po. 181027 México 11-801 D.F.

In this work we describe the development of an array of 7 neutron detectors through the use of solid plastic scintillating material and photomultiplier tubes. Furthermore the nuclear reaction d(D,^{3}He)n produced with a Tandem accelerator at ININ was used to measure the energy dependent neutron detection efficiency of our detectors with the associated particle technique.

We built our detectors using plastic scintillator material from Bicron (BC 408). The voltage divider to couple the tubes to the photomultipliers to the detector were built following closely a previous design [3].

The geometry chosen is hexagonal prisms 30 cm long with 5 cm apothem. Because of its large dimensions this detector has a large probability of detecting neutrons, while at the same time, allows to vary the solid angle coverage by placing it at different distances from the neutron source [5].

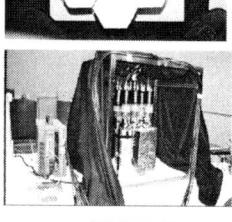
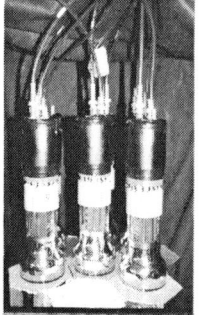
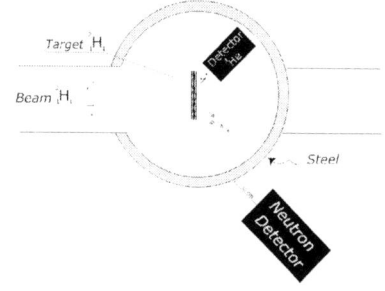

FIGURE 1. Detector array   FIGURE 2. Experimental setup

To measure the detection efficiency, we prepared the detector, in an experimental setup where it was bombarded by neutrons from the nuclear reaction d(D,^{3}He)n with neutron energies from 2.56 to 6 MeV. ^{3}He particles where detected by a tightly collimated solid state detector in the reaction chamber and identified by its energy. The neutron detector is placed outside the chamber where the cinematically correlated neutron should be detected. Because of the very thin wall (0.8 mm steel) neutron losses through absorption or dispersion can be neglected. This is the so called associated particle technique. The detector's neutron detection efficiency is then obtained simply through the ratio of detected neutrons to ^{3}He at

CP726, Nuclear Physics, Large and Small: International Conference on
Microscopic Studies of Collective Phenomena, edited by R. Bijker, R. F. Casten, and A. Frank
© 2004 American Institute of Physics 0-7354-0207-8/04/$22.00

each energy. Drogs' Analytical expression (including multiple scattering [1]) for this efficiency has been computed and compared to our data as shown in the figure.

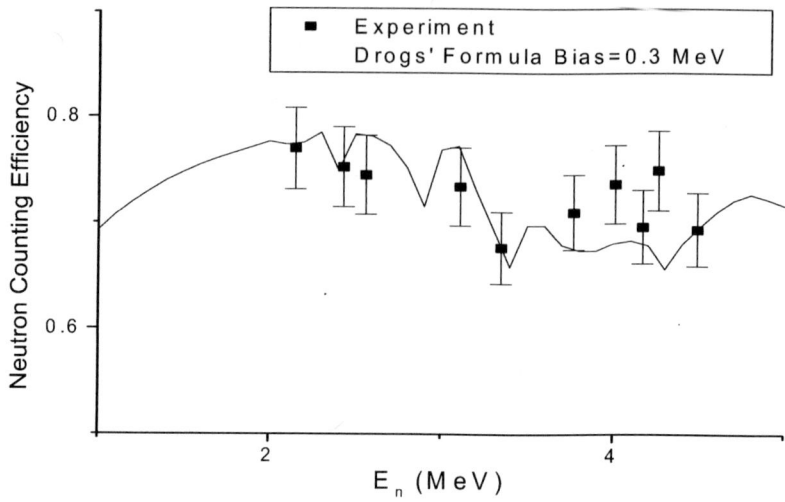

FIGURE 3. Absolute neutron detection efficiency as a function of neutron energy.

The line shows the result of the calculation of the geometrical efficiency using Drogs' multiple scattering expression. The dots, and error bars, represent our measurements.

Oscillations in the calculation come from the use of true experimental neutron capture and scattering cross sections. It is very important to stress that this efficiency has a strong dependency of the electronic bias [1][2].

A neutron detection array (hodoscopes) made of 7 hexagonal prisms of plastic (Bicron 408) scintillators, has been designed and constructed, together with the photomultipliers and voltage dividers.

The detection efficiency has been determined for an individual element of our array, as a function of neutron energy in the range 2.56 to 4.49 MeV for a given value of bias. The maximum efficiency was near 80% around 2.56 as can be see in Graph 1, close to the predicted value of the corresponding model calculation showed in the same graph.

This work was possible thanks to the help of the IFUNAM's workshop. DGAPAUNAM: IN114896 and IN111102. We also like to thank P. Villaseñor and D. G. Linarte B. operators of the tandem at ININ.

REFERENCES

1. J.L. Fowler *et al*."Efficiency Calibration of Scintillation Detectors in The Neutron Energy Range 1.5-25 MeV By The Associated Particle Technique. NIM 175 (1980) 449-463
2. E. Chávez *et al*. "Efficiency and Light Response Measurement of a Neutron" Detector Using the AP Technique – Mex.Nucl. (May 2002)
3. J.M Villarreal, "Establecimiento de un sistema de detección de neutrones"Tesis, F de Ciencias, UNAM. 1995, pp. 12,53.
4. http://www.detectors.saint-gobain.com/Data/Element/BC-408/NE 105
5. G.Knoll Radiation Detection and Measurement 3rd ed. 1999 pp 224-227

Shape coexistence in the neutron-deficient Pt isotopes in a configuration mixing IBM

Irving O. Morales*, Carlos E. Vargas* and Alejandro Frank†

*Facultad de Física e Inteligencia Artificial, Universidad Veracruzana
Sebastián Camacho No. 5, Centro, CP 91000, Xalapa, Ver. México
†Instituto de Ciencias Nucleares, UNAM,
Apdo. Postal 70-543, 04510 México, D.F. México

Abstract. The recently proposed matrix-coherent state approach for configuration mixing IBM is used to describe the evolving geometry of the neutron deficient Pt isotopes. It is found that the Potential Energy Surface (PES) of the Platinum isotopes evolves, when the number of neutrons decreases, from spherical to oblate and then to prolate shapes, in agreement with experimental measurements. Oblate-Prolate shape coexistence is observed in ^{194,192}Pt isotopes.

The shape coexistence phenomenon [1] observed in the atomic nuclei has been a topic which has produced much excitement in recent years [2].

While in the lead isotopes [3] there is a clear dominance of spherical (regular N bosons) configurations, in the platinum isotopes it is found that both the regular and intruder configurations, become very rapidly deformed (prolate or oblate). A simple alternative to describe the geometric properties of the Pt isotopes, as well as their evolution with neutron number, is given by the Interacting Boson Model [4] in conjunction with the matrix coherent-state approach [3]. For any IBM Hamiltonian, a potential energy surface can be generated that depends on the quadrupole-shape variables β and γ, which caracterize the deviation from spherical and axial symmetry, respectively. Details can be found elsewhere [3].

The model space is built from N and N+2 boson configurations. The Hamiltonian is

$$\hat{H} = \hat{H}_{0p-0h} + \hat{H}_{2p-2h} + \hat{H}_{mix} \ , \tag{1}$$

where $\hat{H}_{ip-ih}$ is the Hamiltonian for the ith configuration, taken in the simplified form

$$\hat{H}_{ip-ih} = \varepsilon_i \hat{n}_d + \kappa_i \hat{Q}_i \cdot \hat{Q}_i + \kappa_i' \hat{L} \cdot \hat{L} \ . \tag{2}$$

This Hamiltonian provides a simple parametrization of the essential features of nuclear structural evolution in terms of a vibrational term $\hat{n}_d$ and a quadrupole interaction.

In the Figure we show the Potential Energy Surfaces in β and γ. In ^{200,198,196}Pt isotopes, the lowest minimum is spherical. In ^{194}Pt the lowest energy minimum has an oblate shape coexisting with a prolate one almost at the same energy. In ^{192}Pt the asolute minimum is prolate coexisting with the oblate, at an exitation energy of 200 keV. In ^{190}Pt and lighter platinum isotopes, the prolate minima dominate and there is no more coexistence, as the oblate minima are predicted at very high energies. This work was

CP726, Nuclear Physics, Large and Small: International Conference on
Microscopic Studies of Collective Phenomena, edited by R. Bijker, R. F. Casten, and A. Frank
© 2004 American Institute of Physics 0-7354-0207-8/04/$22.00

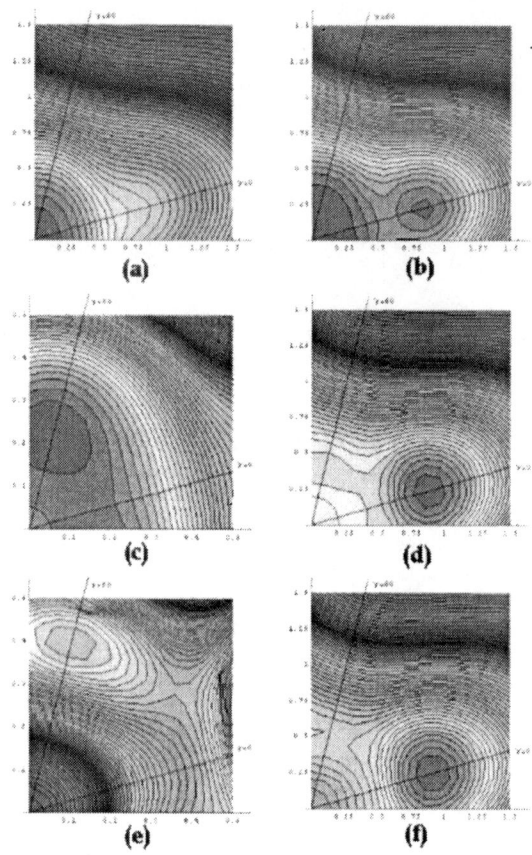

FIGURE 1. Potential Energy Surfaces, insert a) in ^{196}Pt, b) and c) in ^{194}Pt, d) and e) in ^{192}Pt, and f) in ^{190}Pt. The parameters of the Hamiltonian were extracted from the Ref. [5]

supported in part by CONACYT (México). I. O. M. and C. E. V. thank the organizing committe the partial support to attend the meeting.

REFERENCES

1. Heyde, K. H., Isacker, P. V., Waroquier, M., Wood, J. L., and Meyer, R. A., *Phys. Rep.*, **102**, 291 (1983).
2. Wood, J. L., Heyde, K., Nazarewicz, W., Huyse, M., and Duppen, P. V., *Phys. Rep.*, **215**, 101 (1992).
3. Frank, A., Isacker, P. V., and Vargas, C. E., *Phys. Rev. C*, **69**, 34323 (2004).
4. Iachello, F., and Arima, A., *The Interacting Boson Model*, Cambridge University Press, Cambridge, 1987.
5. Harder, M. K., Tang, K. T., and Isacker, P. V., *Phys. Lett. B*, **405**, 25–30 (1997).

Distribution of the GT strength starting from the ground state of ^{14}N

A. Negret[*], T. Adachi[†], C. Bäumer[**], A.M. van den Berg[‡], G.P.A. Berg[‡], P. von Brentano[§], D. Frekers[**], D. De Frenne[*], K. Fujita[¶], Y. Fujita[†], E.-W. Grewe[**], P. Haefner[**], K. Hatanaka[¶], M. Hunyadi[‡], M. A. de Huu[‡], H. Johansson[‖], E. Jacobs[*], Y. Kalmykov[††], K. Kawabata[‡‡], A. Korff[**], K. Nakanishi[¶], P. von Neumann-Cosel[††], T. Ogama[†], L. Popescu[*], S. Rakers[**], A. Richter[††], N. Ryezayeva[††], Y. Sakemi[¶], A. Shevchenko[††], Y. Shimbara[†], Y. Shimizu[¶], A. Tamii[‡‡], M. Uchida[‡‡], H. J. Wörtche[‡] and M. Yosoi[‡‡]

[*]Vakgroep Subatomaire en Stralingsfysica, Universiteit Gent, B-9000 Gent, Belgium
[†]Department of Physics, Osaka University, Toyonaka, Osaka 560-0043, Japan
[**]Institut für Kernphysik, Westfälische Wilhelms-Universität Münster, D-48149 Münster, Germany
[‡]Kernfysisch Versneller Instituut, Rijksuniversiteit Groningen, NL-9747 AA Groningen, The Netherlands
[§]Institut für Kernphysik,Universität zu Köln, 50937 Köln, Germany
[¶]Research Center for Nuclear Physics, Osaka University, Ibaraki, Osaka 567-0047, Japan
[‖]Gesellschaft für Schwerionenforschung mbH, Darmstadt, Germany
[††]Institut für Kernphysik, Technische Universität Darmstadt, D-64289 Darmstadt, Germany
[‡‡]Departament of Physics, Kyoto University, Sakyo, Kyoto 606-8224, Japan

Abstract.
The Gamow-Teller (GT) strength distribution in the β^- and β^+ directions starting from the ground state of ^{14}N has been investigated. A (d,^{2}He) experiment has been performed at KVI, Groningen with E_d=170 MeV and a (^{3}He,t) experiment at RCNP, Osaka with $E_{^3He}$=420 MeV. It is found that, in both cases, the GT mechanism is populating mainly the 2^+ excited states in the final nuclei, while the ground state transitions are strongly suppressed. The analysis is ongoing.

This study is addressing the long-standing problem of the suppression of the Gamow-Teller (GT) transitions from the ground states of ^{14}C and ^{14}O to the ground state of ^{14}N [1]. Although all the quantum numbers involved in the beta decay of ^{14}C and ^{14}O to the ground state of ^{14}N (J^π=1$^+$, T=0) suggest a GT character, these transitions are strongly hindered. Aiming to understand this anomaly, we investigated the GT strength distribution from the ground state of ^{14}N to levels in ^{14}C and ^{14}O. Theoretically [2], the structure of the ground state of ^{14}N explaining the hindrance of the GT decays from the ground states of ^{14}C and ^{14}O and the ground state of ^{14}N seems to favor the transition to a single 2^+ state rather than to the 0^+ or 1^+ excited states in the final nuclei.

In order to overcome the Q-value limitation of the energy range that can be investigated in the β-decay studies, one has to employ charge-exchange reactions of (n,p) and (p,n) type in order to determine the GT strength distribution in the final nuclei. It is accepted that, although a different interaction mediates the β-decay and the charge-exchange reaction, there is still a proportionality between the B(GT) of the β-decay and

CP726, Nuclear Physics, Large and Small: International Conference on
Microscopic Studies of Collective Phenomena, edited by R. Bijker, R. F. Casten, and A. Frank
© 2004 American Institute of Physics 0-7354-0207-8/04/$22.00

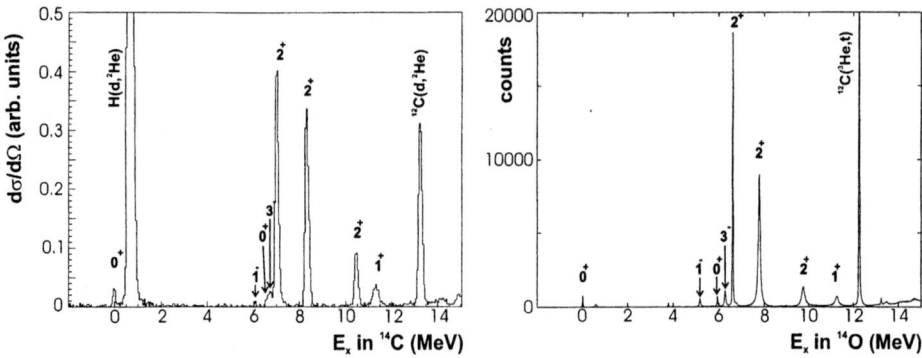

FIGURE 1. The 0° spectra for the $^{14}N(d,^2He)^{14}C$ and $^{14}N(^3He,t)^{14}O$. A melamine ($C_3H_6N_6$) target has been used. The cross section of different transitions is proportional with the B(GT)s. The similarities between the two spectra are revealing the good isospin symmetry.

the cross section of the corresponding reaction at 0° [3].

To investigate the β^+ direction, we used the $^{14}N(d,^2He)^{14}C$ reaction at KVI, Groningen. The AGOR cyclotron delivered the beam at E_d=170 MeV. The 2 protons (which are considered as ^2He if they are coupled into a $^1S_0$ state) were detected simultaneously with the Big Bite Spectrometer and the Eurosupernova detector [4]. The spin and isospin selectivity of the $(d,^2He)$ reaction makes it an ideal filter for GT transitions. The achieved energy resolution was about 150 keV. For the β^- direction we performed a $^{14}N(^3He,t)^{14}O$ experiment at RCNP, Osaka with E_{3He}=420 MeV. A very good energy resolution of about 35 keV can be obtained with the Grand Raiden spectrometer at the WS beamline of the Ring Cycloton [5]. This allowed, as a supplementary result, also the determination of the natural widths of several excited states of ^{14}O (S_p=4.628 MeV).

Figure 1 displays the spectra. Indeed, in both cases, the g.s. - g.s. transitions are strongly hindered and the strength is going mainly to three 2^+ excited states. However, an open question for the theory remains the fragmentation of the GT strength over three 2^+ excited states. The striking similarities of the two resulting spectra observed in these two mirror experiments illustrate a good isospin symmetry.

REFERENCES

1. Ajzenberg-Selove, F., *Nuclear Physics A*, **523**, 1–196 (1991).
2. S. Aroua, e. a., *Nuclear Physics A*, **720**, 71–83 (2003).
3. C.D. Goodman, e. a., *Physical Review Letters*, **44**, 1755–1759 (1980).
4. S. Rakers, e. a., *Nuclear Instruments and Methods A*, **481**, 253–261 (2002).
5. Y. Fujita, e. a., *Nuclear Physics A*, **687**, 311–320 (2001).

A Toy Model for QCD: Hadrons, Penta- and Heptaquarks

M. Nuñez[1], P. O. Hess[1], O. Civitarese[2] and M. Reboiro[2]

[1] *Instituto de Ciencias Nucleares, UNAM, A.P. 70-543, 04510 México D.F., Mexico*
[2] *Departamento de Física, Univ. Nac. La Plata, c.c.67 1900, La Plata, Argentina*

Abstract. A toy model for QCD is presented and applied to the hadron spectrum. As a byproduct the structure of penta- and hepta-quarks is obtained. A complete classification of the states is given. One essential feature of the model is the non-conservation of particle number.

The quark-antiquark part is given by a two-level system (Lipkin model) with a degeneration of each level given by 18 (2 spin times 3 flavor times 3 color degrees of freedom). The levels are filled by quarks. Holes represent antiquarks. The levels are at an energy of ± 0.33 GeV. The quarks levels are coupled to a boson level at 1.6 GeV, which represents pairs of gluons coupled to spin-color zero. [1].

The quark-antiquark pairs are coupled to color zero, spin (S) 0 and 1 and to flavor $((\lambda,\lambda))$, (0,0) and (1,1). The pairs are treated as bosons, i.e., there are four different types. Due to the boson mapping, care has to be taken due to the appearence of unphysical states [2]. One can deal with them in an approximate way [1] or matching the states in the boson space to the microscopic space. Therefore, a complete classification of the fermion and gluon states is important [1].

The Hamiltonian of the system is given by

$$
\begin{aligned}
H = \ & 2\omega_f n_f + \omega_b n_b + \sum_{\lambda S} V_{\lambda S} \left\{ \left[(b_{\lambda S}^\dagger)^2 + 2b_{\lambda S}^\dagger b_{\lambda S} + (b_{\lambda S})^2 \right] (1 - \frac{n_f}{2\Omega}) b + \right. \\
& \left. b^\dagger (1 - \frac{n_f}{2\Omega}) \left[(b_{\lambda S}^\dagger)^2 + 2b_{\lambda S}^\dagger b_{\lambda S} + (b_{\lambda S})^2 \right] \right\} + \\
& n_{(0,1)0} \left(D_1 n_b + D_2(b^\dagger + b) \right) + n_{(2,0)1} \left(E_1 n_b + E_2(b^\dagger + b) \right) \quad ,
\end{aligned}
$$

(1)

where $(b_{\lambda S}^\dagger)^2 = (b_{\lambda S}^\dagger \cdot b_{\lambda S}^\dagger)$ is a short hand notation for the scalar product [1]. The factor $(1 - \frac{n_f}{2\Omega})$ simulates the effect of the terms which would appear in the exact boson mapping of the quark-antiquark pairs [2]. The operators $b^\dagger$ and b are boson creation and annihilation operators of the gluon pairs with spin and color zero. The $2\Omega = 18$ is the

---

[1] Similarly for $(b_{\lambda S})^2$ and $(b_{\lambda S}^\dagger b_{\lambda S})$

CP726, Nuclear Physics, Large and Small: International Conference on Microscopic Studies of Collective Phenomena, edited by R. Bijker, R. F. Casten, and A. Frank
© 2004 American Institute of Physics 0-7354-0207-8/04/$22.00

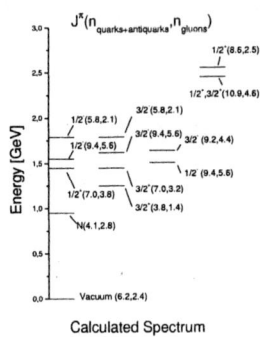

FIGURE 1. Nucleon resonances (first group of levels), Δ resonances (second group), pentaquarks (third group) and heptaquarks (fourth group). On the right side of each level are given the assigned spin and parity (J^π), and the total quark and anti-quark ($n_q + n_{\bar{q}}$) and gluon (n_g) contents (see the text)

degeneration of each level and $b^{\dagger}_{\lambda S}$ refer to the quark-antiquark pairs with spin S and flavor (λ, λ).

In Ref. [1] the classification for meson and baryon like states is given. Meson like states are presented by 18 quarks. With no interaction, all the 18 quarks are in the lower level. Meson like states are given by particle-hole excitations ($(q\bar{q})^n$). With interaction the physical meson states are a mixture of particle-hole states and gluons. In average, the mesons in this toy model consist of about 20% to 30% of gluons. The baryon like states are presented by $q^3(q\bar{q})^n$, where the quarks are distributed in the two levels. In case of no interaction, the lowest level is occupied and there are three valence quarks in the upper level. When the interaction is turned on, the hadrons contain besides the three valence quarks also a sea of quark-antiquark pairs and gluons. For the proton, the gluon content turns out to be about 41%, in close agreement to the experiment.

The parameters of the theory were adjusted to the meson spectrum (four parameters), to the nucleon resonances (two parameters) and to the Δ resonances (two parameters). For details, please consult [1]. As a byproduct we obtain the position of the first penta- and heptaquarks with unusual flavor combinations. The lowest pentaquarks is predicted to have spin $\frac{1}{2}$ with negative parity. The lowest heptaquark has positive parity. In Fig. 1 the spectrum of the lowest lying baryon states is depicted. Note, that the roper resonance (first excited nucleon resonance) is well reproduced. It exhibits a stronger collective nature as the nucleon state, as can be appreciated by the total number of quark plus antiquark and gluon content. The latter results were send for publication.

The work was supported financially by DGAPA (in119002), CONACyT and SNI.

REFERENCES

1. S. Lerma, S. Jesgarz, P. O. Hess, O. Civitarese and M. Reboiro, Phys. Rev. **C 67** (2003), 055209.
2. A. Klein and E. R. Marshalek, Rev. Mod. Phys. **63** (1991), 375.

Isospin purity in the A=42 isobars

J.N. Orce*, P. Petkov[†]**, C.J. McKay*, S.N. Choudry*, S.L. Lesher*, M. Mynk*, D. Bandyopadhyay*, S.W. Yates* and M.T. McEllistrem*

*University of Kentucky, Lexington, Ky 40506-0055, USA
[†]Institute for Nuclear Research and Nuclear Energy, Sofia 11784, Bulgaria
**Institut für Kernphysik der Universität zu Köln, 50937 Köln, Germany

Abstract. The lifetime of the first $2^+_{T=1}$ state in ^{42}Sc has been measured as 74(16) fs. This result gives a value for the isoscalar matrix element of M_0=6.63(76). From the mirror nuclei, ^{42}Ca and ^{42}Ti, the isoscalar matrix element is given as 7.15(48) W.u., confirming isospin purity in the A=42 isobars.

Following the theoretical work by Bernstein, Brown and Madsen [1] and the experimental work of Cottle et al. [2, 3], the study of the N=Z nucleus ^{42}Sc may lead to an accurate experimental test of isospin purity for the A=42 isobaric analogues. For this lifetime determination, the ^{40}Ca(^{3}He,pγ)^{42}Sc reaction has been used at beam energies of 4.2 and 5.1 MeV at the University of Kentucky. The Ca targets were prepared by evaporating natural Ca onto a Ta foil. The thicknesses of the Ca targets were 0.98 and 0.62 mg/cm^2, and argon gas was used to minimize the oxidation of the Ca target while mounting it in the chamber. The recoiling ^{42}Sc nuclei are predicted to stop inside the Ca target in 0.11 mg/cm^2, and after about 1.4 ps. Beam pulses were separated by 533 ns and bunched to about 1.5 ns, allowing the use of the time-of-flight technique, where background contamination is suppressed by gating on appropiate time windows. The γ rays deexciting the recoiling nuclei, with an initial recoil velocity of v/c=0.004, were collected using a BGO Compton-suppressed HPGe detector with 2 keV resolution. DSAM measurements were carried out at angles of 0°, 20°, 37°, 54°, 129° and 147° for the 4.2 MeV test, and 55° and 125° for 5.1 MeV. The 4.2 MeV beam energy was chosen with the purpose of both having enough statistics for line-shape analysis and minimizing side-feeding. At this beam energy only 11% of the observed intensity is feeding the first $2^+_{T=1}$ state from higher levels. The experiments were run for 14 days to provide adequate statistics. Combining the 4.2 and 5.1 MeV measurements, it is expected that a more sensible idea of the physical problems involved in the line-shape analysis, such as side-feeding and consequent proton emission effects would be resolved.

Line-shape analysis [4, 5, 6] with the inclusion of the calculated proton emission effect [7] for the (^{3}He,pγ) reaction has been used for determining lifetimes. This method gives a realistic description of the stopping process of the recoiling ^{42}Sc nuclei in the ^{40}Ca target. Figure 1 shows the fit to the line-shape of the 975 keV transition depopulating the first $2^+_{T=1}$ state at 1586 keV, giving a lifetime of 74(16) fs, which corresponds to a value for the isoscalar matrix element of M_0=6.63(76) W.u. These values had been previously determined as 100(30) fs and M_0=5.70(90) W.u. [8, 9]. A comparison of

CP726, Nuclear Physics, Large and Small: International Conference on
Microscopic Studies of Collective Phenomena, edited by R. Bijker, R. F. Casten, and A. Frank
© 2004 American Institute of Physics 0-7354-0207-8/04/$22.00

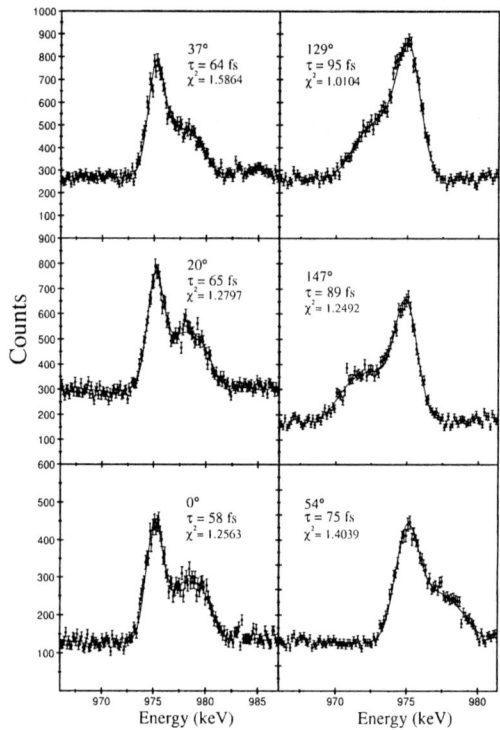

FIGURE 1. Line-shape analysis of the 975 keV transition from the $2^+_{T=1}$ at 1586 keV in ^{42}Sc.

the isoscalar matrix element obtained in the current work with that given by the mirror nuclei [2] indicates that isospin purity is conserved in the region. However, it is expected that a future experiment measuring the proton distribution of the (^{3}He,p) reaction would decrease the uncertainty.

REFERENCES

1. A. M. Bernstein, V.R. Brown, and V.A. Madsen, Phys. Rev. Lett. **42**, 425 (1979)
2. P.D. Cottle *et al.*, Phys. Rev. **C 60**, 031301 (1999)
3. P.D. Cottle *et al.*, Phys. Rev. Lett. **88**, 172502 (2002)
4. G. Böhm *et al.*, Nucl. Instrum. Methods **A 329**, 428 (1993)
5. P. Petkov *et al.*, Nucl. Phys. **A 640**, 293 (1998)
6. D. Tonev *et al.*, Phys. Rev. **C 65**, 034314 (2002)
7. *Projection angular-momentum evaporation code for fusion-evaporation reaction.* U. of Liverpool
8. N.R. Roberson and G. van Middelkoop, Nucl. Phys. **A 176**, 577 (1971)
9. D.P. Balamuth, G.P. Anastassiou, and R.W. Zurmuhle, Phys. Rev. **C 2**, 215 (1970)

Study of Gamow-Teller Transition Strengths in fp-shell Nuclei Using the ^{64}Ni$(d,^2$He$)^{64}$Co Reaction

L. Popescu[*], C. Bäumer[†], A.M. van den Berg[**], D. Frekers[†], D. de Frenne[*], Y. Fujita[‡], E.W. Grewe[†], P. Haefner[†], M. Hunyadi[**], M. de Huu[**], E. Jacobs[*], H. Johansson[§], A. Korff[†], A. Negret[*], P. von Neumann-Cosel[¶], S. Rakers[†], N. Ryezayeva[¶], A. Shevchenko[¶], H. Simon[§] and H.J. Wörtche[**]

[*]Vakgroep Subatomaire en Stralingsfysica, Universiteit Gent, B-9000 Gent, Belgium
[†]Institut für Kernphysik, Westfälische Wilhelms-Universität Münster, D-48149 Münster, Germany
[**]Kernfysisch Versneller Instituut, Rijksuniversiteit Groningen, NL-9747 AA Groningen, The Netherlands
[‡]Department of Physics, Osaka University, Toyonaka, Osaka 560-0043, Japan
[§]Gesellschaft für Schwerionenforschung mbH, D-64291 Darmstadt, Germany
[¶]Institut für Kernphysik, Technische Universität Darmstadt, Germany

Abstract. The ^{64}Ni$(d,^2$He$)^{64}$Co reaction was studied at the AGOR cyclotron of the KVI, Groningen, with the Big-Bite Spectrometer and the EuroSuperNova detector using a deuteron beam of 170 MeV. An energy resolution of about 100 keV was achieved. In addition to the ground-state with $J^\pi = 1^+$, several other 1^+ states could be identified and the strengths of the corresponding Gamow-Teller transitions determined.

To study the spin-isospin response of nuclei charge-exchange reactions play an important role. For $(d,^2$He$)$ experiments it is shown that, as for the (n,p) reaction, there is a proportionality between the reaction cross section and the Gamow-Teller transition strength (B$^+_{GT}$) [1]. We have studied the ^{64}Ni$(d,^2$He$)^{64}$Co reaction at the KVI, Groningen, using a 170 MeV deuteron beam of the AGOR accelerator. The ejectiles were momentum analized with the Big-Bite Spectrometer (BBS) and the EuroSuperNova detection system [2].

The (n,p)-type reaction $(d,^2$He$)$ shows several advantages over the (n,p) reaction: the use of a primary beam provides the possibility to improve the energy resolution, down to about 100 keV with the KVI setup as used in this experiment. In addition, $(d,^2$He$)$ reaction selects spin-flip transitions as the ground-state (gs) of the deuteron is mainly a ^{3}S$_1$ state and the ^{2}He is in a ^{1}S$_0$ state. So, for low momentum transfer, the $(d,^2$He$)$ spectra will be dominated by Gamow Teller transitions. The difficulty in the study of $(d,^2$He$)$ reaction comes from the fact that ^{2}He is an unbound system and as a consequence we have to detect the two protons from the ^{2}He system at the same time. In addition, the deuteron breakup reactions will increase the background in the forward direction. However, the coincident detection will reduce the phase space of the two protons and limit their internal energy below about 1 MeV. Under these conditions the original 2 proton system is mainly in a ^{1}S$_0$ state and only a few percent higher order

CP726, *Nuclear Physics, Large and Small: International Conference on Microscopic Studies of Collective Phenomena*, edited by R. Bijker, R. F. Casten, and A. Frank
© 2004 American Institute of Physics 0-7354-0207-8/04/$22.00

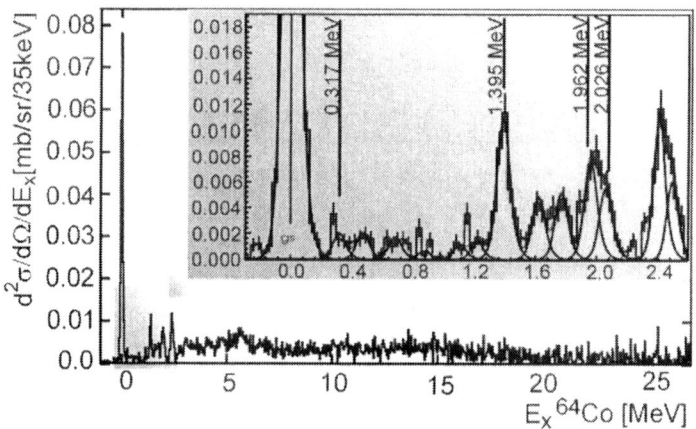

FIGURE 1. ^{64}Ni(d,^{2}He)^{64}Co spectrum for a scattering angle in the interval (0°, 1°) after the background subtraction. A binning of 35 keV/channel has been used.

components contribute [3].

The transition from the ^{64}Co gs to the ^{64}Ni gs is the fastest β^- transition to the gs of a stable nucleus in the fp-shell region (log ft = 4.3 ; B$_{GT}^+$=0.627±0.036). As a consequence, the target nucleus ^{64}Ni is the only fp-shell nucleus that enables the direct determination of the unit cross section for the (d,^{2}He) reaction, the unit cross section being defined as the proportionality factor between the extrapolated cross section to zero momentum transfer and the B$_{GT}^+$. The value that we obtained for the unit cross section in this experiment is 0.56±0.05 mb/sr providing an important calibration standard for the fp-shell mass region. To this error an estimated systematic inaccuracy of 20% should be added.

In Figure1, the ^{64}Ni(d,^{2}He)^{64}Co spectrum for a scattering angle in the interval (0°, 1°) is shown. Apart from the gs, several other states fed by Gamow Teller transitions could be identified in the ^{64}Co excitation energy region 0-2.5 MeV. Above 2.5 MeV the density of the levels becomes too high to separate them. However, in that energy region bump-like structures can easily be observed, indicating the presence of giant resonances. The analysis of these structures is still ongoing. The neutron separation energy is at 6.02 MeV and the so called quasi free continuum region was subtracted by using the semiphenomenological function given by Erell et al. [4].

REFERENCES

1. S. Rakers, e. a., *Physical Review C*, **65**, 044323 (2002).
2. S. Rakers, e. a., *Nuclear Instrumentations and Methods A*, **481**, 252 (2002).
3. S. Kox, e. a., *Nuclear Physics A*, **556**, 621–640 (1993).
4. A. Erell, e. a., *Physical Review C*, **34**, 1822 (1986).

Population of ^{195}Os via a deep-inelastic reaction

J.J. Valiente-Dobón*†, C. Wheldon**†, P.H. Regan†‡, C.Y. Wu§, D. Cline§,
C. Andreoiu*, R. Chapman¶, P. Fallon‖, S.J. Freeman††, A. Hayes§,
H. Hua§, S.D. Langdown†‡, I.Y. Lee‖, X. Liang¶, A.O. Macchiavelli‖,
C.J. Pearson†, Zs. Podolyák†, G. Sletten‡‡, J.F. Smith††, C.E. Svensson*,
R. Teng§, D. Ward‖, D.D. Warner§§ and A.D. Yamamoto†‡

*Department of Physics, University of Guelph, Guelph N1G 2W1, Ontario, Canada.
†Department of Physics, University of Surrey, Guildford GU2 7XH, UK.
**SF7, Hahn-Meitner-Institut, Glienicker Straße 100, D-14109 Berlin, Germany.
‡Wright Nuclear Structure Laboratory, Yale University, New Haven, CT 06520-8124, USA.
§Department of Physics, University of Rochester, NY 14627, USA.
¶Department of Electronic Engineering and Physics, University of Paisley, Paisley PA1 2BE, UK.
‖Lawrence Berkeley National Laboratory, Berkeley, California 94720, USA.
††Department of Physics and Astronomy, Schuster Laboratory, University of Manchester,
Manchester M13 9PL, UK.
‡‡The Niels Bohr Institute, University of Copenhagen, Blegdamsvej 17. 2100 Copenhagen,
Denmark.
§§CCLRC Daresbury Laboratory, Warrington WA 4AD, UK.

Abstract.
The present work reports on the $^{195}_{76}$Os isotope, which is the most neutron-rich osmium isotope for which transitions have been measured. It has been populated following a multi-nucleon transfer reaction between a thin $^{198}_{78}$Pt target and an 850-MeV $^{136}_{54}$Xe beam. Evidence from γ-ray coincidences has been found for an $I^\pi = \left(\frac{27}{2}^-\right)$ isomeric state with a measured half-life of $26 \pm 9ns$.

The osmium-tungsten isotopes are a perfect playground for understanding the shape transition from a prolate deformed shape in the lighter mass isotopes to an oblate deformed shape in heavier isotopes going via a γ-soft potential and ending up in a spherical shape for isotopes near the closed-shell $N = 126$. However due to experimental constraints it is very difficult to access these exotic neutron-rich nuclei. Some progress has been made to populate these exotic systems using relativistic energy projectile fragmentation reactions [1], but these yield only limited spectroscopic information. Multi-nucleon transfer reactions are known to populate neutron-rich nuclei [2] and have been used to populate two of the heaviest known neutron-rich osmium isotopes, ^{194}Os [3] and ^{195}Os [4, 5]. Here, we present spectroscopic information on the ^{195}Os isotope obtained via a multi-nucleon transfer reaction.

The reaction used to populate the nucleus of interest was an 850 MeV $^{136}_{54}$Xe beam provided by the 88" cyclotron incident on a self-supporting, 420 μg.cm^{-2} $^{198}_{78}$Pt target at LBNL. GAMMASPHERE plus CHICO [6] were used to detect deexcitation γ rays plus kinematically coincident reaction products; see Ref. [4].

Figure 1 (left) shows the delayed γ rays from the deexcitation of the $26 \pm 9ns$ isomer

CP726, Nuclear Physics, Large and Small: International Conference on
Microscopic Studies of Collective Phenomena, edited by R. Bijker, R. F. Casten, and A. Frank
© 2004 American Institute of Physics 0-7354-0207-8/04/$22.00

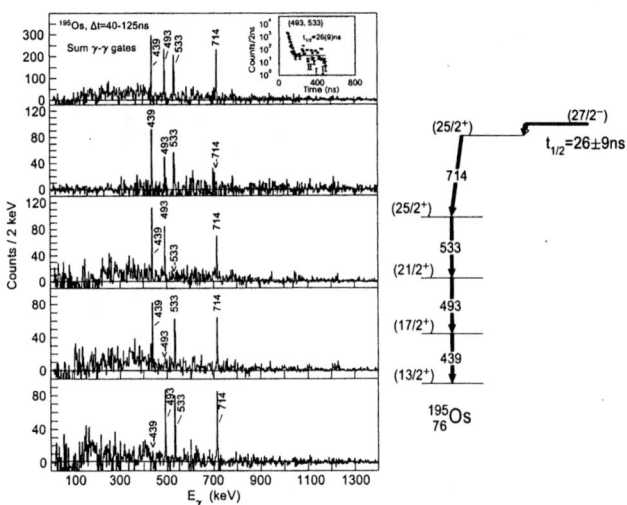

FIGURE 1. Background subtracted delayed γ-ray spectra for ^{195}Os (left). The inset shows the fitted half-life curve ($t_{1/2} = 26 \pm 9ns$). The level scheme deduced from γ-γ coincidences is shown on the right.

in ^{195}Os. The half-life (see inset) has been obtained by fitting the double gated (493 and 533 keV gates) time spectra with an exponential curve and a constant background. The γ-ray transitions from the decay of the isomer 439, 493, 533 and 714 keV, have been found to be in coincidence, the ordering is based on increasing γ-ray energies. The transition that directly deexcites the isomer $I^\pi = \left(\frac{27}{2}^-\right)$ could not be observed in the current data due to its presumed low energy and therefore high conversion coefficient. The energy limit of this transition has been discussed in Ref. [7], though prior to the current work, no coincidence data were available. Spin and parities can not be assigned experimentally for delayed γ rays, but are estimated from comparisons with Blocked BCS calculations [7]. The nucleus ^{195}Os is expected to have a triaxial, γ-soft shape, as it is adjacent to the predicted transition from proplate to oblate shapes at ^{196}Os [3].

This work has been supported by NSERC (Canada), Premier's Research Excellence Award from the Ontario government, EPSRC (UK), the US Department of Energy and the National Science Foundation.

REFERENCES

1. Podolyak, Zs., et al., *Nucl. Phys.* **A 722**, 2003, pp. 273–278.
2. Takai, H., et al., *Phys. Rev.*, **C 38**, 1988, pp. 1247–1261.
3. Wheldon, C., et al., *Phys. Rev.*, **C 63**, 2000, pp. 011304-1–011304-5.
4. Valiente-Dobón, J. J., et al., *Phys. Rev.*, **C 69**, 2004, pp. 024316-1–024316-13.
5. Regan, P. H., et al., *Laser Physics Letters*, in press, 2004, volume 1.
6. Deleplanque, M. A. and Diamond, R. M. (Eds.), *Gammasphere Proposal*, LBNL Report 5202 (1988); Simon, M. W., et al., *Nucl. Intr. and Meth.*, **A 452**, 2000, pp. 205–222.
7. Caamaño, M., Ph.D. thesis, University of Surrey, Guildford, (2002).

Excited bands in even-even rare-earth nuclei

Carlos E. Vargas* and Jorge G. Hirsch†

*Facultad de Física e Inteligencia Artificial, Universidad Veracruzana
Sebastián Camacho No. 5, Centro, CP 91000, Xalapa, Ver. México
†Instituto de Ciencias Nucleares, UNAM,
Apdo. Postal 70-543, 04510 México, D.F. México

Abstract. The energetics of states belonging to normal parity bands in even-even dysprosium isotopes, and their B(E2) transition strengths, are studied using an extended pseudo-SU(3) shell model. States with pseudospin 1 are added to the standard pseudospin 0 space, allowing for a proper description of known excited normal parity bands.

The nuclear shell model [1] provides a detailed microscopic description of a number of properties of atomic nuclei. Nevertheless, a shell model description of heavy nuclei requires further assumptions, being of particular relevance the systematic and proper truncation of the Hilbert space. The SU(3) shell model [2] has been successfully applied in light nuclei. In heavy nuclei the strong spin-orbit interaction renders the SU(3) model useless, while at the same time pseudospin emerges as a good symmetry. The success of the pseudo-SU(3) model [3] lies on the goodness of this symmetry.

The first applications of the pseudo-SU(3) model considered it as a dynamical symmetry. Further developments enabled mixed-representation calculations. A realistic Hamiltonian including SU(3) symmetry-breaking terms could be diagonalized. A fully microscopic description of many rotational bands and electromagnetic transition strengths in both even-even [4] and odd-A [5, 6] heavy deformed nuclei emerged. The inclusion of states with pseudospin 1 and 3/2 (in addition to those with $\tilde{S} = 0$ and $1/2$) for protons and neutrons allowed to describe up to eight rotational bands in odd-mass nuclei [7]. Scissors M1 excitations in odd-mass heavy nuclei [8] were also described. In this work we present the results for excited bands in even-even dysprosium isotopes.

The basis states are built with proton and neutron states with pseudospin 0 and 1. The Hamiltonian has a *principal* part H_0:

$$H_0 = \sum_{\alpha=\pi,\nu} \{H_{sp,\alpha} - G_\alpha H_{pair,\alpha}\} - \frac{1}{2} \chi \, \tilde{Q} \cdot \tilde{Q} . \tag{1}$$

Added to them are four 'rotor-like' terms that are diagonal in the SU(3) basis:

$$H = H_0 + a K_J^2 + b J^2 + c \tilde{C}_3 + A_{sym} \tilde{C}_2 . \tag{2}$$

A detailed analysis of each term of this Hamiltonian and its parametrization can be found in Ref. [9].

The Figure shows the yrast, γ, β and five other excited bands in ^{162}Dy. Experimental data (see the NNDC page at http://www.nndc.bnl.gov) are plotted on the left

CP726, *Nuclear Physics, Large and Small: International Conference on Microscopic Studies of Collective Phenomena*, edited by R. Bijker, R. F. Casten, and A. Frank
© 2004 American Institute of Physics 0-7354-0207-8/04/$22.00

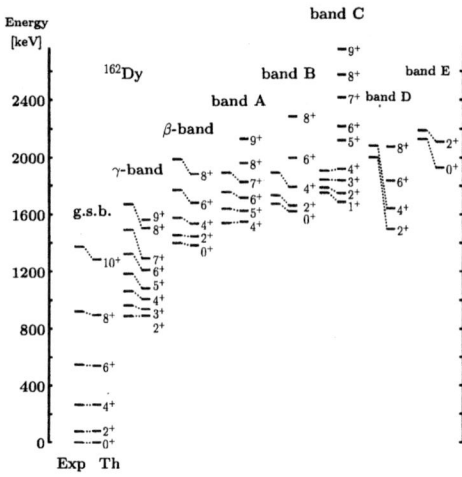

FIGURE 1. Positive parity bands in ^{162}Dy. The labels indicate the total angular momentum and parity of each level. Experimental data are plotted on the left-hand side of each column and theoretical ones on the right-hand side.

hand side of each column and represent nearly all measured normal parity bands (eight), while those obtained with the model are shown on the right hand side. In the previous pseudo SU(3) study of this nuclei only four bands were described. The new bands are being described thanks to the inclusion in the Hilbert space of $\tilde{S} = 1$ proton or neutron states. The agreement is very good for the eight rotational bands, not only for the energies but also from the B(E2) transition strengths [9]. This work was supported in part by CONACYT (México).

REFERENCES

1. Mayer, M. G., *Phys. Rev.*, **75**, 1969 (1949).
2. Elliott, J. P., *Proc. Roy. Soc. London Ser. A*, **245**, 128 (1958).
3. Ratna Raju, R. D., Draayer, J. P., and Hetch, K. T., *Nucl. Phys. A*, **202**, 433 (1973).
4. Beuschel, T., Hirsch, J. G., and Draayer, J. P., *Phys. Rev. C*, **61**, 54307 (2000).
5. Vargas, C. E., Hirsch, J. G., and Draayer, J. P., *Nucl. Phys. A*, **673**, 219 (2000).
6. Vargas, C. E., Hirsch, J. G., Beuschel, T., and Draayer, J. P., *Phys. Rev. C*, **61**, 31301 (2000).
7. Vargas, C. E., Hirsch, J. G., and Draayer, J. P., *Phys. Rev. C*, **66**, 64309 (2002).
8. Vargas, C. E., Hirsch, J. G., and Draayer, J. P., *Phys. Lett. B*, **551**, 98 (2003).
9. Vargas, C. E., and Hirsch, J. G., *Submitted to Phys. Rev. C*, **0**, 0 (2004).

Two-dimensional Fourier analysis of the nuclear mass errors

Víctor Velázquez*, Juan C. López-Vieyra[†], Alejandro Frank[†], Jorge G. Hirsch[†] and José Barea[†]

*Departamento de Física, Facultad de Ciencias, Universidad Nacional Autónoma de México, A.P. 70-348, 04511 México DF, México.
[†]Instituto de Ciencias Nucleares, Universidad Nacional Autónoma de México, A.P. 70-543, 04510 México DF, México

Abstract. Differences between measured and calculated masses are studied using two-dimensional Fourier analysis. Low frequency components are associated with residual correlations.

INTRODUCTION

The prediction of nuclear masses is very important for a complete understanding of several processes in the universe. There are some successful mass formulae which are commonly used. The simplest is the liquid drop model (LDM), but there are others which consider shell model corrections, like the finite range droplet model (FRDM) [1], and Duflo-Zuker (DZ) [2]. In addition, Garvey-Kelson (GK)[3] can be used and involve statistics systematics.

In previous works [4, 5] it was noted the presence of correlations in the difference $M_{exp} - M_{the}$ for the three first models, where a one-dimensional Fourier analysis was developed. Here a 2-D Fourier analysis is performed in order to get an unique sight of the correlations phenomena in the Z-N plane.

FOURIER TRANSFORMATION

A two-dimensional plot of mass difference $\Delta M(N,Z) = M_{exp}(N,Z) - M_{the}(N,Z)$ is presented in Fig.1 of [6]. Their two dimensional Fourier transform (2-DFT) is calculated as:

$$F_{k,l} = \frac{1}{\sqrt{N_{max}Z_{max}}} \sum_{Z=0}^{Z_{max}} \sum_{N=0}^{N_{max}} \Delta M(N,Z) e^{-2\pi i (Zk/Z_{max} + Nl/N_{max})} \tag{1}$$

In the Fig. 1 we show the magnitude of the 2-DTF for each model. The different degrees of correlation can be reflected with the position of the largest 2-DFT components. For the LDM the peaks at low frequencies (in corner) show a high degree of correlation. The migration of main peaks to high frequencies are associated with a loss of correlation. The extreme case corresponds to GK where the peaks appears at high frequencies and

CP726, Nuclear Physics, Large and Small: International Conference on
Microscopic Studies of Collective Phenomena, edited by R. Bijker, R. F. Casten, and A. Frank
© 2004 American Institute of Physics 0-7354-0207-8/04/$22.00

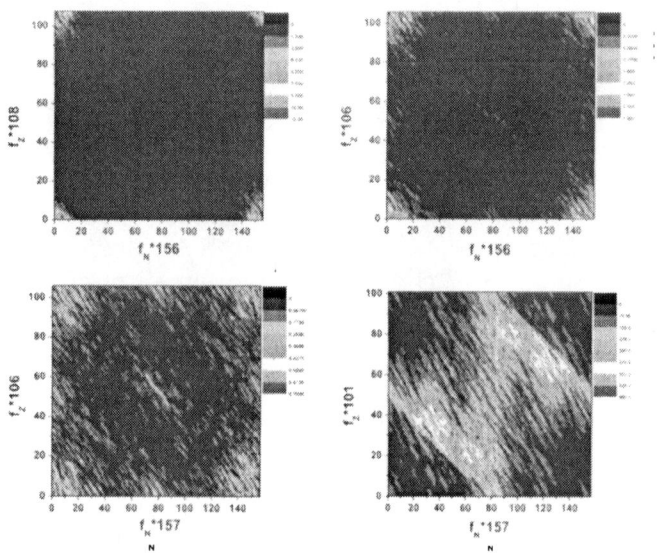

FIGURE 1. 2-DF transformation for LDM (top-left), FRDM (top-right), Dufb-Zuker (bottom-left), and Garvey-Kelson (bottom-right)

the correlations is such that the r.m.s. is equal to 0.189 MeV, in comparison with the 3.49 MeV in the LDM.

The diagonal patterns in all visible 2-DFT analysis of the nuclear data set, reflect the central distribution of measured masses, which is not rectangular in the N-Z plane.

REFERENCES

1. Moller P., Nix J. R., Myers W. D., and Zwiatecky W. J., Atomic Data and Nuclear Data Tables, **59** 185(1995).
2. Dufb J. and Zuker A. P., Phys. Rev. **C52** R23(1995).
3. Garvey G. T. and Kelson I., Phys. Rev. Lett. **16** (1966)197.
4. Hirsch J.G., Frank A., and Velázquez V., Phys. Rev. **C69** 37304(2004).
5. Velázquez V., Frank A., and Hirsch J. G., 'Systematic correlations and chaos in mass formulae.' in *Computational and Group Theoretical Methods in Nuclear Physics*, edited by J. Escher, O. Castaños, J. Hirsch, G. Stoicheva, and S. Pittel, World Scientifi c c, Singapore, 2004.
6. Hirsch J.G., Frank A., Barea J., Van Isacker P.,and Velázquez V.: this proceedings.

LIST OF PARTICIPANTS

Elí Aguilera	ININ, Mexico	efar@nuclear.inin.mx
Ani Aprahamian	Notre Dame, USA	aapraham@nd.edu
José Arias	Sevilla, Spain	pepe@nucle.us.es
Akito Arima	Tokyo, Japan	
José Barea	ICN-UNAM, Mexico	barea@nuclecu.unam.mx
Bruce Barrett	Arizona, USA	bbarrett@physics.arizona.edu
Noemie Benczer-Koller	Rutgers, USA	nkoller@physics.rutgers.edu
Roelof Bijker	ICN-UNAM, Mexico	bijker@nuclecu.unam.mx
María Ester Brandan	IF-UNAM, Mexico	brandan@fisica.unam.mx
Peter von Brentano	Köln, Germany	brentano@ikp.uni-koeln.de
Mark Caprio	Yale, USA	mark.caprio@yale.edu
David Cardamone	Arizona, USA	dmcard@physics.arizona.edu
Rick Casten	Yale, USA	rick@riviera.physics.yale.edu
Jacek Dobaczewski	Warsaw, Poland	jacek.dobaczewski@fuw.edu.pl
Jerry Draayer	LSU, USA	draayer@sura.org
Kalin Drumev	LSU, USA	kalin@epscor.phys.lsu.edu
Jorge Dukelsky	Madrid, Spain	dukelsky@iem.cfmac.csic.es
Tomas Dytrych	LSU, USA	tdytrych@nuclear.phys.lsu.edu
Jon Engel	UNC, USA	engelj@physics.unc.edu
Ruben Fossion	Ghent, Belgium	ruben.fossion@ugent.be
Alejandro Frank	ICN-UNAM, Mexico	frank@nuclecu.unam.mx
Denis de Frenne	Ghent, Belgium	denis.defrenne@ugent.be
Moshe Gai	Yale/UConn, USA	moshe.gai@yale.edu
Tom Gaisser	Bartol, USA	gaisser@bartol.udel.edu
Homero García Martínez	ININ, Mexico	homero_gm@yahoo.com.mx
Joaquín Gómez-Camacho	Sevilla, Spain	gomez@nucle.us.es
Peter Hess	ICN-UNAM, Mexico	hess@nuclecu.unam.mx
Jorge Hirsch	ICN-UNAM, Mexico	hirsch@nuclecu.unam.mx
Francesco Iachello	Yale, USA	francesco.iachello@yale.edu
Omar Irving Morales	Veracruzana, Mexico	irvingfisica@hotmail.com
Piet van Isacker	GANIL, France	isacker@ganil.fr
Etienne Jacobs	Ghent, Belgium	etienne.jacobs@ugent.be
Jan Jolie	Köln, Germany	j.jolie@ikp.uni-koeln.de
Shelly Lesher	Kentucky, USA	slesher@pa.uky.edu
Geza Levai	ATOMKI, Hungary	levai@atomki.hu
Sean Liddick	MSU, USA	liddick@nscl.msu.edu

Alexander Lisetskiy	MSU, USA	lisetski@nscl.msu.edu
Kim Lister	Argonne, USA	lister@anl.gov
René Luna	CINVESTAV, Mexico	rluna@fis.cinvestav.mx
Elizabeth McCutchan	Yale, USA	elizabeth.ricard-mccutchan@yale.edu
Ricardo Monroy	IF-UNAM, Mexico	rmonroy@fisica.unam.mx
Marcos Moshinsky	IF-UNAM, Mexico	moshi@fisica.unam.mx
Witek Nazarewicz	Oak Ridge, USA	witek@utk.edu
Alexandru Negret	Ghent, Belgium	alexandruliviu.negret@ugent.be
Maribel Núñez	ICN-UNAM, Mexico	mnunez@nuclecu.unam.mx
Nico Orce	Kentucky, USA	jnorce@uky.edu
María Esther Ortiz	IF-UNAM, Mexico	ortiz@fisica.unam.mx
Takaharu Otsuka	Tokyo, Japan	otsuka@phys.s.u-tokyo.ac.jp
Elizabeth Padilla	ICN-UNAM, Mexico	padilla@nuclecu.unam.mx
Thomas Papenbrock	Oak Ridge, USA	papenbro@phy.ornl.gov
Norbert Pietralla	Stony Brook, USA	npietralla@notes.cc.sunysb.edu
Stuart Pittel	Bartol, USA	pittel@bartol.udel.edu
Lucia Popescu	Ghent, Belgium	LuciaAna.Popescu@ugent.be
Patrick Regan	Surrey, UK	P.Regan@surrey.ac.uk
Peter Ring	München, Germany	ring@physik.tu-muenchen.de
Gergana Stoitcheva	Oak Ridge, USA	stoitchevags@ornl.gov
Mario Stoitsov	Oak Ridge, USA	stoitsovmv@ornl.gov
Igal Talmi	Weizmann, Israel	igal.talmi@weizmann.ac.il
José Javier Valiente Dobón	Guelph, Canada	valiente@physics.uoguelph.ca
Carlos Vargas Madrazo	Veracruzana, Mexico	cavargas@uv.mx
Víctor Velázquez	FC-UNAM, Mexico	vmva@hp.fciencias.unam.mx
Tamas Vertse	ATOMKI, Hungary	vertse@atomki.hu
Dave Warner	Daresbury, UK	d.d.warner@dl.ac.uk
Victor Zamfir	Yale, USA	zamfir@galileo.physics.yale.edu

StuFiesta

Nuclear Physics, Large and Small
Microscopic Studies of Collective Phenomena

PROGRAM

MONDAY April 19

17:00   Welcome

Chair: **Igal Talmi** (Weizmann)

17:15   **Jorge Dukelsky** (CSIC, Madrid)
New generalizations of the Richardson-Gaudin models
17:45   **Witek Nazarewicz** (ORNL)
Description of weakly bound or unbound nuclear states
18:15   **Jon Engel** (UNC)
Time-reversal violation in nuclei and electric-dipole moments of atoms
18:45   **Marcos Moshinsky** (IF-UNAM)
Derivation of the pseudo-SU(3) symmetry of heavy nuclei in the shell model picture

19:30   Barbecue

TUESDAY April 20

Chair: **Maria Ester Brandan** (IF-UNAM)

09:00 **Peter von Brentano** (Köln)
Identical bands, pseudospin symmetry and supersymmetry
09:30 **José Barea** (ICN-UNAM)
New correlations induced by nuclear supersymmetry
10:00 **Bruce Barrett** (Arizona)
The ab initio large-basis no-core shell model
10:30 **Takaharu Otsuka** (Tokyo)
Perspectives in the shell model

11:00 Coffee

Chair: **Elí Aguilera** (ININ)

11:30 **Jacek Dobaczewski** (Warsaw)
Energy density functionals in nuclei
12:00 **Mario Stoitsov** (ORNL)
HFB with Skyrme forces and exact particle number projection for
deformed nuclei
12:30 **Gergana Stoitcheva** (ORNL)
Shifted-contour method in nuclear structure for shell model Monte Carlo
calculations
13:00 **Thomas Papenbrock** (ORNL)
Nuclear shell model frontiers
13:30 **Akito Arima** (Tokyo)
Shell model calculations with random interactions and binding energies

14:00 Lunch

16:30 Poster session

19:00 Conference Banquet (La Trapiche)

WEDNESDAY April 21

Chair: **Peter Hess** (ICN-UNAM)

09:00   **Noemie Benczer-Koller** (Rutgers)
        Stuart Pittel and the $f_{7/2}$ shell revisited: magnetic moments of the Ca
        isotopes
09:30   **Jorge Hirsch** (ICN-UNAM)
        Regularities vs. chaos in nuclear masses
10:00   **Joaquín Gómez-Camacho** (Sevilla)
        Probing additional dimensions in the universe with neutron experiments
10:30   **Tom Gaisser** (Bartol)
        Nuclear abundances in the high-energy cosmic radiation

11:00   Coffee

Chair: **Moshe Gai** (Yale/UConn)

11:30   **Francesco Iachello** (Yale)
        Quantum phase transitions in nuclei and other systems
12:00   **Alejandro Frank** (ICN-UNAM)
        Geometry of coexistence in the interacting boson model
12:30   **José Arias** (Sevilla)
        Phase transitions in the interacting boson model
13:00   **Norbert Pietralla** (Stony Brook)
        Analytic X(5)-to-rigid rotor transition in the confined beta-soft rotor model
13:30   **Jan Jolie** (Köln)
        Future investigations in nuclear structure

14:00   Lunch

THURSDAY April 22

Chair: **Maria Esther Ortiz** (IF-UNAM)

09:00 **Kim Lister** (Argonne)
Neutron-proton correlations along the N=Z line
09:30 **Patrick Regan** (Surrey)
Structural evolution in the (N, Z, I) coordinate frame for $A \sim 100$
10:00 **Jerry Draayer** (LSU)
Extended pairing model for well-deformed nuclei
10:30 **Rick Casten** (WNSL-Yale)
Alternative interpretation of E0 transitions in transitional nuclei

11:00 Coffee

Chair: **Ani Aprahamian** (Notre Dame)

11:30 **Roelof Bijker** (ICN-UNAM)
Spectroscopy of pentaquarks
12:00 **Victor Zamfir** (WNSL-Yale)
Evolution of nuclear observables in the spherical-deformed phase
transition and the interacting boson model
12:30 **Tamas Vertse** (ATOMKI, Hungary)
A shell model representation with antibound states
13:00 **Peter Ring** (München)
Symmetry conserving relativistic mean-field theory and applications
for halo phenomena at the neutron drip line
13:30 **David Warner** (Daresbury)
Neutrons and protons - Vive la différence!

264